AF361934

Meiofauna Biodiversity and Ecology

Meiofauna Biodiversity and Ecology

Editors

Federica Semprucci
Roberto Sandulli

MDPI • Basel • Beijing • Wuhan • Barcelona • Belgrade • Manchester • Tokyo • Cluj • Tianjin

Editors
Federica Semprucci
Università degli Studi di Urbino Carlo Bo
Italy

Roberto Sandulli
"Parthenope" University of Naples
Italy

Editorial Office
MDPI
St. Alban-Anlage 66
4052 Basel, Switzerland

This is a reprint of articles from the Special Issue published online in the open access journal *Diversity* (ISSN 1424-2818) (available at: https://www.mdpi.com/journal/diversity/special_issues/ meiofauna_biodiversity_ecology).

For citation purposes, cite each article independently as indicated on the article page online and as indicated below:

LastName, A.A.; LastName, B.B.; LastName, C.C. Article Title. *Journal Name* **Year**, *Article Number*, Page Range.

ISBN 978-3-03943-132-8 (Hbk)
ISBN 978-3-03943-133-5 (PDF)

Cover image courtesy of Yours Alexei.

Contents

About the Editors

Federica Semprucci is a researcher at the Department of Biomolecular Sciences of the University of Urbino (Italy) with a current didactic activity in Developmental Biology. She graduated in Natural Sciences and obtained her PhD in Environmental Sciences in 2007 and in "Mechanisms of cell regulation: morpho-functional and evolutionary aspects" in 2011, both at the University of Urbino. Her research activity tries to better understand the meiofaunal biodiversity patterns and their drivers in coastal systems from temperate to tropical regions. She is a specialist of the Nematoda phylum and she is exploring new possible ways to utilize nematodes as a tool in the Ecological Quality (EcoQ) assessment. Furthermore, she has published several papers on nematode taxonomy, describing new species from the Maldives and Hawaii and giving some re-descriptions of species belonging to Cyatholaimidae family. She is currently involved in several research activities in Italy, Tunisia, Malaysia, Gulf of Mexico and South Korea.

Roberto Sandulli is Full Professor of Zoology and Marine Biology at Parthenope University of Naples (Italy). He has shared joint research programs with the Department of Biology at the University of Gent (Belgium), the Marine Laboratory of Aberdeen (UK), the Zoology Department at the University of Aberdeen (UK), the Institute of Marine Biology of Crete (Greece), the Department of Biology of Columbia University (USA), the Biology Department of Jacksonville State University (USA), the Plymouth Marine Laboratory (UK), the Stazione Zoologica "A. Dohrn" of Naples (Italy), IFREMER, Brest (F), and several Italian Universities and Research Centers. He is reference lecturer of UNESCO, and a previous president of the Benthos Committee of the Italian Society of Marine Biology (SIBM, 2002–2004, 2010–2015). Since 2016, he has been appointed President of the Coastal Zone Management committee of the Italian Society of Marine Biology (SIBM, 2016–2018), and since January 2019, he has been a member of the Directive Committee of SIBM. He is the author of over 170 scientific publications concerning marine biology, the ecology of marine meiobenthos, the effect of pollution on benthic communities, marine biodiversity, allochthonous species invasion, bioconstructions, red coral distribution, MPAs management, the spatial microdistribution of meiofauna, and seagrass reimplantation.

Preface to "Meiofauna Biodiversity and Ecology"

Meiofauna are small organisms ranging 30–500 μm in body size, inhabiting marine sediments and other substrata all over the world, even the most extreme ones. We can find many different meiofaunal species in a very small handful of sediment, with the most varied and curious shapes, that share peculiar lifestyles, ecological relationships, and evolutionary traits. They contribute significantly to the processes and functioning of marine ecosystems, thanks to their high abundance and taxonomical diversity, fast turnover and metabolic rates. Some meiofaunal taxa have also revealed their considerable utility in the evaluation of the ecological quality of coastal marine sediments in accordance with European Directives. Therefore, understanding the distribution patterns of their biodiversity and identifying the factors that control it at a global level and in different types of habitats is of great importance. Due to their very small morphological characteristics utilized for the taxonomical identification of these taxa, the suite of necessary skills in taxonomy, and the general taxonomic crisis, many young scientists have been discouraged to tackle meiofauna systematics. The papers collected in this book, however, bring together important themes on the biology, taxonomy, systematics, and ecology of meiofauna, thanks to the contribution of researchers from around the world from the USA, Brazil, Costa Rica, Mexico, Cuba, Italy, Belgium, France, Denmark, Russia, Kuwait, Vietnam, and South Korea. This was certainly an additional opportunity to build a more solid network among experts in this field and contribute to increasing the visibility of these tiny organisms.

A special thanks to Prof. Wonchoel Lee for the wonderful taxonomic drawings of the species described in this volume that contribute to make our cover unique.

Federica Semprucci, Roberto Sandulli
Editors

 *diversity*

Editorial

Editorial for Special Issue "Meiofauna Biodiversity and Ecology"

Federica Semprucci [1,2,*] **and Roberto Sandulli** [3,4]

[1]   Department of Biomolecular Sciences (DiSB), University of Urbino "Carlo Bo", 61029 loc. Crocicchia, Italy

[2]   Fano Marine Center, The Inter-Institute Center for Research on Marine Biodiversity, Resources and Biotechnologies, 61032 Fano, Italy

[3]   Department of Sciences and Technologies (DiST), Parthenope University, 80143 Naples, Italy; roberto.sandulli@uniparthenope.it

[4]   Consorzio Nazionale, Interuniversitario per le Scienze del Mare (CoNISMa), 00196 Rome, Italy

*   Correspondence: federica.semprucci@uniurb.it

Received: 15 June 2020; Accepted: 17 June 2020; Published: 19 June 2020

Abstract: Meiofauna are a component of aquatic environments from polar to tropical regions. They may colonize all types of habitats and include very enigmatic and exclusive taxa. The biodiversity of this component in marine ecosystems is far from being accurately estimated, but this would be a new challenge given the importance that meiofaunal components may play in marine ecosystem functioning and processes. This Special Issue collects many interesting topics in research on meiofauna contributing to plugging a gap on several key issues in their biodiversity, distribution, and ecology, from numerous regions that include the USA, Brazil, French Guiana, Costa Rica, Mexico, Cuba, Italy, Kuwait, Vietnam, Madagascar, the Maldives, and South Korea.

Keywords: biodiversity; ecology; taxonomy; DNA barcoding; new species; epibiosis; β-diversity; biological traits; bioindicators; meiofauna paradox

---

Meiofauna are small organisms (body size range: 30–500 µm) that inhabit seabeds all over the world, even the most extreme ones [1–3]. They live in and on all types of marine substrata as well as on other living organisms from invertebrates to vertebrates. Twenty-four of the 35 animal phyla have at least one representative within meiofauna. In a handful of sediment or in a few square centimeters, we can find many different meiofaunal species, with the most varied and curious forms, that share peculiar lifestyles, ecological relationships, and evolutionary traits [1]. Thanks to their remarkable abundance and biodiversity, their rapid life cycle, and their metabolic rate, they contribute significantly to the processes and functioning of marine ecosystems [4]. Some taxa have also revealed considerable utility in the evaluation of the ecological quality of coastal marine sediments in accordance with European Directives 2000/60/CE and MSFD 2008/56/CE, e.g. [5,6]. Therefore, understanding the distribution patterns of their biodiversity and identifying the factors that control it at a global level is of great importance.

Current estimates of the meiofaunal biodiversity level are significantly lacking. There are authors who even hypothesize that some meiobenthic phyla could have the same level of biodiversity magnitude as insects in terrestrial habitats [7]. Unfortunately, the minute morphological characterizations utilized for the taxonomical identification of these taxa, the suite of necessary classification skills, and the general increase in taxonomic crisis that still discourages many young researchers from field taxonomy have notably hampered advances in this field of interest. This Special Issue also assumes additional considerable importance if we consider the crucial role that the terrestrial nematode *Caenorhabditis elegans* has held in the history of biology in the last few decades. In fact, this benthic species could represent an important tool not only for the monitoring and conservation of marine ecosystems, but also as a hidden treasure trove of new natural products that could represent an advance in the biomedical sector.

Accordingly, the purpose of this Special Issue was to bring together researchers from around the world to share their most recent studies on some important themes in meiofaunal biodiversity and ecology. Many people replied to the Special Issue call from the USA, Brazil, Costa Rica, Mexico, Cuba, Italy, Belgium, France, Denmark, Russia, Kuwait, Vietnam, and South Korea. This was certainly an additional opportunity to build a more solid network among experts in this field and contribute to increasing the visibility of these tiny organisms.

The manuscripts published in this Special Issue cover not only topics in the traditional morphological taxonomy of meiofaunal groups (Gastrotricha: [8]; Nematoda: [9]; Copepoda: [10,11]), but also studies combining both morphological and molecular approaches (Gastrotricha: [12]; Nematoda: [13]).

The occurrence of meiofaunal taxa in a wide spatial range is often a mystery because they do not have pelagic larvae. Some hypotheses have been formulated in the past see [1] and references therein, but Ingels and co-authors raise some new and interesting insights to explain the so-called "meiofauna paradox" thanks to a study on the meiofaunal epibionts of loggerhead sea turtles [14]. However, meiofaunal organisms may be themselves a substratum for other smaller benthic groups (i.e., bacteria or ciliates); in simple terms, biodiversity within biodiversity! In this respect, an extensive review is included on the ciliate and nematode epibiosis phenomenon with a description of three new epibiont species and an updated distribution of all the records of nematode-suctoria association around the world [15]. As reported above, it is fundamental to identify the factors that control meiofaunal distribution patterns in marine ecosystems under both natural and anthropogenic gradients. Indeed, the relationship between organic matter, prokaryotes, and meiofauna across a river-lagoon-sea gradient is investigated in [16]. The role of habitat on the diversity patterns of nematodes in the Cuban archipelago is also evaluated, taking into consideration not only β-diversity but also biological traits [17]. Biological traits, if adequately addressed, could represent an additional approach for the detection of environmental changes [18]. Hard substrata may host a highly diversified meiobenthic community, but overall, a limited number of papers are present on this habitat type see [19] and references therein. Gallucci et al [20] further demonstrate, based on work in south-eastern Brazil, that substrate identity and the surrounding environment are important in structuring smaller meiofauna, particularly the nematodes. In the Mekong delta system (Vietnam), biochemical component changes due to dam construction have been investigated revealing a nematode assemblage that has adapted well to organic enrichment, heavy metal accumulation, and oxygen depletion, but the dam located in the Ba Lai estuary may potentially continue to drive this ecosystem to its tipping point, underlining the need for further investigations [21]. Foraminifera may become a consistent part of meiobenthic communities in marine and transitional environments [22]. Al-Enezi et al [23] document for the first time the biodiversity pattern of benthic foraminifera from Kuwait Bay and the northern islands in this area.

Acknowledgments: As guest editors for this Special Issue, we would like to express our gratitude to all the authors and anonymous reviewers who have contributed new ideas and modern perspectives to meiofauna research. We also would like to warmly thank the staff members at the MDPI editorial office for their support during the editorial process.

Conflicts of Interest: The authors declare no conflict of interest.

References

1. Giere, O. *Meiobenthology: The Microscopic Motile Fauna of Aquatic Sediments*, 2nd ed.; Springer: Berlin, Germany, 2009; p. 527.
2. Sandulli, R.; Semprucci, F.; Balsamo, M. Taxonomic and functional biodiversity variations of free living nematodes across an extreme environmental gradient: A study case in a Blue Hole cave. *Ital. J. Zool.* **2014**, *81*, 508–516. [CrossRef]

3. Baldrighi, E.; Zeppilli, D.; Appolloni, L.; Donnarumma, L.; Chianese, E.; Russo, G.F.; Sandulli, R. Meiofaunal communities and nematode diversity characterizing the Secca delle Fumose shallow vent area (Gulf of Naples, Italy). *PeerJ* **2020**, *8*, e9058. [CrossRef] [PubMed]

4. Danovaro, R.; Gambi, C.; Dell'Anno, A.; Corinaldesi, C.; Fraschetti, S.; Vanreusel, A.; Vincx, M.; Gooday, A.J. Exponential decline of deep-sea ecosystem functioning linked to benthic biodiversity loss. *Curr. Biol.* **2008**, *18*, 1–8. [CrossRef] [PubMed]

5. Semprucci, F.; Losi, V.; Moreno, M. A review of Italian research on free-living marine nematodes and the future perspectives on their use as Ecological Indicators (EcoInds). *Mediterr. Mar. Sci.* **2015**, *16*, 352–365. [CrossRef]

6. Semprucci, F.; Balsamo, M.; Appolloni, L.; Sandulli, R. Assessment of ecological quality status along the Apulian coasts (Eastern Mediterranean Sea) based on meiobenthic and nematode assemblages. *Mar. Biodiv.* **2018**, *48*, 105–115. [CrossRef]

7. Feist, S.W.; Mees, J.; Reimer, J.D.; Gittenberger, A.; Bruce, N.L.; Walter, M.E.; Rocha, J.R.; Berta, M.A.; Molodtsova, T.N.; Hopcroft, R.R.; et al. The Magnitude of Global Marine Species Diversity. *Curr. Biol.* **2012**, *22*, 2189–2202.

8. Todaro, M.A.; Sibaja-Cordero, J.A.; Segura-Bermúdez, O.A.; Coto-Delgado, G.; Goebel-Otárola, N.; Barquero, J.D.; Cullell-Delgado, M.; Dal Zotto, M. An Introduction to the Study of Gastrotricha, with a Taxonomic Key to Families and Genera of the Group. *Diversity* **2019**, *11*, 117. [CrossRef]

9. Tchesunov, A.; Jeong, R.; Lee, W. Two New Marine Free-Living Nematodes from Jeju Island Together with a Review of the Genus *Gammanema* Cobb 1920 (Nematoda, Chromadorida, Selachinematidae). *Diversity* **2020**, *12*, 19. [CrossRef]

10. Jeon, D.; Lee, W.; Soh, H.Y.; Eyun, S.-I. A New Species of Monstrillopsis Sars, 1921 (Copepoda: Monstrilloida) with an Unusually Reduced Urosome. *Diversity* **2020**, *12*, 9. [CrossRef]

11. Lee, W. Doolia, A New Genus of Nannopodidae (Crustacea: Copepoda: Harpacticoida) from off Jeju Island, Korea. *Diversity* **2020**, *12*, 3. [CrossRef]

12. Campos, A.; Todaro, M.A.; Garraffoni, A.R.S. A New Species of *Paraturbanella* Remane, 1927 (Gastrotricha, Macrodasyida) from the Brazilian Coast, and the Molecular Phylogeny of Turbanellidae Remane, 1926. *Diversity* **2020**, *12*, 42. [CrossRef]

13. Martínez-Arce, A.; De Jesús-Navarrete, A.; Leasi, F. DNA Barcoding for Delimitation of Putative Mexican Marine Nematodes Species. *Diversity* **2020**, *12*, 107. [CrossRef]

14. Ingels, J.; Valdes, Y.; Pontes, L.P.; Silva, A.C.; Neres, P.F.; Corrêa, G.V.V.; Silver-Gorges, I.; Fuentes, M.M.; Gillis, A.; Hooper, L.; et al. Meiofauna Life on Loggerhead Sea Turtles-Diversely Structured Abundance and Biodiversity Hotspots That Challenge the Meiofauna Paradox. *Diversity* **2020**, *12*, 203. [CrossRef]

15. Baldrighi, E.; Dovgal, I.; Zeppilli, D.; Abibulaeva, A.; Michelet, C.; Michaud, E.; Franzo, A.; Grassi, E.; Cesaroni, L.; Guidi, L.; et al. The Cost for Biodiversity: Records of Ciliate–Nematode Epibiosis with the Description of Three New Suctorian Species. *Diversity* **2020**, *12*, 224. [CrossRef]

16. Bianchelli, S.; Nizzoli, D.; Bartoli, M.; Viaroli, P.; Rastelli, E.; Pusceddu, A. Sedimentary Organic Matter, Prokaryotes, and Meiofauna across a River-Lagoon-Sea Gradient. *Diversity* **2020**, *12*, 189. [CrossRef]

17. Armenteros, M.; Pérez-García, J.A.; Marzo-Pérez, D.; Rodríguez-García, P. The Influential Role of the Habitat on the Diversity Patterns of Free-Living Aquatic Nematode Assemblages in the Cuban Archipelago. *Diversity* **2019**, *11*, 166. [CrossRef]

18. Semprucci, F.; Cesaroni, L.; Guidi, L.; Balsamo, M. Do the morphological and functional traits of free-living marine nematodes mirror taxonomical diversity? *Mar. Environ. Res.* **2018**, *135*, 114–122. [CrossRef]

19. Losi, V.; Sbrocca, C.; Gatti, G.; Semprucci, F.; Rocchi, M.; Bianchi, C.N.; Balsamo, M. Sessile macrobenthos (Ochrophyta) drives seasonal change of meiofaunal community structure on temperate rocky reefs. *Mar. Environ. Res.* **2018**, *142*, 295–305. [CrossRef]

20. Gallucci, F.; Christofoletti, R.A.; Fonseca, G.; Dias, G.M. The Effects of Habitat Heterogeneity at Distinct Spatial Scales on Hard-Bottom-Associated Communities. *Diversity* **2020**, *12*, 39. [CrossRef]

21. Yen, N.T.M.; Vanreusel, A.; Lins, L.; Thai, T.T.; Nara Bezerra, T.; Quang, N.X. The Effect of a Dam Construction on Subtidal Nematode Communities in the Ba Lai Estuary, Vietnam. *Diversity* **2020**, *12*, 137. [CrossRef]

22. Balsamo, M.; Semprucci, F.; Frontalini, F.; Coccioni, R. Meiofauna as a tool for marine ecosystem biomonitoring. In *Marine Ecosystems*; Cruzado, A., Ed.; InTech: Rijeka, Croatia, 2012; Volume 4, pp. 77–104.
23. Al-Enezi, E.; Khader, S.; Balassi, E.; Frontalini, F. Modern Benthic Foraminiferal Diversity: An Initial Insight into the Total Foraminiferal Diversity along the Kuwait Coastal Water. *Diversity* **2020**, *12*, 142. [CrossRef]

 *diversity*

Article

The Cost for Biodiversity: Records of Ciliate–Nematode Epibiosis with the Description of Three New Suctorian Species

Elisa Baldrighi [1], Igor Dovgal [2], Daniela Zeppilli [3], Alie Abibulaeva [2], Claire Michelet [4], Emma Michaud [4], Annalisa Franzo [5], Eleonora Grassi [6], Lucia Cesaroni [6], Loretta Guidi [6], Maria Balsamo [6], Roberto Sandulli [7,*] and Federica Semprucci [6]

[1] Institute for Biological Resources and Marine Biotechnologies (IRBIM), Italian National Research Council (CNR), Via Pola 4, 71010 Lesina, Italy; elisabaldrighi82@gmail.com
[2] A.O. Kovalevsky Institute of Biology of the Southern Seas of RAS, 299011 Sevastopol, Russia; dovgal-1954@mail.ru (I.D.); abibulaeva@ibss-ras.ru (A.A.)
[3] IFREMER, Centre Brest, REM/EEP/LEP, ZI de la pointe du diable, CS10070, F-29280 Plouzané, France; daniela.zeppilli@ifremer.fr
[4] LEMAR, Université Brest, CNRS, IRD, IFREMER, F-29280 Plouzane, France; claire.s.michelet@gmail.com (C.M.); emma.michaud@univ-brest.fr (E.M.)
[5] Oceanography Section, Istituto Nazionale di Oceanografia e di Geofisica Sperimentale, OGS I-34151 Trieste, Italy; afranzo@inogs.it
[6] Department of Biomolecular Sciences (DiSB), University of Urbino 'Carlo Bo', loc. Crocicchia, 61029 Urbino, Italy; e.grassi5@campus.uniurb.it (E.G.); lucia.cesaroni@uniurb.it (L.C.); loretta.guidi@uniurb.it (L.G.); maria.balsamo@uniurb.it (M.B.); federica.semprucci@uniurb.it (F.S.)
[7] Department of Science and Technology (DiST), Parthenope University Naples, Consorzio Nazionale Interuniversitario per le Scienze del Mare (CoNISMa), 00196 Rome, Italy
* Correspondence: roberto.sandulli@uniparthenope.it; Tel.: +39-347-4274046

http://zoobank.org/urn:lsid:zoobank.org:pub:D2927905-34B3-402B-AA81-0C13D12D1D81
Received: 4 May 2020; Accepted: 1 June 2020; Published: 4 June 2020

Abstract: Epibiosis is a common phenomenon in marine systems. In marine environments, ciliates are among the most common organisms adopting an epibiotic habitus and nematodes have been frequently reported as their basibionts. In the present study, we report several new records of peritrich and suctorian ciliates-nematode association worldwide: from a deep-sea pockmark field in the NW Madagascar margin (Indian Ocean), from a shallow vent area in the Gulf of Naples (Mediterranean, Tyrrhenian Sea), in a MPA area in the Gulf of Trieste (Mediterranean, Adriatic Sea), from a mangrove system in French Guiana (South America, Atlantic Ocean), and from the Maldivian Archipelago. In addition, three new species of Suctorea from the Secca delle Fumose shallow vent area (Gulf of Naples) were described: *Loricophrya susannae* n. sp., *Thecacineta fumosae* n. sp. and *Acinetopsis lynni* n. sp. In the light of these new records and data from the existing literature, we discuss the suctorian–nematode epibiosis relationship as a lever to biodiversity.

Keywords: epibiosis; ciliophora; suctorea; nematoda; meiofauna; biodiversity

---

1. Introduction

Epibiosis (greek *epi* "on top" and *bios* "life") is a facultative spatial association between two organisms: the epibiont and the basibiont [1]. Epibionts are organisms that, during their sessile phase, remain attached to the surface of a living substratum, while the basibiont provides the support for the epibiont. Both concepts suggest ecological functions [2,3]. Epibiosis is a common phenomenon in marine systems and can be considered a direct consequence of surface limitation and/or a wave turbulence effect that obliges many lightweight organisms to evolve attachment systems to adhere to hard and relatively stable surfaces (e.g., of other living organisms; [4]). Epibiosis is the evolutionary result of an interaction between environmental factors and benthic life forms; it is a dynamic process and the ecological consequences for the basibiont and the colonizer (e.g., bacteria, fungi, algae and protozoans) can be of different nature (i.e., positive, negative or without effects for the host) depending on the environmental conditions and on the epibiotic assemblage composition and density [1,3,5]. Direct and indirect interactions among epibionts and with the host, and changing environmental conditions drive the dynamics of the epibiotic community [1]. Epibiosis can be temporary, i.e., linked to the seasonal presence of the basibiont and /or epibiont or it can represent a temporary colonization due to a decrease in basibiont defenses or to its fitness. Epibiosis may modify a number of interactions between the basibiont and the biotic and abiotic components of the environment [3,5,6]. This is the reason why epibiosis may act as an ecological lever by modifying and greatly amplifying or buffering biotic and abiotic stresses [5]. In some cases, epibionts are considered as commensals (e.g., [7]) because they are not harmful to the hosts; however, some of them can indirectly influence growth, survival rate and reproductive capability of basibionts, showing a negative impact on their fitness [5,8].

Epibiotic assemblages are rarely species-specific [2], and many colonizers are substratum generalists. Different basibiont species may also host different epibiotic communities (e.g., [9,10]). In most investigations, less than 20% of epibionts were reported as restricted to this mode of life, and less than 5% occur exclusively on one basibiont species [11,12]. Nevertheless, some exceptions were documented in previous studies (e.g., [7,13,14]) indicating species-specific host-epibiont relationships. In general, the epibiont must be able to cope with the basibiont lifestyle and its surface properties. The properties of the basibiont surface, i.e., its consistency, surface ornamentation, the presence of previous settlers (e.g., biofilms) and the deployment of defenses, determine which of the available potential epibionts will successfully settle and grow when a suitable substratum becomes available. Indeed, many basibionts have developed a variety of defense mechanisms to prevent epibiosis or to remove epibionts: these span from mechanical defenses (e.g., mucus secretion, burrowing behavior, movement in narrow caves for abrasion and epibiont elimination) to chemical methods (e.g., secretion of secondary metabolites such as antibacterial or antifungal compounds), if the nature of the relation is disadvantageous [3,15].

An important component of epibiont communities are ciliated protozoans. These organisms also constitute a significant component of the overall marine and freshwater ecosystems, and play an important role in the food chain [16]. Suctorian ciliates, together with peritrichs, are the most species-rich groups of Ciliophora. They live in all types of water bodies and they are epibionts on a wide diversity of hosts and substrates. Some species are ectoparasitic or endoparasitic species, but many of these ciliates are commensals of aquatic invertebrates or vertebrates [17]. Suctorian ciliates are quite selective by feeding principally on small ciliates, flagellates and amoebae that are captured by tentacles [17,18].

Many meiofaunal organisms such as Copepoda Harpacticoida, Ostracoda, Halacarida, Tanaidacea, Kinorhyncha and free-living Nematoda were found to be common basibionts for suctorian and peritrich ciliates and prevalent across estuarine to marine ecosystems [14,19–21]. However, many aspects of this relationship need to be clarified: the criteria for the host selection; adhesion mechanisms; the role of environmental variables in influencing the distribution and diversity of ciliates adhered to meiofauna, and the ecological significance of epibiont–basibiont interactions across different habitats [21,22].

Nematoda is the most abundant, ubiquitous and diverse meiofaunal marine phylum [23] and they cover a key ecological role in the ecosystem processes [24]. Thanks to their cuticle characteristics, often made by a thick and multi-layered collagenous covering, they are ideal basibionts for many suctorian ciliates (e.g., [22,25]). In particular, nematodes of the families Desmodoridae and Desmoscolecidae have found to be largely colonized due to the well-developed cuticular ornamentation that favors the adhesion of epibionts (e.g., [26]).

In a recent study based on published records, Chatterjee et al. [27] provided a checklist of suctorian epibionts on meiobenthic marine nematodes. Despite the amount of data presented from different geographical zones and types of environments, this phenomenon is still largely underestimated and the nematode-ciliate association might be more common than it actually appears to be. This is mainly due to three reasons: (i) in papers concerning nematode taxonomy and/or ecology the presence of epibionts was often overlooked or simply reported without a description of the ciliate(s), their number and distribution on the basibiont body surface; (ii) the methodology used for nematode extraction from the sediment (i.e., centrifugation) may induce the loss of some epibionts; (iii) specialists of ciliate or nematode taxonomy work separately and their focus of research is usually on the taxonomy and ecology of only one of the two groups. All these aspects have largely hampered a clear comprehension of this phenomenon.

In the present paper, we reported some new finds of Suctorea from the Secca delle Fumose shallow vent area (Naples, Italy) and we described three new species. Secca delle Fumose belongs to the degassing structure offshore of the Campi Flegrei caldera and its biology and ecology has been investigated only recently [28,29]. We reported also several new records of peritrich and suctorian ciliates-nematode association worldwide, providing an update of the check-list presented by Chatterjee et al. [27]. In the light of these new records and the literature data, we discussed the suctorian–nematode epibiosis relationship as a lever to biodiversity.

2. Material and Methods

2.1. Research Areas and Sampling Strategy

Sediment samples were collected from five different areas located worldwide: a deep-sea pockmark field in the northwestern Madagascar margin (Indian Ocean), a shallow vent area in the Gulf of Naples (Tyrrhenian Sea), a MPA area in the Gulf of Trieste (Adriatic Sea), a mangrove system in French Guiana (South America, Atlantic Ocean), and a coral reef system in the Maldivian Archipelago (Figure 1). Samples were collected either by a multi-corer (MUC), a manual corer or by SCUBA divers with the help of manual cylindrical corers (Table 1 for details). Hereafter, we briefly report the main characteristics of each study area.

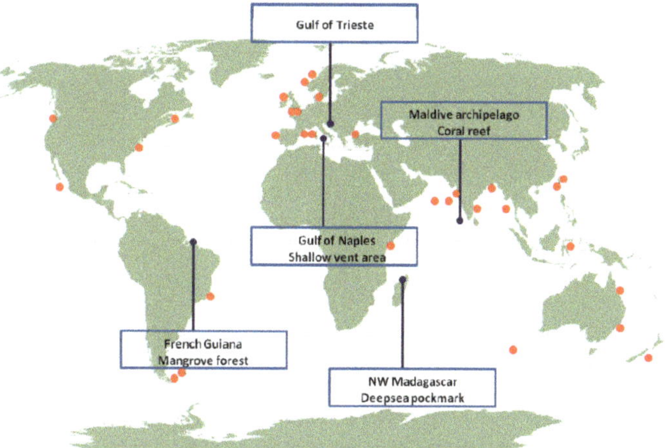

Figure 1. Sampling locations in the present study (blue dots) and locations in which nematode–ciliate associations were reported from the available literature (red dots).

Table 1. Geographical location and study sites, sediment features and methods used for each sampling and analysis activities.

System Investigated	Deep-Sea Pockmark	Shallow Water HV	Nearby MPA	Mangrove Forest	Coral Reef
Study site	NW Madagascar margin	Gulf of Naples	Gulf of Trieste	French Guiana	Archipelago of Maldives
Geographical area	Indian Ocean	Tyrrhenian Sea	Adriatic Sea	Atlantic Ocean	Indian Ocean
Sampling depth	528–775 m	9–14 m	18 m	0–50 cm	19–63 m
Sediment tipology	Muddy	Sandy	Sandy-mud	Muddy	Sandy (in reef); silty (out reef)
Sampling period	Sept.–Oct. 2014	November 2017	September 2011	November 2017	May 2013
N. of stations sampled	2 (Site 1; Site 2)	4 (H; G; CN and CS)	1 (St. C1)	3 (St. 1; St.2 and St.3)	4 (M1, M4—in; M8, M9—out reef)
N. of stations with infested nematodes	1 (Site 1—inside pockmark)	3 (G; CN and CS)	1 (St. C1)	2 (St. 1 and St.2)	4 (M1; M4; M8 and M9)
Sampling technique	MUC	Manual corer	Manual corer	Manual corer	Manual corer
Sediment layer(s) considered	0–1 cm	0–1; 1–3; 3–5; 5–10 cm	0–10 cm	0–2; 2–10; 10–16 cm	0–5 cm
Sive sized used	1 mm–32μm	1 mm–32 μm	1 mm–38 μm	1 mm–32 μm	0.5 mm–42 μm
Extraction method	Ludox colloidal silica	Ludox colloidal silica	Ludox colloidal silica	Ludox colloidal silica	Ludox colloidal silica
Reference	Sanchez et al. (under review)	[29]	[30]	Michelet et al. (under review)	[31]

2.1.1. Deep-Sea Pockmark: Madagascar Margin

This study was conducted on the northwestern part of the Madagascar along the Mahavavy slope to collect samples within a pockmark area (Site 1) and along the Betsiboka slope to collect samples outside the pockmark (Site 2) (PAMELA-MOZ01 cruise; [32]).

Nematodes with epibiont ciliates were found only at Site 1, which exhibited higher total sulfur concentrations (up to 4.7%), a lower dissolved oxygen penetration and the presence of CH_4 (<1 µM) [33] compared to Site 2 outside the pockmark. Overall, higher sedimentation rates were observed at Site 1, with two or three main input events over the last 60 years [34], a period also characterized by a very high accumulation of total sulfur.

2.1.2. Secca delle Fumose Shallow Vent: Gulf of Naples

The study area of Secca delle Fumose (SdF) is located in the northwestern side of the Gulf of Naples. SdF is a submarine relief consisting of a network of ancient Roman pillars, among which thermal vents releasing hot gas-rich hydrothermal fluids (9–14 m water depth range) occurred. In this study, we selected four sampling sites: one diffusive emission site (H) characterized by the presence of white microbial mats covering the soft bottom; one geyser site (G) at 65 m distance from the H site, with surrounding rocky substrate covered by yellow sulphur deposits and with hot water emissions reaching 80 °C at the sediment surface; two inactive sites (CN and CS) located at distance of 100 m from the active sites H and G. From site H we reported the highest sediment temperature (37.5 °C) and lowest pH value (7.56); site G was characterized by the presence of sulphur ion S^{2-}, a pH of 8 and a temperature of 29.1 °C. From the inactive sites, CN and CS, we detected temperature (21.8 °C) and pH (8.1) values comparable to the background. Nematode with ciliates were only found at sites G, CN and CS [29].

2.1.3. Marine Protected Area: Gulf of Trieste

The study was carried out at the station C1, which is located ca. 200 m offshore nearby the outer border of the Marine Protected Area (MPA) of Miramare in the Gulf of Trieste (North Adriatic Sea). The Gulf is characterized by annual fluctuations of temperature (from 5 °C to ≥24 °C at the surface and from 6 °C to ≥20 °C at the bottom) and in summer the water column is usually stratified. Sedimentation is mainly controlled by river inputs rather than by marine currents [30]. Only one individual of nematode with ciliates was found from the sampling station C1.

2.1.4. Mangrove Forests: French Guiana

The study area was located in the vicinity of the Cayenne estuary, French Guiana (South America). Sediment samples were taken from three stations characterized by the presence of mangrove forests and situated on river edges in the polyhaline zone at an increasing distance from the Cayenne city. Station 1 was located near to a wastewater treatment plant which drained the waters from industrial, commercial and urban areas. Station 2 was located at the intersection between two rivers Cayenne and Montsinnery and Station 3 was 10 km from the estuary mouth and from the agricultural and urban environments. 4. Overall, Station 3 in downstream of the Cayenne estuary appeared to be more preserved from anthropogenic effects than the other two stations. Basibiont nematodes were found at St. 1 and St. 2 [Michelet et al., under review].

2.1.5. Coral Reefs: Archipelago of Maldives

The archipelago of Maldives is in the Indian Ocean, in the central part of the Chagos-Maldives-Laccadive Ridge [35]. The archipelago is formed by a single atoll chain in the northern and southern areas, and by a double atoll chain in the central area. All the Maldivian sediments are of coralline origin with a range of grain size very heterogeneous and poorly correlated to the level of hydrodynamic conditions [31]. This is mainly due to the short transport underwent by the sediments that are deposited almost immediately after their erosion by the reefs [36]. The archipelago is dominated by monsoons with southwest to northwestern winds (~225°–315°) from April to November (namely westerly monsoon); and northeast–eastern winds (~45°–90°) prevailing from November to March (e.g., northeastern monsoon) [35]. The samples were collected in the South and North Malé, and Felidhoo atolls (see [37] for details). A total of 20 sites were investigated: 10 outer (ocean-facing sides situated on the atoll rim) and 10 inner (lagoon sides of the atoll rim) reefs. Suctorians were found in the nematodes of the following locations: North Malé atoll, stations M1 and M4 located in two inner reefs at depths of 40 and 19 m, respectively; South Malé atoll, stations M8 and M9 located in two outer reefs at depths of 21 and 63 m of depths, respectively.

2.2. Samples Processing

Meiofauna organisms were extracted from the sediments by centrifugation with Ludox colloidal silica HS-40 [38]. From each sample, the first 100–110 nematodes encountered in the cuvette were hand-picked, transferred from fixative to glycerol through a series of ethanol-glycerol solutions and finally mounted on slides in anhydrous glycerin [39]. All nematodes on permanent slides were identified at the genus level using the pictorial keys of Platt and Warwick [40,41] and Warwick et al. [42], as well as the original species descriptions and identification keys available through NeMys [43]. An estimation of the percentage of nematode basibionts colonized was calculated based on the total nematode density in each sample.

The specimens showing epibionts attached to the body wall were isolated for an in-depth identification of the ciliate species. Measurements of ciliates were made using the program Toup View 3.7 for digital camera.

The terminology and systematic position of suctorian ciliates follow Dovgal [17,44].

A Permanent microscopic slide of the nematode specimen with the new suctorian species is deposited at REM/EEP/LEP of Ifremer Centre Brest, Brittany, France.

3. Results

3.1. New Records of Associations between Nematodes and Ciliates from Deep-Sea and Shallow Water Systems

The list of new records found in the present study is presented in Table 2; we also reported an estimated percentage of colonized nematodes at each sampling site, aware that the values may be underestimated in reality. A total of six genera and twelve species of epibiont ciliates were found on fifteen nematode genera and four families from five different habitats (Table 2 and Figure 1). Three ciliate species were recognized as new: *Loricophrya susannae* n. sp., *Thecacineta fumosae* n. sp., and *Acinetopsis lynni* n. sp.

Table 2. Summary of data reported in the present study. Study sites and stations; estimated percentage of colonized nematodes (No. IN) on total nematodes found at each sampling site; B gender = gender of the basibiont; BG = basibiont genus; ES = epibiont species; No. E/B = number of epibiont found on basibiont; juv. = juvenile.

Study Site	Station	No. IN	B Gender	BG	ES	No. E/B	Position on the Basibiont Body
NW Madagascar margin-deep sea pockmark	St.1	15	female (13)-male (3)	*Desmodora*	*Paracineta homari*	1–4	Middle body and tail region
	St.1	17	female (12)-male (4)	*Desmodora*	*Trematostoma rotunda*	1–4	Middle body and head region
	St.1	17 15%	female (10)-male (6)	*Desmodora*	*Paracineta homari*	1–4	Cloaca region
Gulf of Naples-shallow vent area Gayser	Site G	25	female (15)-male (6)-juv. (4)	*Desmodora*	*Trematostoma rotunda*	1–8	Along the entire body length
	Site G	14	female (6)-male (5)-juv. (3)	*Paradesmodora*	*Paracineta homari*	1–6	Along the entire body length
	Site G	2	male (2)	*Polysigma*	*Thecacineta calix*	2–5	Along the entire body length
	Site G	2	male (2)	*Pseudochromadora*	*Paracineta homari*	1	Tail region, before anus
	Site G	8	female (4)-male (4)	*Pseudodesmodora*	*Thecacineta calix*	2–12	Along the entire body length
	Site G	2	male (2)	*Desmodoridae*	*Loricophrya bosporica*	1–2	Tail region
	Site G	1 5%	female	*Desmodora*	*Acinetopsis lynni* sp. n.	4	Middle body and tail region
Gulf of Naples-shallow vent area Inactive site	Site CN	9	female (2)-male (3)-juv. (4)	*Desmodora*	*Trematostoma rotunda*	1–5	Along the entire body length
	Site CN	7	female (1)-male (4)-juv. (2)	*Pseudodesmodora*	*Thecacineta calix*	1–5	Along the entire body length
	Site CN	1 2%	male	*Sygmophoranema*	*Thecacineta calix*	1	Tail region
Gulf of Naples-shallow vent area Inactive site	Site CS	6	female (1)-male (4)-juv. (1)	*Desmodora*	*Paracineta homari*	1–5	Along the entire body length
	Site CS	3	Juvenile	*Paradesmodora*	*Paracineta homari*	1	Middle body
	Site CS	1	Juvenile	*Perspiria*	*Loricophrya susannae* n. sp.	1	Middle body
	Site CS	2	male (1)-juv. (1)	*Pseudodesmodora*	*Thecacineta calix*	1–6	Along the entire body length
	Site CS	3	female (2)-juv. (1)	*Chromaspirina*	*Loricophrya susannae* n. sp.	1	Middle body
	Site CS	2	male	*Polysigma*	*Thecacineta calix*	1–2	Along the entire body length
	Site CS	1	female	*Prochaetosoma*	*Thecacineta calix*	1	Middle body
	Site CS	1 2%	female	*Perepsilonema*	*Thecacineta fumosae* n. sp.	1	Middle body
Gulf of Trieste	St. C1	1 0.2%	juvenile	*Pseudochromadora*	phoronts of Apostomatia	19	Along the entire body length
French Guiana-Mangrove forest Polluted area	St. 1	4	male (1)-female (3)	*Desmodora*	*Paracineta homari*	5–6	Along the entire body length
	St. 1	2 0.5%	female	*Desmoscolex*	*Loricophrya bosporica*	1	Tail region
French Guiana-Mangrove forest Medium polluted area	St. 2	15	male	*Desmodora*	*Loricophrya bosporica*	6	Along the entire body length
	St. 2	1 2.7%	female	*Spirinia*	*Cothurnia* sp.	2	Tail region
Archipelago of Maldives - coral reef	M1	1	male	*Paradesmodora*	*Loricophrya sivertseni and L. stresemanni*	2	After cloaca opening
Inner reef	M4	1	female	*Croconema*	*Thecacineta calix*	5	Intestine region and caudal region before anus
	M4	1 1%	male	*Paradesmodora*	*Thecacineta cothurnioides*	1	Second half of the body, before cloaca
Outer reef	M8	1	juvenile	*Desmodorella*	*Paracineta* sp.	1	First half of pharingeal region
	M9	1 1%	female	*Echinodesmodora*	*Thecacineta cothurnioides*	1	Second half of intestine region

3.1.1. Deep-Sea Pockmark: Madagascar Margin

The highest percentage (15%) of colonized nematodes were reported from the deep-sea pockmark and only from the station inside the pockmark. All basibiont nematodes belonged to the genus *Desmodora*, the most abundant genus found at that station [Sanchez et al., unpublished data] and most of them were females (35 on a total of 48 specimens). *Trematosoma rotunda* and *Paracineta homari* were the two epibiont suctorian species on *Desmodora* sp., they were found attached to the middle and tail/cloaca region (*P. homari*) and in the head region (*T. rotunda*) from 1 to 4 individuals. *T. rotunda* was already reported as epibiont on *Desmodora scaldensis* (Supplementary Material, Table S1A; [45]) inhabiting a mangrove system in India, while the suctorian species *Paracineta homari* (Supplementary Material, Figure S1A) was reported for the first time as ciliate epibiont on nematodes [27].

3.1.2. Secca delle Fumose Shallow Vent: Gulf of Naples

We found from 2% (inactive sites CN and CS) to 5% (active site G) of nematodes with epibionts from the shallow vent area Secca delle Fumose (Table 2). Both females and males were found to be basibionts for ciliates and a minor number of juveniles (16 on a total of 90 nematodes). Basibionts belonged to ten different nematode genera and three families (i.e., Desmodoridae, Draconematidae and Epsilonematidae). Desmodoridae was the most represented family accounting for eight genera (*Chromaspirina, Desmodora, Paradesmodora, Perspiria, Polysigma, Pseudochromadora, Pseudodesmodora* and *Sygmophoranema*), followed by Draconematidae (*Prochaetosoma*) and Epsilonematidae (*Perepsilonema*) with only one genus. Seven species of suctorian ciliates were identified, and three of them are new species, *Loricophrya susannae* n. sp., *Thecacineta fumosae* n. sp., and *Acinetopsis lynni* sp. n. (Table 2, see below for taxonomic descriptions). From Secca delle Fumose we also found new associations between nematodes and suctorians, indeed we reported for the first time *Thecacineta calix* epibiont on four nematode genera *Polysigma, Pseudodesmodora, Sygmophoranema* and *Prochaetosoma* (Supplementary Material, Figure S1B–E) and *Thecacineta fumosae* n. sp. on *Perepsilonema*. The number of epibionts on the basibiont varied from a minimum of 1 to a maximum of 12 and they were usually attached along the entire body length of the nematodes or in the middle–posterior part of the body. Interestingly, we found nematodes with ciliates from all investigated sediment layers from the top 1 cm until 10 cm depth and in particular at site G the highest number of basibiont nematodes from the deepest layer were reported.

3.1.3. Marine Protected Area: Gulf of Trieste

In the Gulf of Trieste, only one nematode (0.2%) showed epibionts along its body (Supplementary Material, Figure S1F). The specimen belonged to the genus *Pseudochromadora* (Desmodoridae) and a total of 19 putative phoronts of Apostomatia (Oligohymenophorea, Ciliophora) were attached to the nematode along the entire length of its body. This was the first finding of phoronts of Apostomatia as epibionts on nematode and neither in the recent review by Chatterjee et al. [27] mentioned the representative of Apostomatia was as a nematode epibiont.

3.1.4. Mangrove Forests: French Guiana

In the mangrove forests of French Guiana, nematodes as basibionts of ciliates ranged from 0.5% (polluted area) to 2.7% (medium polluted area) of total abundance. Two suctorian (*Paracineta homari* and *Loricophrya bosporica*) and one peritrich (*Cothurnia* sp.) ciliates were found on specimens of three nematode genera: *Desmodora, Spirinia* (Desmodoridae) and *Desmoscolex* (Desmoscolecidae) (Table 2; Supplementary Material, Figure S2A) in a number of 1 to 6 along the entire body length or on the tail region. The associations *Desmoscolex–L. bosporica* and *Spirinia–Cothurnia* were not new (Supplementary Material Table S1; [27]). Differently, the associations *Desmodora–L. bosporica* (Supplementary Material, Figure S2B) and *Desmodora–P. homari* (see above) were reported for the first time in the present study.

3.1.5. Coral Reefs: Archipelago of Maldives

Five nematode specimens (1%) belonging to four different genera (*Paradesmodora, Croconema, Desmodorella* and *Echinodesmodora*) and one family (Desmodoridae) were found as basibionts for suctorian ciliates (five different species) inhabiting the coral reefs of Maldives (Table 2). Also in this study case, the majority of basibiont–epibiont associations were new and never reported before in the literature (Supplementary Material Table S1; [20,27]): *Loricophrya sivertseni* [46] and *L. stresemanni* [46] were found as epibionts on the same specimen of *Paradesmodora* (Supplementary Material, Figure S2C); *Paradesmodora/Echinodesmodora–Thecacineta cothurnioides* (Supplementary Material, Figure S2D) and *Desmodorella–Paracineta* sp. The number of epibionts hosted varied from 1 to 5 and they were mainly attached to the posterior part of the body of adults.

3.2. Taxonomic Account of Ciliates: Systematic Position

Class SUCTOREA Claparède et Lachmann, 1859
Subclass EXOGENIA Collin, 1912
Order METACINETIDA Jankowski, 1978
Family PARACINETIDAE Jankowski, 1975
Genus *Loricophrya* Matthes, 1956
Loricophrya susannae n. sp. (Figure 2A,B)

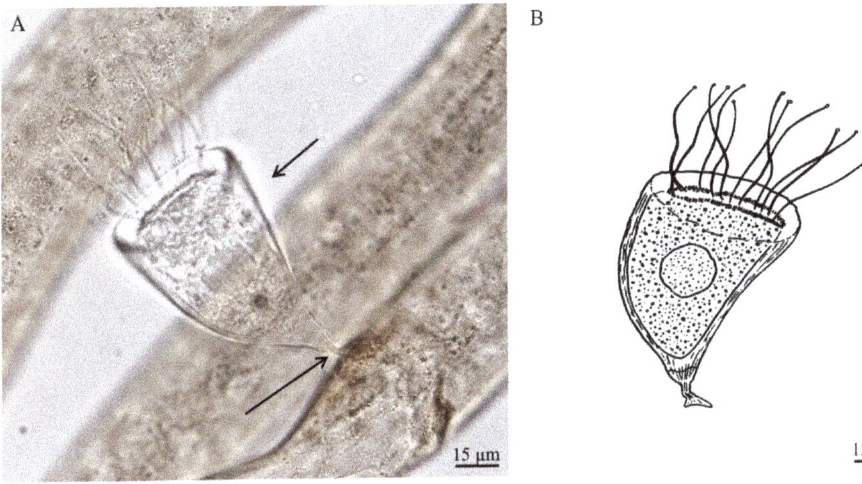

Figure 2. (**A**) Bright field microscopy image of *Loricophrya susannae* n. sp. on *Chromaspirina* from the shallow vent area of Secca delle Fumose (Gulf of Naples, Italy); (**B**) drawing of *Loricophrya susannae* n. sp. (present study). The black arrows indicating the thin, striated pseudostyle (below) and conical, bent stylotheca (above).

Etymology: The specific name is in memoriam of Prof Susanna De Zio for her basic contribution to the taxonomy of marine meiobenthic Tardigrada.

Diagnosis: Suctorian ciliate covered with conical, smooth, transparent, weakly bent, stylotheca. Pseudostyle short, thin, curved, longitudinally striated (Figure 2A). The cell body entirely covered by stylotheca, attached to the lorica in the mouth area. There are about 15 thin, flexile tentacles. Macronucleus ovoid, centrally located.

Morphological description: Suctorian ciliate covered with conical, smooth, transparent, weakly bent, lorica (stylotheca 70 × 44 µm). The stalk-like protuberance of lorica (pseudostyle 10 µm in length) is shorter than the lorica itself, thin, curved, provided with longitudinally striae, expanded in plate in zone of contact with substrate (Figure 2A,B). The cell body unflattened, entirely covered by stylotheca, attached to the lorica in the area of aperture. The cytoplasm colorless, contains some dark inclusions (Figure 2A). There are about 15 thin, flexile tentacles (22–57 µm in length) with characteristic terminal knobs. Tentacles are evenly distributed at the apical body surface. Macronucleus relatively large, spherical (18 × 15 µm), centrally located. Reproduction not observed.

Measurements, based on two individuals (in µm): Stylotheca length 70, maximal stylotheca width 44, pseudostyle length 10, pseudostyle diameter 1, epicone length 6, width 6, body length 52, body width 36, macronucleus length 18, width 15, tentacle length 22–57.

Differential diagnosis: Genus *Loricophrya* Matthes, 1956 include 11 species (12, [47]), however the new species is relative to *L. parva* (Schulz, 1932), from which differs by presence of thin, striated pseudostyle and conical (nor urn-like), bent, stylotheca. In addition, in all known representatives of the genus the cell body is not entirely covered by stylotheca.

Type material: Permanent microscopic slides of nematodes with the new suctorian species were deposited at REM/EEP/LEP of Ifremer Centre Brest, Brittany, France.

Type locality: Secca delle Fumose, Gulf of Naples, Italy.

Type host: *Chromaspirina* sp. and *Perspiria* sp.

Order VERMIGEMMIDA Jankowski, 1973
Family THECACINETIDAE Matthes, 1956
Genus *Thecacineta* Collin, 1909
Thecacineta fumosae n. sp. (Figure 3A,B)

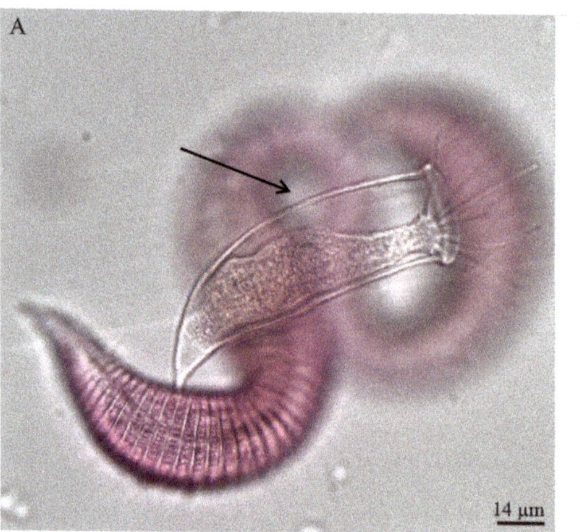
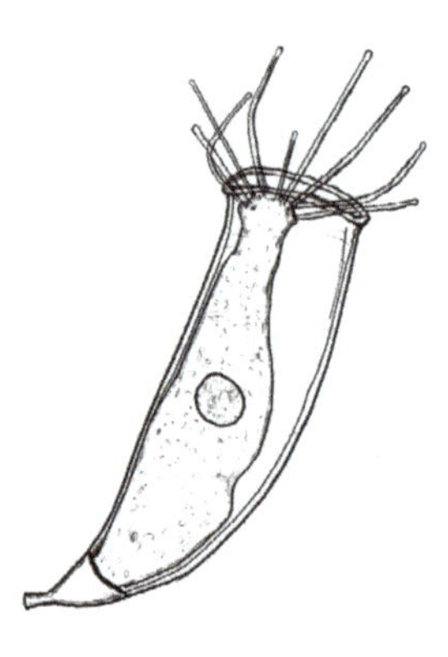

Figure 3. (**A**) Bright field microscopy image of *Thecacineta fumosae* n. sp. on *Perepsilonema* from the shallow vent area of Secca delle Fumose (Gulf of Naples, Italy); (**B**) drawing of *Thecacineta fumosae* n. sp. (present study). The black arrow indicating the transparent lorica.

Etymology: The species name refers to the locality name, Secca delle Fumose.

Diagnosis: Marine loricate suctorian. Cell body attached to the bottom of lorica. The apical part of body narrowed, not protruded from lorica aperture. Up to 12 capitate tentacles placed at apical body surface. Macronucleus spherical, located in the middle of body. Lorica slightly curved, smooth, transparent, without any ribs. The mouth of lorica is some wider than the rest. Stalk short, curved, with good developed, conical epicone, which have the same width as bottom of lorica (Figure 3A). Reproduction not observed.

Morphological description: Marine. Cell body granulated, colorless, attached to the bottom of lorica. The basal part of two thirds of the cell body flared, then body to become narrow, but with weakly enlarged apical part, bearing tentacles. The cell body not protruded from lorica aperture. Up to 12 capitate tentacles (23–33 µm in length) evenly distributed at apical body surface. Macronucleus ovoid, located in the middle of body (Figure 3B). Lorica (82 µm in length and 27 µm in width) slightly curved, smooth, transparent, without any ribs or striae (Figure 3A). The aperture of lorica (29 µm in diameter) is some wider than the rest, with somewhat arched annular edge. The stalk (14 µm in length) is clearly delimited from lorica, short, weakly curved, without any folds or striae, with good developed, conical epicone, which have the same width as bottom of lorica in the area of contact with them (Figure 3B).

Measurements, based on one individual (in µm): Lorica length 82, lorica width 27, lorica aperture diameter 29, body length 78, body width 20, stalk length with epicone 14, diameter 11, maximal epicone diameter 33, length of tentacles 23–33.

Differential diagnosis: The new suctorian species differs from the relative species as *Thecacineta cothurnioides* (Collin, 1909) recorded on harpacticoid copepod *Cletodes longicaudatus* from Banyuls-sur-Mer at Mediterranean coast of France [48] and nematode *Tricoma* sp. from Ratnagiri, west coast of India, Indian Ocean [49] and *Thecacineta urceolata* [26] found on nematode *Desmodora pontica* from Ludao, Taiwan by its curved lorica, stalk with wide, conical epicone. From other species of the genus the new species differs by its transparent lorica.

Type material: Permanent slide of the nematode with the new suctorian species was deposited stored at REM/EEP/LEP of Ifremer Centre Brest, Brittany, France.

Type locality: Secca delle Fumose, Gulf of Naples, Italy.

Type host: *Perepsilonema* sp.

Subclass ENDOGENIA Collin, 1912
Order ACINETIDA Raabe, 1964
Family ACINETOPSIDAE Jankowski, 1978
Genus *Acinetopsis* Robin, 1879

Acinetid ciliates with tentacles of two types: hypertrophied, agile, prehensile ones, and regular feeding (sucking) ones [50]. The body is trapezium-like, laterally flattened, loricate and stalked. The macronucleus is spherical or ovoid.

In accordance with Dovgal [17], there are three species of the genus: *Acinetopsis rara* [51] (type species), *A. tentaculata* [52] and *A. elegans* [53].

However, the species found of Swarczewsky [53] on gills of amphipod crustacean *Carinurus solskii* from Baikal Lake were not provided with lorica and thus must be excluded from the genus *Acinetopsis*. It is Jankowski's opinion [54] that the species is representative of genus *Tokophrya* Bütschli, 1899. *A. rara* (Figure 4(1)) was found [51] on hydroids from genus *Sertularia* collected near Concarneau (France). The ciliate is from 70 to 90 µm (in accordance with [51]), covered with stalked lorica, the height of which is one third less than its width. The stalk is very thin, 100 µm long. The body is uniformly granular, greyish, with a small contractile vacuole, a flat apical surface, from the centre of which one contractible tentacle extends. The body does not reach the bottom of the lorica. As observed by Grell and Meister [55] the *A. rara* feeds on representatives of genus *Ephelota* that is often far greater than the predator. The sucking tentacles of *A. rara* are much smaller and bear knobs devoid of haptocysts. The prehensile tentacles are ordinary in structure but, on the contrary, gigantic, very lively and enriched with haptocysts [55]. The original figure of Batisse [56] illustrates that, in addition to the central trapping tentacle, there are two groups of 12–13 thin, short sucking tentacles.

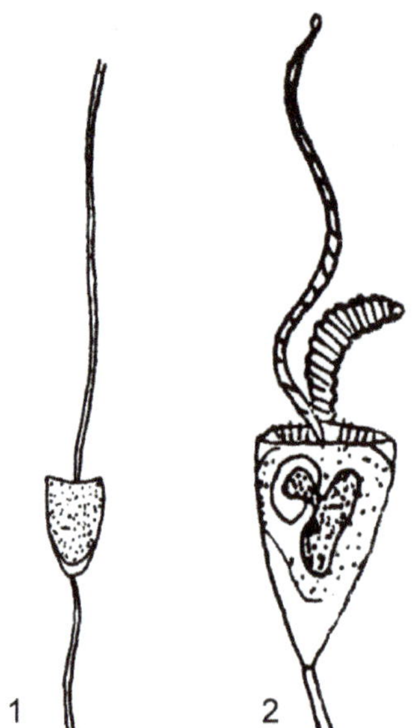

Figure 4. *Acinetopsis rara* (**1**) (modified from Robin, 1879) and *Acinetopsis tentaculata* (**2**) (modified from Root, 1922).

Acinetopsis tentaculata [52] (Figure 4(2)) has a body enclosed in a flattened, cup-shaped lorica, borne on a slender stalk, of which some are longer than the lorica. The body is irregularly flattened-ovoid in shape, bearing one or two agile prehensile tentacles and two groups of small sucking tentacles on its apical surface. The macronucleus is ovoid; there are also one or more micronuclei. A single contractile vacuole is located near base of the body. Reproduction occurs by endogenous budding.

Measurement (in μm, in accordance with Root, [52]): Lorica length 187, width 105, body length 138, width 100, thick 73, length of stalk 287, extended prehensile tentacle length 500.

Locality: Woods Hole, USA.

Host: *Ephelota coronata* Wright, 1858 found on hydroids *Obelia commissuralis* and *O. geniculata*.

Jankowski [54] believed that the *A. tentaculata* is a younger synonym of *A. rara*.

Acinetopsis lynni n. sp. (Figure 5A–D).

Etymology: The specific name is in honor of outstanding protistologist Denis Lynn (1947–2018).

Diagnosis: Suctorian ciliate enclosed in a flattened, narrow, elongate, smooth lorica with short extended upward stalk. Body attached to the bottom of lorica and fills about two thirds or three fourths of it. There are two agile hunting tentacles and from three to four short, contractile sucking tentacles. The macronucleus is ovoid, positioned near foot of the body. Reproduction occurs by endogenous budding with formation of single protomite (Figure 5A,B).

Morphological description: Marine suctorian ciliate enclosed in stalked lorica. The cell body is attached to the bottom of lorica, which fills about two thirds or three fours of it. Cytoplasm colorless and faintly granulated, with numerous inclusions (Figure 5B,C). The apical body surface, which bearstentacles, is noticeably concave downward. Lorica elongate (49–62 μm × 17–19 μm), somewhat expanded, slightly flattened, smooth, but with two or three annular striae on the inside of walls in the top. Stalk short (5–7 μm × 2–4 μm), straight, some extended upward, very weakly longitudinally striated, with adhesive disc. There are two extremely long (about the same length as body), agile hunting tentacles (27–46 μm) and from three to four short, contractile sucking tentacles (5–10 μm × 1–2 μm). Macronucleus ovoid, positioned near foot of the body. Reproduction by endogenous budding with the formation of single protomite. Protomite relatively large (24 μm × 11 μm), its length makes up more than half of the body length (Figure 5B–D).

Measurement, based on four individuals (in μm): Lorica length 49–62, lorica width 17–19, lorica aperture diameter 12–20, body length 36–41, body width 14–18, macronucleus diameter 7–9, stalk length 5–7, stalk diameter 2–4, prehensile tentacle length 27–46, sucking tentacle length 5–10, sucking tentacle diameter 1–2, protomite length 24, width 11.

Differential diagnosis: The new species differs from other representatives of the genus by narrow, elongated lorica, short stalks, and the presence of only three to four contractile sucking tentacles. In addition, the cell body of the new species does not fill all lorica.

Type material: Permanent slide of the nematode with the new suctorian species was deposited and stored at REM/EEP/LEP of Ifremer Centre Brest, Brittany, France

Type locality: Secca delle Fumose, Gulf of Naples, Italy.

Type host: *Desmodora* sp.

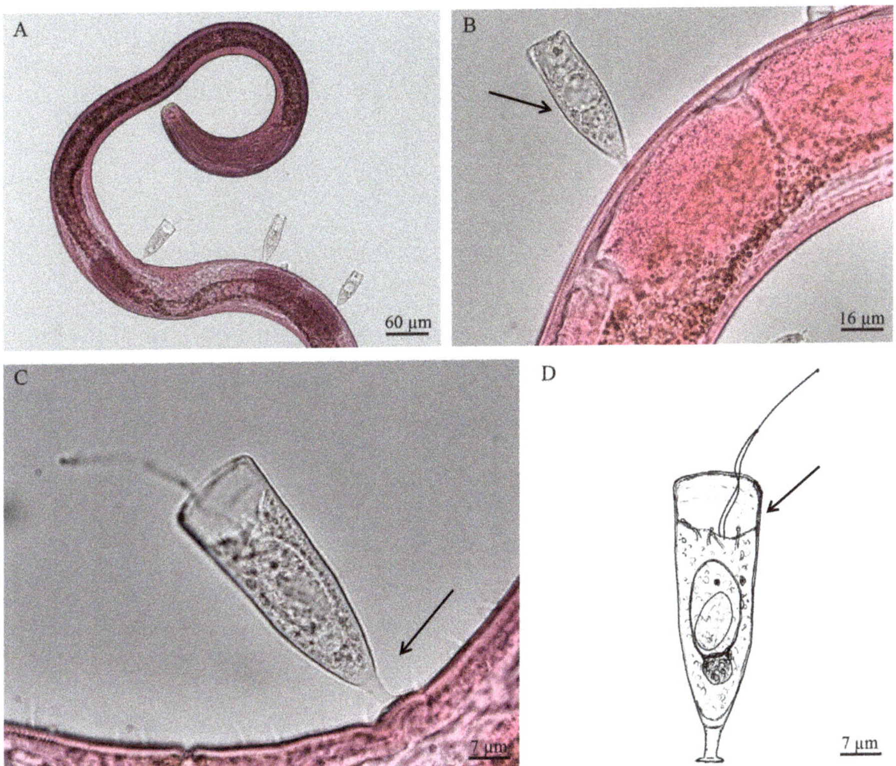

Figure 5. (**A–C**) Bright field microscopy images of *Acinetopsis lynni* n. sp. on *Desmodora* from the shallow vent area of Secca delle Fumose (Gulf of Naples, Italy); (**D**) drawing of *Acinetopsis lynni* n. sp (present study). The black arrows indicating the narrow, elongated lorica (**B**), the short stalk (**C**) and presence only a three to four contractile sucking tentacles (**D**).

3.3. Nematode-Ciliate Association: An Analysis

A total of 22 species of epibionts—among these 20 suctorian ciliates, 1 peritrich ciliate and 1 apostome ciliate—have been reported until now to infest different species of nematodes inhabiting deep-sea and shallow systems worldwide (Figure 1; Supplementary Material, Tables S1 and S2). Figure 6 documents the number of nematode basibionts (n. Genus and n. Species) that each epibiont is able to infest along with the number of habitats (e.g., mangrove forests, coral reef, meadows, open slope systems, sandy beach, etc.) where those associations were recorded. The graph was built up considering all the available literature and the new records from the present study. *Thecacineta calix* was the suctorian ciliate with the wider distribution both in term of basibiont genera/species and kind of habitats, followed by *Trematosoma rotunda*, *Thecacineta cothurnioides*, *Loricophrya bosporica* and *Paracineta homari*. All the other ciliates were characteristic of 1 to 2 nematode species or genera and recovered from 1 to 2 habitats at most maximum.

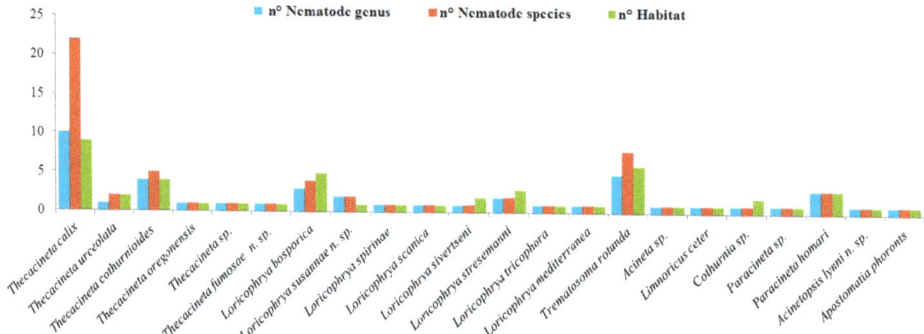

Figure 6. Number of nematode basibionts (n. Genera and n. Species) for each epibiont ciliate and number of habitats (n. Habitat) from where these associations have been reported (data from the literature and present study).

A total of 33 identified species, 23 genera and 5 families of nematodes, have been reported to be ideal basibionts for ciliates (Supplementary Material, Tables S1 and S3). The family with the highest number of nematode basibionts was that of Desmodoridae, accounting for 27 identified species, followed by Epsilonematidae (3), Desmoscolecidae (2), Comesomatidae (1) and Draconematidae (1). Figure 7 analyzes the nematode–ciliates association, considering the number of epibiont species per basibiont and the number of habitats where the association was recorded. There were some nematode genera/species that clearly constituted a more attractive alive substrate for the epibionts (i.e., they can be colonized by different ciliate species), such as *Desmodora* sp., *Paradesmodora* sp., *Spirinia parasitifera*, *Chromaspirina* sp., *Pseudochromadora* sp., and *Tricoma* sp. All the other nematodes were found to be colonized by 1 to 2 ciliate species. The nematode genera that could be basibiont for the higher number of epibiont species, were reported also from a wide range of habitat types. Moreover, if we grouped all the habitats in which these associations have been reported in three main categories: deep-sea/extreme environments, polluted/impacted and pristine environments (see also Table 2 and Table S1 for details), most of the records were found in pristine areas (e.g., sandy beaches, coastal waters, seagrass meadows, coral reefs) and deep-sea/extreme environments (e.g., hypoxic/anoxic sediments, seamounts, open slopes, pockmark areas, hydrothermal vents, deep sediments from Antarctica). Only a minority of cases were reported from polluted systems (e.g., mangrove forests, sandy beaches near domestic sewage). The graphs presented in Figures 6 and 7 were realized according to the available literature and our new records.

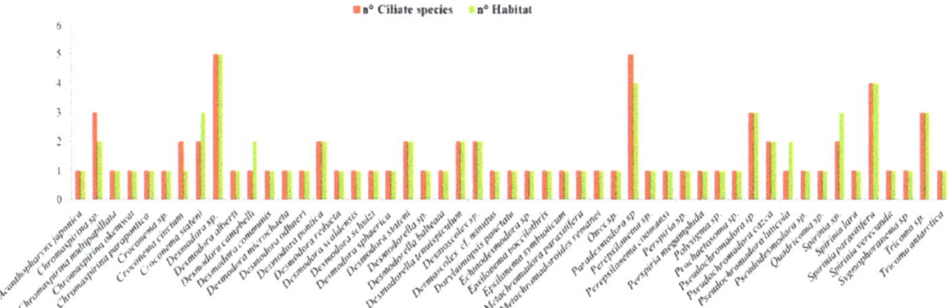

Figure 7. Number of epibiont species for each basibiont and the habitat distribution (n. Habitat) from where these associations were reported (data from the literature and present study).

4. Discussion

4.1. Nematodes as 'Promoters' of Biodiversity

In the present study, we reported three new suctorian species—i.e., *Acinetopsis lynni* n. sp., *Loricophrya susannae* sp. n. and *Thecacineta fumosae* sp. n.—as epibionts of four different basibiont nematode genera (i.e., *Desmodora*, *Perspiria*, *Chromaspirina* and *Perepsilonema*). Moreover, we also reported many new interactions between nematodes and ciliates never documented before [27]. In detail, we found the suctorian ciliates *Paracineta homari* and *Paracineta* sp. as epibionts of nematodes from different environments. Until now, *P. homari* and species of the genus *Paracineta* were reported as epibionts of crustaceans such as pagurid crabs [50] and copepods [57]. The same for the representative of subclass Apostomatia, reported here for the first time as epibiont of *Pseudochromadora* from the Gulf of Trieste, but until now found mainly on crustaceans [58].

We also documented the presence of well-known ciliate epibionts of some nematode genera and/or species colonizing new nematode basibionts. This was the case of the widespread suctorian *Thecacineta calix* [20], found here as epibiont on four new nematode basibionts, i.e., *Polysigma*, *Pseudodesmodora*, *Sygmophoranema* and *Prochaetosoma* inhabiting the shallow vent area in the Gulf of Naples. This finding confirmed the ubiquitous nature of this ciliate adapted to different environments and its ability to colonize many hosts. *Loricophrya bosporica*, previously reported on the nematodes *Desmoscolex* cf. *minutus* and *Metachromadoroides remanei* [59,60] inhabiting the polluted anoxic Black Sea sediments and methane seeps, was reported here on *Desmodora* from the polluted mangrove forests in French Guiana. *Loricophrya bosporica* defined as extremophile [27], showed a wider range of distribution and number of suitable nematode basibionts. The same consideration could be made for two other species of *Loricophrya*: *L. sivertseni* and *L. stresemanni*. We found both species sharing the same micro-niche on the tail of one specimen of *Paradesmodora* from the Maldivian coral reef. These two ciliates were reported from Norwegian fjords on *Spirinia parasitifera* (*L. stresemanni*) and on other nematodes that were not identified (*L. sivertseni*) [25,27].

Thecacineta cothurnioides was previously reported on *Chromaspirina* sp., *C. parapontica* and *Tricoma* sp. nematodes [21,22,53] from shallow continental shelf and polluted mangrove systems and on copepods from the Mediterranean coast of France [20]. In the present study, *T. cothurnioides* was found as epibiont on *Paradesmodora* and *Echinodesmodora* inhabiting the Maldivian coral reef broadening the geographical range of distribution of this ciliate and their suitable hosts.

Previous findings from the existing literature and new records from the present study, showed evidences that nematodes are 'promoters' of biodiversity for ciliates, a biodiversity of which we have only a small perception since it remains largely underestimated. Sartini et al. [61] suggested that freshwater snails could be regarded as a source of biodiversity in limnic environments, since with their shells they offer a range of microhabitats that peritrich ciliates can occupy. Moreover, freshwater gastropods show tolerance to a wide range of environmental conditions and they are commonly found inhabiting many freshwater systems. Similarly, nematodes constitute a suitable substratum colonized by new ciliates species and in addition to that, they support new nematode–ciliate associations in different habitats worldwide. Nematodes are good basibiont candidates, being among the most abundant, diversified, and ubiquitous meiofauna groups, able to colonize many environments, from shallow to deep-sea waters and survived in adverse and extreme conditions [23].

It is known that the basibiont highly enhances epibiont dispersal, being a motile living substratum for sessile marine organisms, and that an epibiont actively selects the colonized species and the attachment site on the basibiont [62]. Indeed, epibiotic associations represent an excellent framework with which to examine diversity patterns among geographical regions on a variety of scales [1].

4.2. Advantages and Disadvantages of Nematode–Ciliate Association

The debate is still open on whether for nematodes to be promoters of biodiversity is more of a cost than a benefit. Previous studies conducted mainly on crustaceans as basibionts for various epibionts (e.g., [1,3,15,63]), analyzed in detail costs and benefits for both the actors involved in this association (Table 3).

Epibiosis has a number of effects on both the epibiont and the basibiont, and these include advantages for the epibiont such as: (i) new available surface to colonize; (ii) dispersal and geographical expansion; (iii) increase in the supply of nutrients; (iv) protection against predation [3]. In general, the epibiont must be able to cope with all aspects of the basibiont lifestyle and surface properties, and, frequently, older individuals or older parts of an individual tend to be more heavily covered by epibionts [1].

On the other hand, epibiosis can be disadvantageous for the epibiont since the basibiont is a biologically unstable, variable, and non-durable living organism [1]. Epibionts can be: (i) removed from the basibiont by abrasion or molt, (ii) exposed to some detrimental host defense, (iii) exposed to inadequate environmental conditions, and (iv) captured by some basibiont predator [1,3,15].

The nature of epibiont impact on the basibiont is variable and strong context-specific [15]. In some way, epibiosis can be beneficial for the basibiont by providing: (i) both mimetic protection and cleansing, (ii) defense mechanisms against predators, and (iii) nutrient flow [3]. On the other hand, epibiosis disadvantages for the basibiont are numerous (Table 3). Epibionts may: (i) restrict the mobility of the basibiont, (ii) affect growth and molting, (iii) affect the functions of several organs (e.g., eyes, gills), (iv) cause an increase in predation risk, (v) compete for nutrients with basibiont, and (vi) mask the chemical identity of the host [1,3,15]. Wahl [5] sustained that epibiosis acts as an ecological lever by amplifying or buffering the basibiont–environment interactions (i.e., amplifying or buffering biotic and abiotic stress). The way the epibiosis influences an interaction may be of two dissimilar kinds: one called *exploitative* (i.e., epibiont exerts a stress on the basibiont) and the other called *interference mediation* (i.e., the susceptibility of a basibiont may be increased or decreased according to the identity or quantity of epibionts) [5]. Harder [15], reporting some examples among basibionts such as algae, molluscs, cnidaria or echinoderms, stated that any potential basibiont must either tolerate epibiosis or employ some sort of defense against this phenomenon.

Could all these costs vs. benefits listed above be applicable to nematode–ciliate association? If we consider the costs for the basibiont, they can be all potentially valid for nematodes particularly when epibiont ciliates exceed a certain number (e.g., Supplementary Material Figure S1F) and/or the size of a single epibiont is almost equal to the size of the nematode (e.g., Supplementary Material Figure S2A). Under those conditions, every vital activity of the basibiont may became difficult, for instance: motility, mating, gas exchange and perception of the external environment, reduction in defense and competition capacities. This is the case of the *exploitative* nature of the epibiosis [5].

Harder [15] suggested a variety of beneficial effects for the basibiont induced by the presence of epibionts (Table 3); however, we think they are hardly applicable in the case of nematodes. The positive effects of camouflage, or of epibiont mechanisms against predators and associational resistance may be effective in the presence of a certain number of epibionts covering the basibiont surface; however, in the case of nematodes (organisms usually smaller than 0.5 mm) higher is the number of epibionts greater is the negative effect for the host (= *interference mediation*; [15]). Our studied case of colonized nematodes from polluted sediments of mangrove system (French Guiana) could be an example of ecological lever. Under an environmental stress condition, the vulnerability of the basibionts favored the epibionts' colonization amplifying the abiotic stress. Ansari and Bhadury [21] reported a high number of infested nematodes inhabiting polluted mangrove systems (India). The authors explained that in such stress conditions nematodes were easily colonized due to a defense lowering, while for the epibionts, this association was even more advantageous for survivors. The overall feeling is that, for nematodes, the cost of epibiosis exceeds the benefits.

Table 3. Costs vs. benefits in the epibiont–basibiont relationship of epibiosis (summarized from Key et al., 1999; Harder, 2008; Wahl, 2009; Fernandez-Leborans, 2010).

(A) Epibiont Effects on Basibiont	
Cost	**Benefit**
Change of properties of basibiont surface into basibiont–epibiont–water interface	Epibiont mechanisms against predators that can serve concurrently for the basibiont
Increase friction with the water = decrease basibiont velocity	Associational resistance
Increase predation risk by modifying chemical signals and by increasing the basibiont weight = motility reduction	Camouflage Nutrient flow from epibiont and vitamins
Decrease the visus of the basibiont and cause wounds on the cuticle	Ciliate epibionts adhere to external cuticular layer of the basibiont with no effect and without causing any arm = neutral effect
A high infestation makes difficult for the host to maintain the position into the water and sediment	
Harmful energetic cost especially when energy conditions are limited	
Decreased reproduction, growth and survivorship: epibiont can be a hindrance to mating	
Starvation for food competition with epibiont in the case of sharing of feeding source(s)	
Increase basibiont vulnerability to infections and energy cost for locomotion	
Diminish gas exchanges if they colonize gas exchange surface	
Epibionts as cause for stress on basibiont by compromising the capacity of defense and competition	
(B) Basibiont Effects on Epibiont	
Cost	**Benefit**
Exposure to inadequate environmental conditions	Available surface to colonize on which to live
Removing of epibiont by abrasion or by moult	Shock-absorbing substratum
Capture of the basibiont by predator	Dispersion of epibiont mobile life stage and gene flow
Competition with other ciliates attached to the same basibiont or with conspecific when abundant	Free transport for the epibiont
Shared doom with basibiont	Protection against predators
Exposure to detrimental host defense	Favorable hydrodynamic conditions Increasing in availability of nutrients for the epibiont

4.3. The Cost for Biodiversity

Previous authors (e.g., [22,26,64]) suggested that the success of nematodes as basibionts was related to some of their morphological and physiological features coupled with their lifestyle, to which the number and type of epibiont ciliates are related.

Among their morphological features, a thick, heavily ornamented or annulated cuticle was reported as the primary characteristic to be a suitable basibiont nematode for ciliates [27,65]. This characteristic presents in certain nematodes body surface shows a similitude with the calcified body surface of many crustaceans, usually reported as good habitat for epibionts [8,16].

Similarly, all nematode genera we found with ciliate epibionts presented the typical thick ornamented cuticle. We reported nematodes usually found as basibionts for ciliates (e.g., *Desmodora, Desmoscolex, Perepsilonema*) and new colonized genera (e.g., *Polysigma, Pseudodesmodora, Sygmophoranema* and *Prochaetosoma*) increasing the number of 'available' nematode basibionts and the perception that this number can be even higher.

The importance of the cuticle for the epibiont settlement was confirmed once we considered the Secca delle Fumose (Gulf of Naples) case study. We reported colonized nematodes from all sites, with the only exception being the active site H, dominated by *Oncholaimus* and followed by *Daptonema*. Both these nematodes presented a thin, smooth or weakly striated cuticle. It is also true that site H presented the highest temperature and lowest pH values, conditions that might be conducive to the survival of ciliates. Nevertheless, is well known the existence of suctorian ciliates that can survive under extreme conditions [27,59]. Indeed, we think that it was the absence of 'good' basibionts, instead of the environmental conditions, that consequently determined the absence of the epibionts [8].

If the properties of the body surface play a crucial role in most interactions between nematodes and ciliates, for nematodes it represents one of the highest prices to pay—particularly when the epibionts are of a conspicuous number. In nematodes, the cuticle is involved in many vital functions: exchange of gasses and nutrients, info-chemicals and defense metabolites, mediation of many processes of recognition of an organism by a partner, a parasite, an epibiont or predator, transmission of many types of biotic and abiotic stress [66]. The properties and functions of this interface may be modified substantially by the presence and activities of epibiotic communities, with consequences for its interaction with the environment and for the relative fitness of the host organism [5]. In our samples, we reported a number of epibionts per nematode ranging from 1 to 19. If we consider that all our nematodes showed a length less than 500 μm and ciliates' length was ≥30–40 μm, it is reasonable to think that in a number of 2–3 epibionts start being a cause of stress for the basibiont. In a number exceeding 5–6 epibionts, we think that the condition of stress was even increased due to a significant increase in the energetic demand of individuals for locomotion [8].

At the cuticle level, there is mucus production that characterizes many of the basibiont nematodes. Mucus production and the release of other physiological secretions/excretions enhance the attachment of bacteria and food particles on the basibiont body surface and ciliates may take advantage of this available direct or indirect food source [25]. For instance, the desmoscolecid nematodes, in adults, possess an annulated body cuticle covered with a fine granular substance (which may be a secretion) and imbedded mineral particles and aggregations of bacteria [67]. Nevertheless, the use of bacteria and particles as food sources for ciliates may be strictly linked to the feeding strategy of the epibiont. Since the majority of ciliates found as epibionts on our nematodes were predator suctorian ciliates, the theory of ciliates feeding on the nematode body surface may be hardly attributed to these epibionts [18,27]. What is certain is that the attachment on a living mobile body surface—like that of nematodes—highly facilitates the ciliate feeding [3,5].

The position of the epibiont may be linked to several factors, such as burrowing behavior, locomotion, the presence of chemical inductors, bacterial exudates, pH and microtexture [5]. Moreover, the area along the body devoted to the secretions/excretions release such as cloaca, anus and vulva regions has been found to attract the epibionts and often their settlement occurred around those body regions (e.g., [58]). It has been hypothesized that these attachment preferences may be connected with the mode of reproduction of some ciliates (i.e., vermigemmic budding) with a migratory stage unable to swim [27,58]. We found that half of the studied basibiont nematodes were colonized on the posterior body region (anus-cloaca) and on the vulva region; however, in all the other cases, ciliates were attached everywhere on the body as previously reported [21,26,64]. We also noticed that the majority of basibionts were adult nematodes, as reported from other authors (e.g., [26,68]), which may be explained by an epibiont strategy to avoid molt and with no distinction between sex. The role of gender and maturity stage (adult vs. juvenile) in the selection criteria by an epibiont remains yet to be clarified [21].

As for the living mode of nematodes and their ability to penetrate into the sediment, Chatterjee et al. [27] reported that available data on epibiont ciliates on nematodes living in the deeper sediment layers are very few. Surprisingly, we found colonized nematodes until the deepest layers (5–10 cm) in a number comparable with colonized nematodes inhabiting the upper sediment layers. However, in this case, we can hypothesize a damage for the epibionts due to abrasion—a mechanical defense [3] probably also adopted by nematodes against the epibionts.

Mikac et al. [7] defined the association between peritrich ciliates and polychaetes as ectocommensalism—where the ciliates have the advantages of increased food availability but the polychaetes do not have any benefits and are not harmed. Authors stated that the major advantage which ciliates gain from being associated with a motile substratum is increased food availability, assured by the free transport to a variety of habitats, and by increased water flow. Similarly, Key et al. [63] defined the relationship between blue-crab and bryozoa as 'phoretic'. Phoresis, or phoresy, it is used to describe a non-permanent, commensalistic interaction in which one organism attaches itself to another (the host) solely for the purpose of travel. Actually, we think that the ciliate–nematode relationship cannot be defined, neither as ectocommensal nor phoretic, since the cost for carrying this diverse population of ciliates appears high for the basibiont. We are aware that further investigations are needed to analyze the effects for the basibionts in depth.

Very little is known about nematode mechanisms of defense. As mentioned above, we can suppose that abrasion may be a strategy to remove epibionts when they move deeper into the sediment. Only recently, new researches have started and are developing, focused on the nematode immune defenses [69]. We know that nematodes can produce several classes of AMPs (antimicrobial peptides) as natural response to fungal, bacteria and yeast attack [70]. Moreover, some AMPs' gene expressions were specifically found at the epidermis level to avoid the infection of certain spores [69]. Many nematodes produce AMPs, but neither their activity nor relative gene expression have been investigated for most of them. Understanding how different nematode species defend themselves against potential pathogens in their environment(s) requires the characterization of the defense molecules from each species [70]. The presence of AMPs involved in the defense against ciliate epibionts is quite probable, but for now with no specific, supporting data.

4.4. Host–Epibiont Species–Specificity and the Environment

Epibiotic relationships are rarely species-specific [1], nevertheless, that is not confirmed for polichaete–peritrich ciliate relationships since species-specific relationships were documented [7,13]. Results of the present study and of the literature highlighted that the majority of ciliates have been reported as epibiont on only one genus and/or species of nematode and with a limited range of distribution (Figure 6). Nevertheless, eight species of ciliates were found colonizing different nematode species, and *Thecacineta. calix* confirmed its ability to colonize a wide number of hosts also among nematodes (22 species of nematodes) and to inhabit different kind of environments [20]. Talking about a host species-specificity may also be premature at the moment in cases of ciliates found only on one nematode, considering the scarcity of data. However, we can confirm that most of the suctorian species we report in this study were found only on nematodes [27], with the only exceptions being *T. calix* and *Paracineta homari*. At the genus level, we reported two ciliates that were also found on other hosts (i.e., crustaceans): *Acineta*, and *Cothurnia* [8] even with different species.

Conversely, looking at a possible epibiont species-specificity (Figure 7) for the basibiont, we reported the same result: most nematodes were found to be basibiont only for one epibiont species and only a few nematodes could recruit different epibionts in different environments. It has been proved that processes of colonization are mediated by specific settlement cues (e.g., surface features) and biogenetic signals from the basibiont [15].

Environmental conditions are also important in favoring (or not favoring) epibiotic relationships [5], since they can determine the presence or absence of the host and/or of the epibiont, as well as enhance the colonization process. From this research, we reported nematode–ciliate relationships from extreme shallow-water and deep-sea environments to impacted and pristine systems, and spanning different geographical areas: the results support the idea that this association might be highly common and diversified in nature. Our study added three new species of suctorian ciliates epibionts on nematodes, pointing out nematodes as a living biotic substrate source of diversity.

In conclusion, our analysis and results suggest that the nematode–ciliate relationship deserves further investigation to elucidate all aspects of this association: diversity, costs and benefits for the basibiont and epibiont, and the ecological meaning of this phenomenon. Subsequently, we can think to apply the epibiosis in the environmental monitoring.

Supplementary Materials: The following materials are available online at http://www.mdpi.com/1424-2818/12/6/224/s1, Figure S1: Bright field microscopy image of (A) *Paracineta homari* on *Desmodora* sp. from deep-sea pockmark (Madagascar margin); (B) *Thecacineta calix* on *Polysigma* sp. from the shallow vent area of Secca delle Fumose (Gulf of Naples, Italy); (C) *Thecacineta calix* on *Pseudodesmodora* sp. from the shallow vent area of Secca delle Fumose (Gulf of Naples, Italy); (D) *Thecacineta calix* on *Sygmophoranema* sp. from the shallow vent area of Secca delle Fumose (Gulf of Naples, Italy); (E) *Thecacineta calix* on *Prochaetosoma* sp. from the shallow vent area of Secca delle Fumose (Gulf of Naples, Italy); (F) *Apostomatia* on *Pseudochromadora* sp. from the Gulf of Trieste. Figure S2: Bright field microscopy image of (A) *Loricophrya bosporica* on *Desmoscolex* sp. from mangrove forests (French Guiana); (B) *Loricophrya bosporica* on *Desmodora* sp. from mangrove forests (French Guiana); (C) *Loricophrya sivertseni* and *L. stresemanni* on *Paradesmodora* sp. from Maldivian coral reefs; (D) *Thecacineta cothurnioides* on *Echinodesmodora* sp. from Maldivian coral reefs. Table S1: Summary of all documented nematode–ciliate associations from the available literature. Table S2: List of epibiont ciliates of nematodes. Listed are the number of nematode genera and species colonized by each ciliate and the number of habitats where these associations were documented. Table S3: List of basibiont nematodes. Listed are the number of ciliate species found to be epibiont for each nematode and the number of habitats where these associations were documented.

Author Contributions: Conceptualization E.B., I.D., F.S.; methodology E.B., I.D., A.A., C.M., A.F., E.G., L.C., F.S.; writing E.B., I.D.; revisions E.B., I.D., D.Z., A.F., E.M., L.G., M.B., R.S., F.S. All authors have read and agreed to the published version of the manuscript.

Funding: Financial support was provided by the project "Prokaryote-nematode Interaction in marine extreme envirONments: a uniquE source for ExploRation of innovative biomedical applications" (PIONEER) funded by the Total Foundation and IFREMER (2016–2019). The work of second (ID) and fourth (AA) authors was made within the framework of research issue of A.O. Kovalevsky Institute of Biology of the Southern Seas #AAAA-A19-119060690014-5. French Guiana sampling and working were funded by "Office de l'Eau de la Guyane" (OEG) and "Agence française pour la biodiversité" (AFB). The authors are grateful to the working group "Mangroves" and the LEEISA laboratory (Cayenne) for its help in the fieldwork made within the European Water Framework Directive.

Acknowledgments: We thank the chief scientists, the scientific party and the crew during the cruises PAMELA-MOZ01 cruise; in the framework of the Passive Margin Exploration Laboratories (PAMELA) project, funded by TOTAL and Ifremer. The authors EB, DZ and RS are grateful to Soprintendenza of the Underwater Archeological Park of Baia (Gulf of Naples) (prot. 5667, 24/10/2016) for the authorization to sampling.

Conflicts of Interest: The authors declare no conflict of interest.

References

1. Ecological Studies. *Marine Hard Bottom Communities*; Wahl, M., Ed.; Springer: Berlin/Heidelberg, Germany, 2009; Volume 206, p. 61. [CrossRef]

2. Wahl, M.; Hay, M.E.; Enderlein, P. Effects of epibiosis on consumer-prey interactions. In *Interactions and Adaptation Strategies of Marine Organisms. Proceedings of the 31st European Marine Biology Symposium, held in St. Petersburg, Russia, 9–13 September 1996*; Kluwer Academic Publishers: New York, NY, USA, 1997; Volume 355, pp. 49–59. [CrossRef]

3. Fernandez-Leborans, G. Epibiosis in Crustacea: An overview. *Crustaceana* **2010**, *83*, 549–640. [CrossRef]

4. Fernandez-Leborans, G. Ciliate—Decapod epibiosis in two areas of the north-west Mediterranean coast. *J. Nat. Hist.* **2003**, *37*, 1655–1678. [CrossRef]

5. Wahl, M. Ecological lever and interface ecology: Epibiosis modulates the interactions between host and environment. *Biofouling* **2008**, *24*, 427–438. [CrossRef] [PubMed]

6. Wahl, M.; Mark, O. The predominantly facultative nature of epibiosis: Experimental and observational evidence. *Mar. Ecol. Prog. Ser.* **1999**, *187*, 59–66. [CrossRef]

7. Mikac, B.; Semprucci, F.; Guidi, L.; Ponti, M.; Abbiati, M.; Balsamo, M.; Dovgal, I. Newly discovered associations between peritrich ciliates (Ciliophora: Peritrichia) and scale polychaetes (Annelida: Polynoidae and Sigalionidae) with a review of polychaete–peritrich epibiosis. *Zoöl. J. Linn. Soc.* **2019**, *188*, 939–953. [CrossRef]

8. Fernandez-Leborans, G.; Chatterjee, T.; Grego, M. New records of epibiont ciliates (Ciliophora) on Harpacticoida (Copepoda, Crustacea) from the Bay of Piran (Gulf of Trieste, Northern Adriatic). *Cah. Biol. Mar.* **2012**, *53*, 53–63.

9. Chiavelli, D.A.; Mills, E.L.; Threlkeld, S.T. Host preference, seasonality, and community interactions of zooplankton epibionts. *Limnol. Oceanogr.* **1993**, *38*, 574–583. [CrossRef]

10. Davis, A.; White, G.A. Epibiosis in a guild of sessile subtidal invertebrates in south-eastern Australia: A quantitative survey. *J. Exp. Mar. Biol. Ecol.* **1994**, *177*, 1–14. [CrossRef]

11. Barnes, D.K.A.; Clarke, A. Epibiotic communities on sublittoral macroinvertebrates at Signy Island, Antarctica. *J. Mar. Biol. Assoc. UK* **1995**, *75*, 689–703. [CrossRef]

12. Cook, J.A.; Chubb, J.C. Veltkamp Epibionts of *Asellus aquaticus* (L.) (Crustacea, Isopoda): An SEM study. *Freshw. Biol.* **1998**, *39*, 423–438. [CrossRef]

13. Magagnini, G.; Verni, F. Epibiosis of *Scyphidia* sp. (Ciliophora, Peritrichida) on *Nerilla antennata* (Archiannelida, Nerillidae). *Boll. Zoöl.* **1988**, *55*, 185–189. [CrossRef]

14. Ansari, K.G.M.T.; Guidi, L.; Dovgal, I.; Balsamo, M.; Semprucci, F. Some epibiont suctorian ciliates from meiofaunal organisms of Maldivian archipelago with description of a new ciliate species. *Zootaxa* **2017**, *4258*, 375. [CrossRef] [PubMed]

15. Harder, T. Marine Epibiosis: Concepts, Ecological Consequences and Host Defence. In *Marine and Industrial Biofouling*; Springer: Berlin/Heidelberg, Germany, 2009; pp. 219–231.

16. Chatterjee, T.; Fernandez-Leborans, G.; Ramteke, D.; Ingole, B.S. New records of epibiont Ciliates (Ciliophora) from Indian coast with descriptions of six new species. *Cah. Biol. Mar.* **2013**, *54*, 143–159.

17. Dovgal, I.V. Evolution, phylogeny and classification of Suctorea (Ciliophora). *Protistology* **2002**, *2*, 194–270.

18. Chatterjee, T.; Fernandez-Leborans, G.; Chan, B.K.K. New record of ciliate *Thecacineta calix* (Ciliophora: Suctorea) epibiont on *Agauopsis halacarid* mite (Acari, Halacaridae) from Taiwan. *Scr. Sci. Nat.* **2012**, *2*, 121–127.

19. Dovgal, I.; Chatterjee, T.; Ingole, B. An overview of Suctorian ciliates (Ciliophora, Suctorea) as epibionts of halacarid mites (Acari, Halacaridae). *Zootaxa* **2008**, *1810*, 60–68. [CrossRef]

20. Chatterjee, T.; Nanajkar, M.; Dovgal, I.; Sergeeva, N.; Bhave, S. New records of epibiont *Thecacineta calix* (Ciliophora, Suctorea) from the Caspian Sea and Angira Bank, Arabian Sea. *Cah. Biol. Mar.* **2019**, *60*, 445–451.

21. Ansari, K.G.M.T.; Bhadury, P. Occurrence of epibionts associated with meiofaunal basibionts from the world's largest mangrove ecosystem, the Sundarbans. *Mar. Biodivers.* **2016**, *47*, 539–548. [CrossRef]

22. Bhattacharjee, D. Suctorian epibionts on *Chromaspirina* sp. (Nematoda: Desmodoridae) from the shallow continental shelf of the Bay of Bengal, northern Indian Ocean. *Mar. Biodivers. Rec.* **2014**, *7*, 1–3. [CrossRef]

23. Zeppilli, D.; LeDuc, D.; Fontanier, C.; Fontaneto, D.; Fuchs, S.; Gooday, A.J.; Goineau, A.; Ingels, J.; Ivanenko, V.N.; Kristensen, R.M.; et al. Characteristics of meiofauna in extreme marine ecosystems: A review. *Mar. Biodivers.* **2017**, *48*, 35–71. [CrossRef]

24. Schratzberger, M.; Ingels, J. Meiofauna matters: The roles of meiofauna in benthic ecosystems. *J. Exp. Mar. Biol. Ecol.* **2018**, *502*, 12–25. [CrossRef]

25. Fernandez-Leborans, G.; Román, S.; Martin, D. A New Deep-Sea Suctorian-Nematode Epibiosis (*Loricophrya-Tricoma*) from the Blanes Submarine Canyon (NW Mediterranean). *Microb. Ecol.* **2017**, *74*, 15–21. [CrossRef] [PubMed]

26. Liao, J.-X.; Dovgal, I. A new *Thecacineta* species (Ciliophora, Suctorea) on *Desmodora pontica* (Nematoda, Desmodorida) from a seagrass bed in Taiwan. *Protistology* **2015**, *9*, 75–78.

27. Chatterjee, T.; Dovgal, I.; Fernandez-Leborans, G. A checklist of suctorian epibiont ciliates (Ciliophora) found on aquatic meiobenthic nematodes. *J. Nat. Hist.* **2019**, *53*, 2133–2143. [CrossRef]

28. Donnarumma, L.; Appolloni, L.; Chianese, E.; Bruno, R.; Baldrighi, E.; Guglielmo, R.; Russo, G.F.; Zeppilli, D.; Sandulli, R. Environmental and Benthic Community Patterns of the Shallow Hydrothermal Area of Secca Delle Fumose (Baia, Naples, Italy). *Front. Mar. Sci.* **2019**, *6*, 685. [CrossRef]

29. Baldrighi, E.; Zeppilli, D.; Appolloni, L.; Donnarumma, L.; Chianese, E.; Russo, G.F.; Sandulli, R. Meiofaunal communities and nematode diversity characterizing the Secca delle Fumose shallow vent area (Gulf of Naples, Italy). *PeerJ* **2020**, *8*, e9058. [CrossRef]

30. Franzo, A.; Guilini, K.; Cibic, T.; Del Negro, P. Interactions between free-living nematodes and benthic diatoms: Insights from the Gulf of Trieste (northern Adriatic Sea). *Mediterr. Mar. Sci.* **2018**, *19*, 538–554. [CrossRef]

31. Semprucci, F.; Balsamo, M. New records and distribution of marine free-living nematodes in the Maldivian Archipelago. *Proc. Biol. Soc. Wash.* **2014**, *127*, 35–46. [CrossRef]

32. Olu, K. PAMELA-MOZ01 *Cruise*. 2014. Available online: https://campagnes.flotteoceanographique.fr/campagnes/14001000/ (accessed on 1 March 2020). [CrossRef]

33. Fontanier, C.; Garnier, E.; Brandily, C.; Dennielou, B.; Bichon, S.; Gayet, N.; Eugene, T.; Rovere, M.; Grémare, A.; Deflandre, B. Living (stained) benthic foraminifera from the Mozambique Channel (eastern Africa): Exploring ecology of deep-sea unicellular meiofauna. *Deep Sea Res. Part I Oceanogr. Res. Pap.* **2016**, *115*, 159–174. [CrossRef]

34. Fontanier, C.; Mamo, B.; Toucanne, S.; Bayon, G.; Schmidt, S.; Deflandre, B.; Dennielou, B.; Jouet, G.; Garnier, E.; Sakai, S.; et al. Are deep-sea ecosystems surrounding Madagascar threatened by land-use or climate change? *Deep Sea Res. Part I Oceanogr. Res. Pap.* **2018**, *131*, 93–100. [CrossRef]

35. Kench, P.S.; Brander, R.W. Wave Processes on Coral Reef Flats: Implications for Reef Geomorphology Using Australian Case Studies. *J. Coast. Res.* **2006**, *221*, 209–223. [CrossRef]

36. Semprucci, F.; Colantoni, P.; Baldelli, G.; Rocchi, M.B.; Balsamo, M. The distribution of meiofauna on back-reef sandy platforms in the Maldives (Indian Ocean). *Mar. Ecol.* **2010**, *31*, 592–607. [CrossRef]

37. Semprucci, F.; Frontalini, F.; Losi, V.; Du Châtelet, E.A.; Cesaroni, L.; Sandulli, R.; Coccioni, R.; Balsamo, M. Biodiversity and distribution of the meiofaunal community in the reef slopes of the Maldivian archipelago (Indian Ocean). *Mar. Environ. Res.* **2018**, *139*, 19–26. [CrossRef] [PubMed]

38. Heip, C.; Vincx, M.; Vranken, G. The ecology of marine nematodes. *Oceanogr. Mar. Biol. Annu. Rev.* **1985**, *23*, 399–489.

39. Seinhorst, J.W. A rapid method for the transfer of nematodes from fixative toanhydrous glicerine. *Nematologica* **1959**, *4*, 67–69. [CrossRef]

40. Platt, H.M.; Warwick, R.M. *Free-Living Marine Nematodes. Part. I. British Enoplids*; Synopses of the British Fauna, Volume 28; Cambridge University Press: Cambridge, UK, 1983; p. 307.

41. Platt, H.M.; Warwick, R.M. *Free-Living Marine Nematodes. Part II. British Chromadorids*; Synopses of the British Fauna, Volume 38; E.J Brill: Leiden, The Netherlands, 1988; p. 502.

42. Warwick, R.M.; Platt, H.M.; Somerfield, P.J. *Free-Living Marine Nematodes. Part III. Monhysterids*; Synopses of the British Fauna, Volume 53; Field Studies Council: Shrewsbury, UK, 1988; p. 296.

43. Bezerra, T.N.; Decraemer, W.; Eisendle-Flöckner, U.; Hodda, M.; Holovachov, O.; Leduc, D.; Miljutin, D.; Mokievsky, V.; Peña Santiago, R.; Sharma, J.; et al. Nemys: World Database of Nematodes. 2019. Available online: http://nemys.ugent.be (accessed on 7 October 2019). [CrossRef]

44. Dovgal, I.V. *Fauna of Ukraine: In 40 Vol. Volume 36: Ciliates—Ciliophora. Issue 1: Class Suctorea*; Naukova Dumka: Kyiv, Ukraine, 2013; p. 267.

45. Ghosh, M.; Mandal, S. Living with Nematode: An Epibiont *Trematosoma rotunda* Associated with Basibiont Desmodora scaldensis from Matla Estuary, Sundarbans, India. *Thalassas* **2019**, *35*, 619–624. [CrossRef]

46. Chen, C.; Golovatch, S.I.; Chang, H.-W. Identity of the east Asian millipede *Habrodesmus inexpectatus* Attems, 1944 (Diplopoda: Polydesmida: Paradoxosomatidae). *J. Nat. Hist.* **2008**, *42*, 2547–2556. [CrossRef]

47. Chatterjee, T.; Nanajkar, M.; Dovgal, I. New record of *Loricophrya stresemanni* (Ciliophora, Suctorea) as epibiont on nematodes from the India Ocean and notes on the genus *Loricophrya*. *Cah. Biol. Mar.* **2019**, *60*, 283–288.

48. Collin, B. Etudes monographiques sur les Acinetiens. II. Morphologie, physiologie, systematique. *Arch. Zool. Exp. Gen.* **1912**, *51*, 1–457.

49. Dovgal, I.; Chatterjee, T.; Ingole, B. New records of *Thecacineta cothurnioides* and *Trematosoma rotunda* (Ciliophora, Suctorea) as epibionts on nematodes from Indian Ocean. *Protistology* **2009**, *6*, 19–23.

50. Lynn, D.H. *The Ciliated Protozoa. Characterization, Classification and Guide to the Literature*, 3rd ed.; Springer: Berlin/Heidelberg, Germany, 2008; p. 605.

51. Robin, M.C. Mémoire sur la structure et la reproduction de quelques infusoires tentaculés suceurs et flagellés. *Journal l'Anatomie Physiologie Normale Pathologiques l'homme Animaux* **1979**, 529–583.

52. Root, F.M. A New Suctorian from Woods Hole. *Trans. Am. Microsc. Soc.* **1922**, *41*, 77. [CrossRef]

53. Swarczewsky, B. Sur Kenntnis der Baikalprotistenfauna. Die an der Baikalgammariden lebeden Infusorien. 1–6. 4. Acinetidae. *Archive Protistenkunde* **1929**, *63*, 362–449.

54. Jankowski, A.V. *Review of Taxa Phylum Ciliophora Doflein, 1901*; Protista: Handbook on Zoology. Volume Part. 2; Alimov, A.F., Ed.; Nauka: St. Petersburg, Russia, 2007; pp. 415–993.

55. Grell, K.G.; Meister, A. Die Ultrastruktur von Acinetopsis rara Robin (Suctoria). I. Tentakeln und Nahrungsaufnahnme. *Protistologica* **1982**, *18*, 67–84.

56. Batisse, A. *Sous-Classe des Suctoria Claparede et Lachmann, 1858*; Traite de Zoologie. Anatomie, Systematique, Biologie. Tome II. Infusoires Cilies. Fascicule 2. Systematique; De Puytorac, P., Ed.; Masson: Paris, France; Milan, Italy; Barcelone, Spain, 1994; pp. 493–563.

57. Sergeeva, N.; Dovgal, I. First finding of epibiont peritrich and suctorian ciliates (Ciliophora) on oligochaetes and harpacticoid copepods from the deep-water hypoxic/anoxic conditions of the Black Sea. *Ecol. Montenegr.* **2014**, *1*, 49–54.

58. Lynn, D.H.; Gómez-Gutiérrez, J.; Strüder-Kypke, M.; Shaw, C.T. Ciliate species diversity and host-parasitoid codiversification in the apostome genus *Pseudocollinia* (Ciliophora, Apostomatia, Pseudocollinidae) that infect krill, with description of *Pseudocollinia similis* n. sp., a parasitoid of the krill *Thysanoessa spinifera*. *Dis. Aquat. Org.* **2014**, *112*, 89–102.

59. Sergeeva, N.; Dovgal, I. *Loricophrya bosporica* n. sp. (Ciliophora, Suctorea) epibiont of *Desmoscolex minutus* (Nematoda, Desmoscolecida) from oxic/anoxic boundary of the Black Sea Istanbul Strait's outlet area. *Zootaxa* **2016**, *4061*, 596. [CrossRef]

60. Ivanova, E.; Dovgal, I.V.; Newton, A. First records of epibiont ciliates (Ciliophora) in methane enriched sediments with species redescriptions. *Ecol. Montenegr.* **2017**, *10*, 51–57.

61. Sartini, B.; Marchesini, R.; D'ávila, S.; D'Agosto, M.; Dias, R.J.P. Diversity and Distribution of Peritrich Ciliates on the Snail *Physa acuta* Draparnaud, 1805 (Gastropoda: Physidae) in a Eutrophic Lotic System. *Zool. Stud.* **2018**, *57*, 2018–2057.

62. Fernandez-Leborans, G.; Gabilondo, R. Inter-annual variability of the epibiotic community on *Pagurus bernhardus* from Scotland. *Estuar. Coast. Shelf Sci.* **2006**, *66*, 35–54. [CrossRef]

63. Key, M.M.; Winston, J.E.; Volpe, J.W.; Jeffries, W.B.; Voris, H.K. Bryozoan fouling of the blue crab *Callinectes sapidus* at Beaufort, North Carolina. *Bull. Mar. Sci.* **1999**, *64*, 513–533.

64. Fisher, R. Ciliate Hitch-hikers—Nematode ecto-commensals from tropical Australian sea grass meadows. *J. Mar. Biol. Assoc. UK* **2003**, *83*, 445–446. [CrossRef]

65. Panigrahi, S.; Bindu, V.K.; Bramha, S.N.; Samantara, M.K.; Mohanty, A.K.; Satpathy, K.K.; Dovgal, I. Report of *Thecacineta calix* (Ciliophora: Suctoria) on nematode *Desmodora* from the intertidal sediments of Southwest Bay of Bengal. *Indian J. Geo-Mar. Sci.* **2015**, *44*, 1840–1843.

66. Decraemer, W.; Coomans, A.; Baldwin, J. Morphology of Nematoda. *Nematoda* **2013**, *2*, 1–60.

67. Riemann, F.; Riemann, O. The enigmatic mineral particle accumulations on the cuticular rings of marine desmoscolecoid nematodes—Structure and significance explained with clues from live observations. *Meiofauna Mar.* **2010**, *18*, 1–10.

68. Ingole, B.; Singh, R.; Sautya, S.; Dovgal, I.; Chatterjee, T. Report of epibiont *Thecacineta calix* (Ciliophora: Suctorea) on deep-sea *Desmodora* (Nematoda) from the Andaman Sea, Indian Ocean. *Mar. Biodivers. Rec.* **2010**, *3*, 3. [CrossRef]

69. Pujol, N.; Davis, P.; Ewbank, J.J. The Origin and Function of Anti-Fungal Peptides in *C. elegans*: Open Questions. *Front. Immunol.* **2012**, *3*, 237. [CrossRef]
70. Tarr, D.E.K. Nematode antimicrobial peptides. *Invert. Surv. J.* **2012**, *9*, 122–133.

Article

Doolia, A New Genus of Nannopodidae (Crustacea: Copepoda: Harpacticoida) from off Jeju Island, Korea

Wonchoel Lee

Department of Life Science, College of Natural Sciences, Hanyang University, Seoul 04763, Korea; wlee@hanyang.ac.kr; Tel.: +82-2-2220-0951

http://zoobank.org/urn:lsid:zoobank.org:pub:EDEFFEEE-4930-4881-9D00-0D88EB6C7496
Received: 20 November 2019; Accepted: 15 December 2019; Published: 18 December 2019

Abstract: A new harpacticoid copepod is described from the waters off Jeju Island, Korea. This species displays a unique set of characteristics including a rostrum that is clearly demarcated from the cephalosome, a setular (spinular) row on the rostrum, a well-developed frill along the posterior margins of each body segment except for the cephalosome, long and cylindrical caudal rami, four segmented female antennules, paired genital apertures in the female, the absence of sexual dimorphism in legs P1–P4, and highly reduced P5 and P6 in the male. This combination of characteristics allocates the specimen to the family Nannopodidae Por, 1986, but the new species belongs to none of the extant genera within the family. A new genus, *Doolia*, is proposed. *Nannopus* is suggested as a sister taxon of the new genus based on shared plesiomorphic characteristics in the maxilliped, legs P1–P4, and P5. *Doolia* gen. nov. is the eighth genus of Nannopodidae, and an amended key for the genus is provided herein.

Keywords: taxonomy; huntermaniidae; cletodidae; rhizotrichidae; *Nannopus*; meiofauna

---

1. Introduction

The family Cletodidae sensu T. Scott, 1904 has been recognized in previous reports as being potentially the most heterogenous harpacticoid family [1]. Lang [2] recognized several lineages within the family, although he did not take any action on the subject. Por [1] revised and redefined the family Cletodidae T. Scott, 1905, establishing four new families: Paranannopidae Por, 1986, Huntemanniidae Por, 1986, Rhizotrichidae Por, 1986, and Argestidae Por, 1986. Among Por's new families, Huys [3] synonimized Huntemanniidae with Nannopodidae Brady, 1880. Consequently, two genera, *Metahuntemannia* Smirnov, 1946 and *Talpina* Dahms & Pottek, 1982 were re-allocated to the subfamily Hemimesochrinae Por, 1986 in the Canthocamptidae Brady, 1880 based on their affinities with genera including *Bathycamptus* Huys & Thistle, 1989, *Micropsammis* Mielke, 1975, *Isthmiocaris* George & Schminke, 2003, and *Perucamptus* Huys & Thistle, 1989 [4].

Currently, Nannopodidae includes about 25 valid species in seven genera: *Nannopus* Brady, 1880, *Pontopolites* T.Scott, 1894, *Huntemannia* Poppe, 1884, *Rosacletodes* Wells, 1985, *Laophontisochra* George, 2002, *Acuticoxa* Huys & Kihara, 2010, and *Talpacoxa* Corgosinho, 2012.

During a study of the harpacticoid copepods in Korean Waters, a nannopodid-like copepod was collected; however, the specimens could not be allocated to any extant genera in the family Nannopodidae. The present study aims to describe this new taxon and establish a new genus based on the current specimens from Korea.

2. Materials and Methods

Specimen Collection and Morphological Examinations

Samples were collected at one site, A5 (34°00′ N, 123°30′ E), among 18 field sites off the coast of Jeju Island, South Korea from 24 September to 3 October 2002 (Figure 1) by the R/V Ara (Jeju National University) [5]. Sediments were collected with a box corer, and five replicates were taken from each box corer. Three replicates were used for analyses of meiofauna. Organisms were extracted from sediments by a Ludox-Am. Copepods were sorted under a dissecting microscope and stored in 70% ethanol. Dissections were applied to the specimens in lactic acid and the dissected appendages were mounted permanently on slides in lactophenol mounting medium. Preparations were sealed with transparent nail varnish. Each appendage was observed and drawn under an Olympus BX51 differential interference contrast microscope using a drawing tube. The descriptive terminology is adopted from Huys et al. [6]. Abbreviations are A1, antennule; A2, antenna; ae, aesthetasc; exp, exopod; enp, endopod; P1–P6, first to sixth thoracopod; exp(enp)-1(2, 3) to denote the proximal (middle, distal) segment of a ramus [6]. Specimens were deposited in the National Institute of Biological Resources (NIBR), Korea, and in the Marine Biodiversity Institute of Korea (MABIK). All scale bars in the figures are in μm.

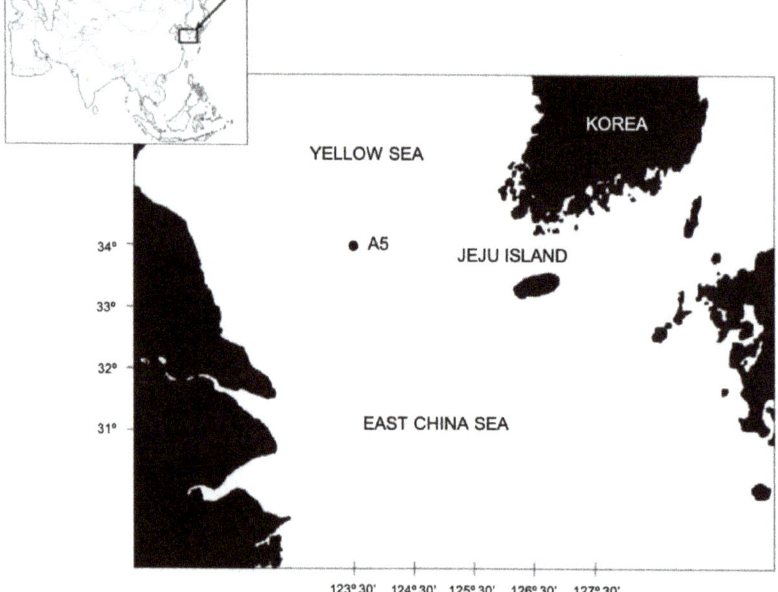

Figure 1. A map of the sampling location (reconstructed from [5]).

3. Systematic Account

Subclass Copepoda Milne Edwards, 1830
Order Harpacticoida Sars, 1903
Family Nannopodidae Por, 1986

3.1. Genus Doolia, gen. nov.

http://zoobank.org/urn:lsid:zoobank.org:act:2D8B250F-51BF-4DA8-AF59-965E25C72E0B

Diagnosis. Cylindrical body densely covered by denticles. Caudal rami cylindrical with six caudal setae and one large pore. Anal operculum well-developed with a spinular row along distal margin.

Rostrum defined at base, with a pair of sensilla and a row of long spinules along anterior margin. Antennule four-segmented. Antenna three-segmented with one-segmented exopod. Mandibular palp well-developed with one segmented exopod and endopod. Maxillary syncoxa with two endites. Maxilliped subchelate. Swimming legs P1–P4 with three-segmented exopods and two-segmented endopods. Female P5 fused medially, with separate exopod. P6 with two naked setae in female. Sexual dimorphism in urosome, antennule, P5, and P6. No sexual dimorphism in swimming legs. Male antennule eight-segmented, subchirocer. Male P5 baseoendopod fused with exopod, and represented by several setae. Male P6 represented by a small plate without ornamentation.

Type species: *Doolia ara* sp. nov. by original designation and monotype.

Etymology. The new genus was named after a famous cartoon character, "Dooly," a baby dinosaur created in 1983 in Korea.

3.2. Doolia ara, sp. nov.

Figures 2–7

http:/zoobank.org/urn:lsid:zoobank.org:act:5CABF7A6-2762-4784-92F3-0E1182D2FCD4

Type locality. Muddy sand substratum, station A5 (34°00′ N, 123°30′ E), off Jeju Island, the south west coast of Korea.

Material examined. Holotype 1♀(NIBRIV0000862381) dissected on seven slides. Paratypes: three♀♀and two♂♂; 2♀♀(MABIK CR00246523 – 4) on five slides, 1♂(MABIK CR00246525) on seven slides, 1♂(NIBRIV0000862382) on six slides, and 1♀(NIBRIV0000862383) in 70% ethanol, all from the type locality, depth 66 m, collected by Y.H. Song and E.J. Nam during 24 September to 3 October 2002.

Etymology. The specific name refers to "Ara", the research vessel of Jeju National University, in appreciation of the crew and Prof. Joon Baek Lee (Jeju National University). Gender, feminine.

Female. Total body length: 239 µm (measured from anterior margin of rostrum to posterior margin of caudal rami). Largest width measured at posterior margin of cephalic shield: 63 µm. Urosome gradually tapering posteriorly (Figure 2A).

Cephalothorax with serrulated posterior margin. Pleural areas rounded without lobate posterolateral angles. Entire surface covered with tiny denticles as illustrated [expressed as dots] in Figure 2A,B. Sensilla and a few pores are present as illustrated in Figure 2A,B.

Rostrum triangular-shaped (Figure 2A,C), broad at base, round anterior margin, clearly separated from cephalosome, and with a pair of sensilla and a row of long spinules along median anterior margin.

Pedigerous somites covered with minute denticles. Prosomites from P2-bearing to P4-bearing with well-developed, comb-shaped hyaline frills.

Urosome (Figures 2A,B and 3B,D) five-segmented, comprising P5-bearing somite, genital double-somite and three free abdominal somites. All urosomites covered with small denticles dorsally and ventrally and with well-developed hyaline frill along the hind margin. Genital double-somite (Figures 2A,B and 3B) without distinct dorsal surface ridge but with a thick internal cuticle layer indicating original segmentation; completely fused internally and externally. Genital field located near anterior margin (Figure 3B) with a large copulatory pore located in a median depression (Figure 3B). P6 with a small protuberance bearing two bare setae, inner seta longer than outer seta, and with a small pore next to Inner seta. Anal somite (Figure 3D) quadrate with well-developed operculum; opercular distal margin with a row of long spinules, and flanked by a pair of sensilla. Caudal rami (Figure 3B,D) long, cylindrical, three times longer than it is wide; each ramus with six setae: seta I bare, shortest; setae II bare and long; seta III absent, or presumably represented by one large pore on dorsal distal surface; setae IV and V separated basally; seta VI bare and small; seta VII tri-articulate at base. Each ramus with spinules on entire surface; additional spinular rows along distal margin.

Antennule (Figure 3A) four-segmented. Segment 1 with a row of long spinules on anterior surface. Segment 2 the largest. Segment 3 with aesthetasc fused basally to one long plumose seta. Armature formula: 1-[1 pinnate], 2-[8 pinnate], 3-[6 pinnate + 2 bare + (1 pinnate + ae)], 4-[4 pinnate + 4 bare + trithek]. Apical trithek consisting of small aesthetasc fused basally to two setae (1 bare, and 1 pinnate).

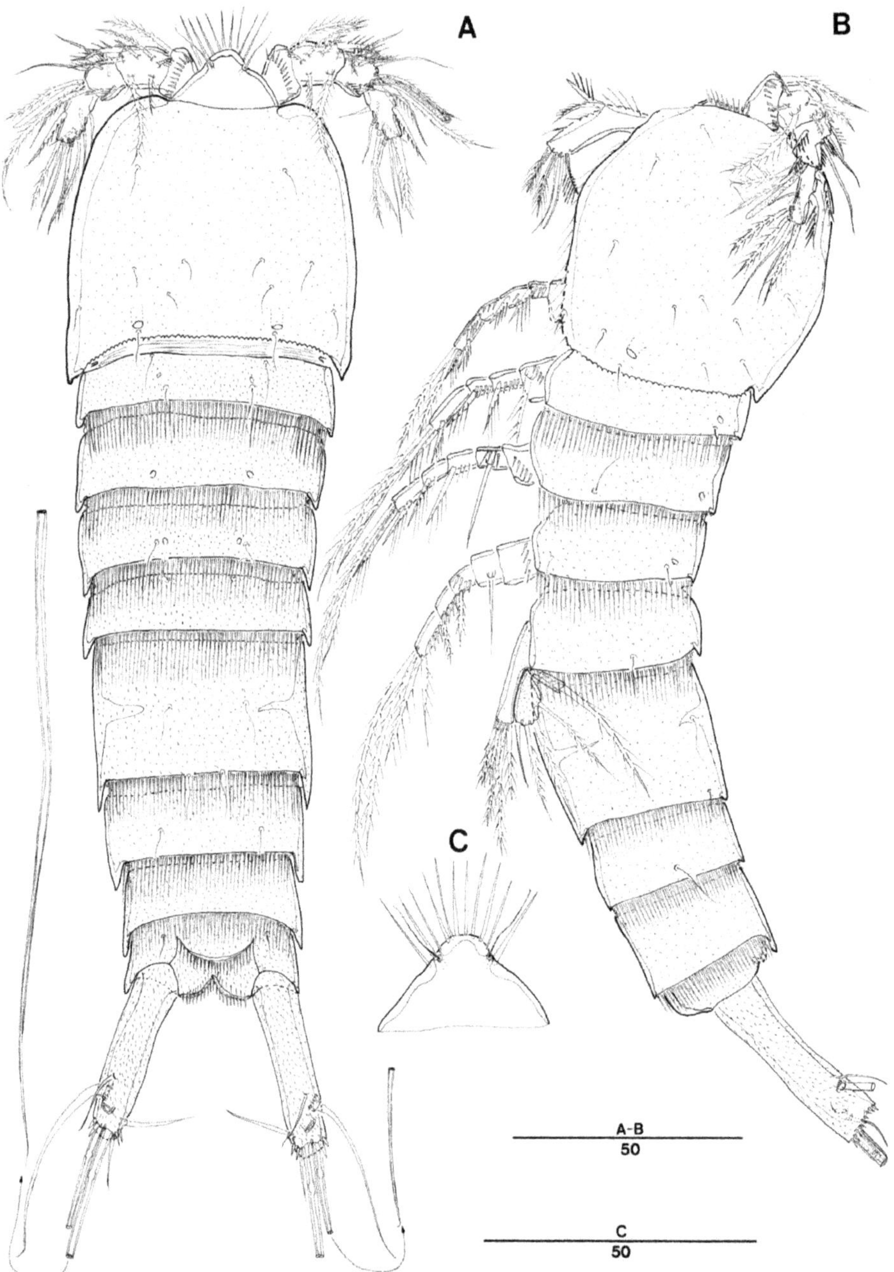

Figure 2. *Doolia ara* gen. et. sp. nov. holotype female. (**A**), habitus, dorsal; (**B**), habitus, lateral; (**C**), rostrum, dorsal.

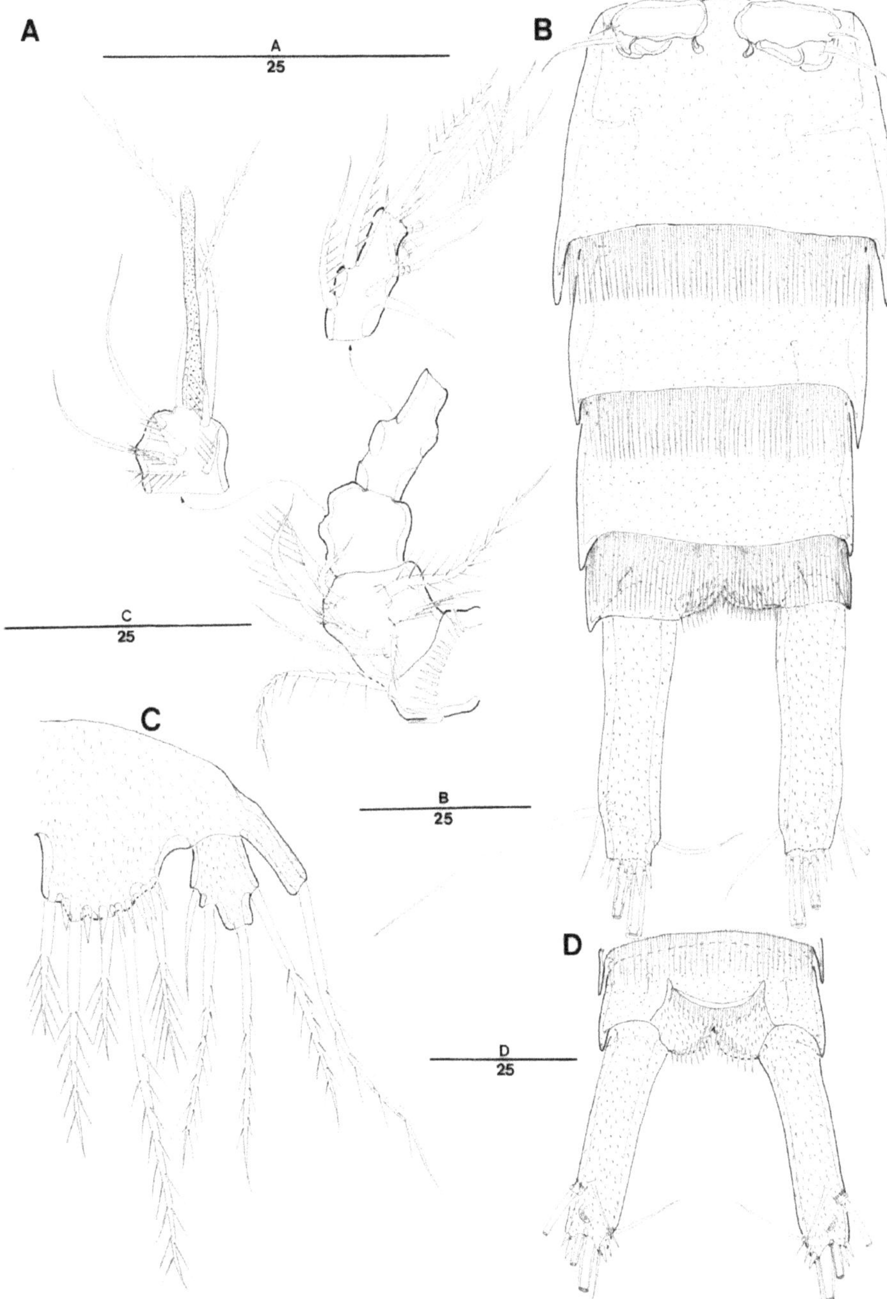

Figure 3. *Doolia ara* gen. et. sp. nov. holotype female. (**A**), antennule; (**B**), urosome (excluding P5 bearing somite), ventral; (**C**), P5, anterior; (**D**), anal segment and caudal rami, dorsal.

Antenna (Figure 4A) three-segmented, comprising coxa, allobasis, free one-segmented endopod and one-segmented exopod. Coxa small (not figured). Allobasis without distinct surface sutures marking original segmentation; with abexopodal pinnate seta on median inner margin and a row of setules along outer margin. Exopod small, about twice as longer as it is wide, with four well-developed pinnate setae and few spinules along outer margin. Endopod slightly shorter than allobasis, lateral armature arising in distal half, consisting of two strong pinnate spines; apical armature consisting of one strong pinnate spine, two pinnate, and two geniculate setae; with two rows of long spinules laterally; apical outer most spine basally fused to one tiny seta.

Mandible (Figure 4B) with well-developed gnathobase bearing several multicuspidate teeth around distal margin and one slender spine at dorsal corner. Palp well-developed. Basis with one plumose seta. Endopod one-segmented, rectangular with one lateral plumose seta, four naked apical setae, and one row of spinules along apical margin. Exopod one-segmented, cylindrical with one naked apical seta.

Maxillule (Figure 4D). Arthrite strongly developed, with two naked setae on anterior surface and eight spines/setae around distal margin and a row of spinules on posterior surface. Coxa with one short cylindrical endite bearing one long naked seta. Endopod and exopod incorporated into the basis, with three naked setae apically, and one plumose and four naked setae laterally.

Maxilla (Figure 4C). Syncoxa with two endites and a row of long spinules along outer margin. Each coxal endite cylindrical; proximal endite with two pinnate spines; distal endite with one naked and two pinnate spines. Allobasis drawn out into strong, slightly curved claw; endopod represented by two naked setae on anterior surface.

Maxilliped (Figure 4E) with one plumose seta and one row of spinules on syncoxa. Basis with one row of tiny spinules along palmar margin. Endopod one-segmented with long, curved, coarsely pinnate claw, and one naked slender seta.

Legs P1–P4 (Figure 5A,B and Figure 6A,B) with well-developed intercoxal plates and praecoxae bearing row of spinules along distal margin near borderline with each related coxa. Coxae and bases with surface ornamentations of spinules as figured. All legs P1–P4 with three-segmented exopods, and two-segmented endopods.

P1 (Figure 5A). Coxa moderate, with long spinular rows as figured. Basis with one strong, bipinnate inner and one stout, unipinnate outer spines. Anterior surface covered with spinules as figured. Exopod longer than endopod; exp-1 and exp-2 with unipinnate spine, respectively; exp-3 longest with two unipinnate spines and two plumose setae. Enp-2 about three times longer than enp-1; enp-1 without spine or seta; enp-2 with row of long setules along outer and inner margins; two terminal plumose setae on enp-2 fused basally, inner one more than three times longer than the outer.

P2–P4 (Figures 5B and 6A,B). Coxae and bases ornamented with spinular rows along outer or distal margins and anterior surface. Outer margin of basis producing a setophore armed with pinnate spine (P2) or naked seta (P3–P4) at each distal end. All segments with rows of spinules along inner and outer margins as figured. Endopod segments of P2–P3 with long setules or spinules alonge inner and outer margins. P4 endopod without setula ornamentations. P2 enp-2 three times longer than enp-1, endopod reaching to proximal area of exp-3, and exp-3 longest. P3 enp-2 2.5 times longer than enp-1, endopod reaching to middle of exp-2, and exp-1 largest. P4 endopod small, enp-2 2.3 times longer than enp-1, endopod reaching to middle of exp-1, and exp-3 longer than exp-1.

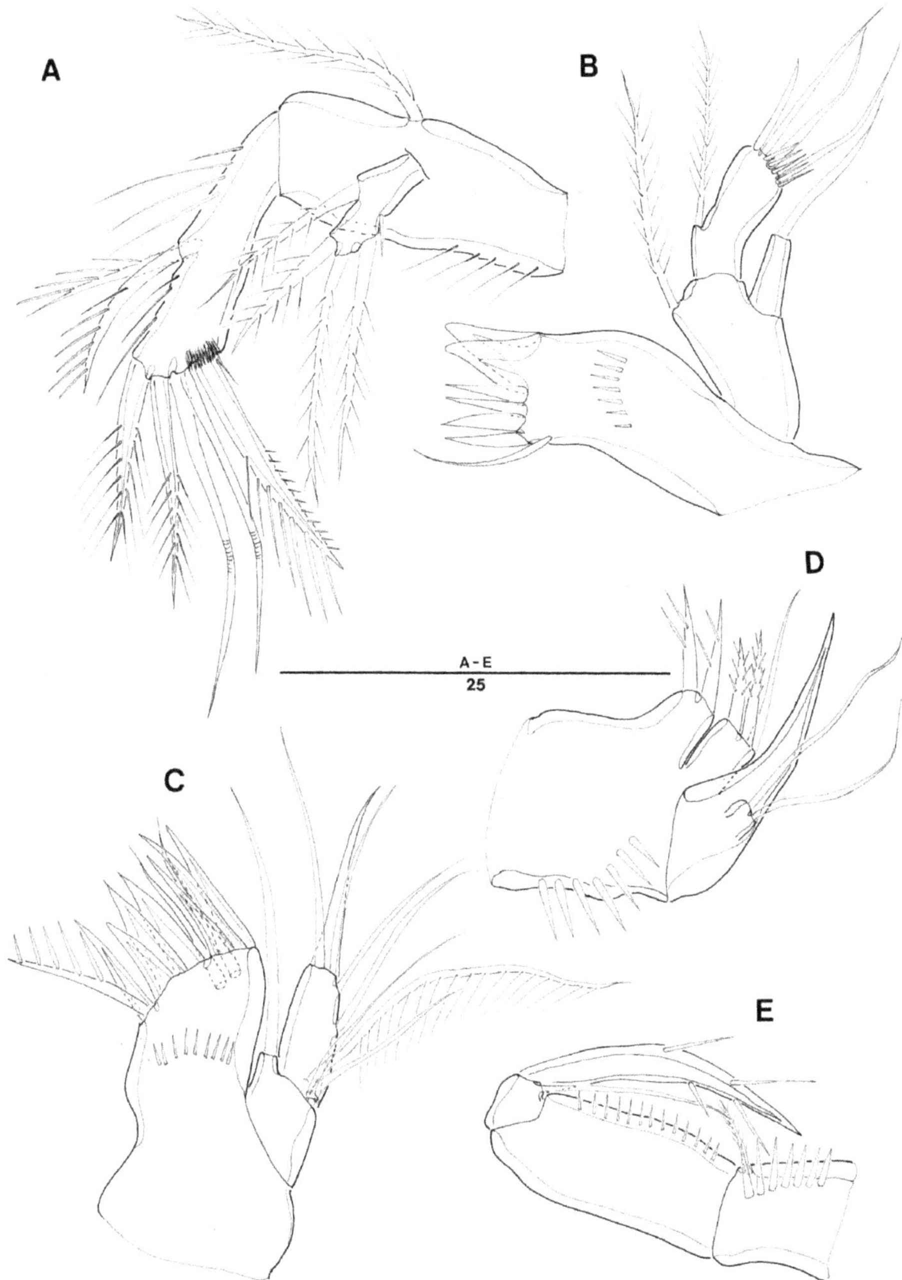

Figure 4. *Doolia ara* gen. et. sp. nov. holotype female. (**A**), antenna; (**B**), mandible; (**C**) maxillule; (**D**), maxilla; (**E**), maxilliped.

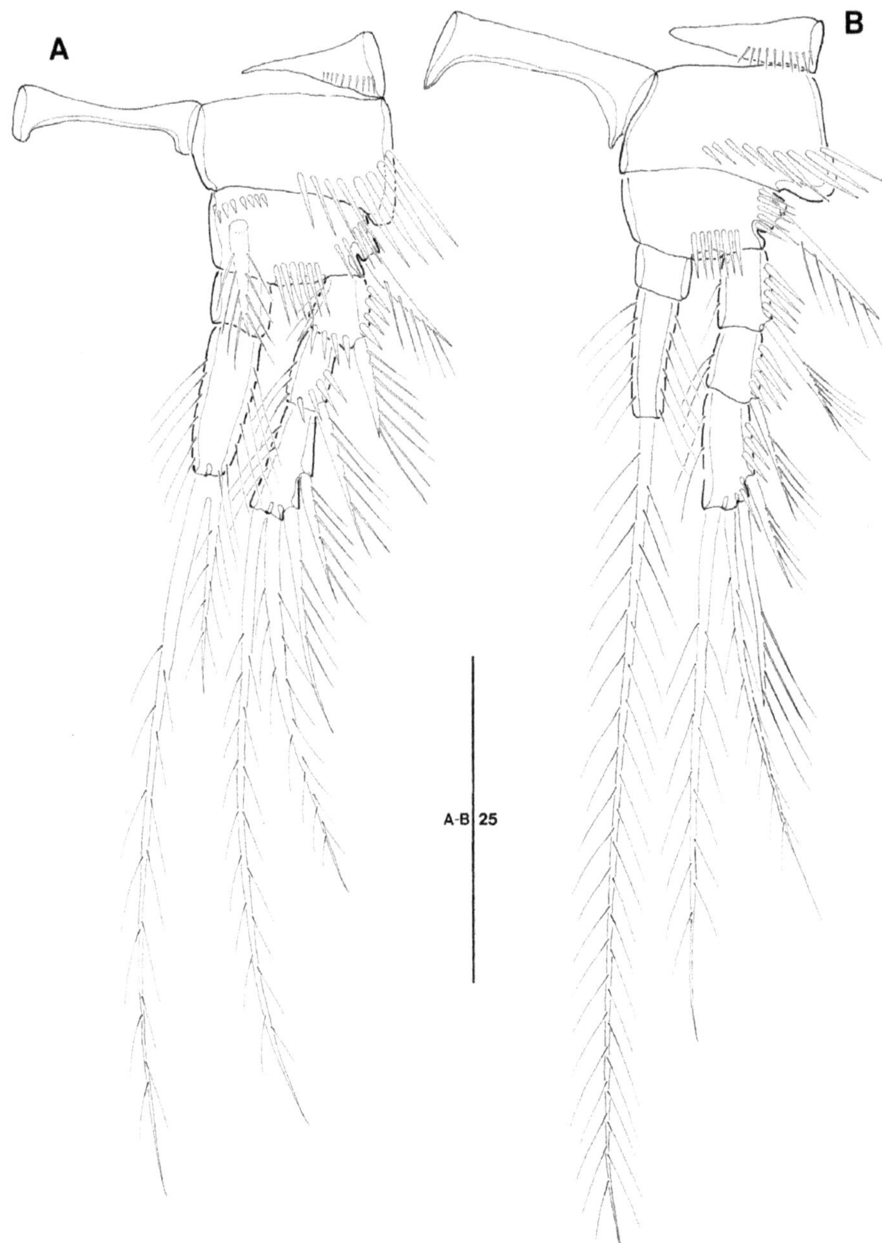

Figure 5. *Doolia ara* gen. et. sp. nov. holotype female. (**A**), P1, anterior; (**B**), P2, anterior.

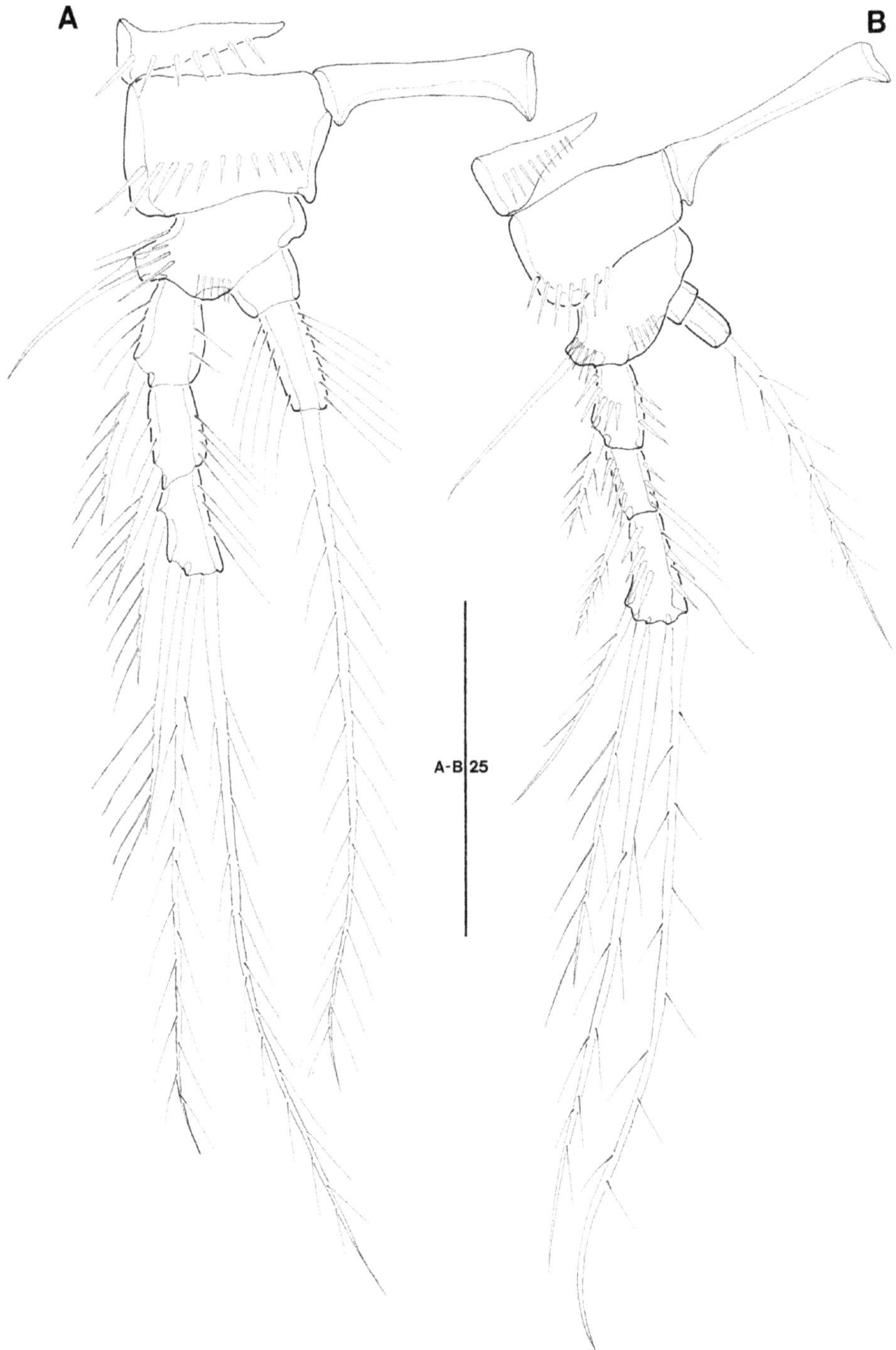

Figure 6. *Doolia ara* gen. et. sp. nov. holotype female. (**A**), P3, anterior; (**B**), P4, anterior.

Spine and setal formulae as follows:

Exopod   Endopod
P1   0.0.022   0.020
P2   0.0.022   0.010
P3   0.0.022   0.010
P4   0.0.022   0.010

P5 (Figure 3C) fused medially, and exopod and baseoendopod separate; each covered with minute spinules as figured. Baseoendopod with a long outer setophore bearing one plumose basal seta. Endopodal lobe only reaching to middle of exopod, with five pinnate setae; second outermost seta longest; a row of spinules along distal margin. Exopod small, slightly longer than wide and with one naked and three pinnate setae.

Male. Body length 243 μm. Largest width measured at about median area of cephalic shield: 61 μm. Urosome gradually tapering posteriorly (Figure 7A).

Cephalothorax with serrulated posterior margin. Pleural areas well-developed and rounded without lobate posterolateral angles. Entire surface covered with tiny denticles as in female. Sensilla present as illustrated in Figure 7A.

Rostrum triangular-shaped (Figure 7A,E), broader than in female, smooth anterior margin, clearly separated from cephalosome, and with a pair of sensilla and a row of long spinules along entire anterior margin (only along median apical margin in female).

Urosome (Figure 7A) six-segmented, comprising P5-bearing somite, genital somite and four abdominal somites. All urosomites covered with small denticles dorsally and ventrally and with well-developed hyaline frill along the hind margin.

Antennule (Figure 7B) eight-segmented and subchirocer with geniculation between segments 5 and 6. Segment 1 with a row of coarse and widely spaced spinules. Segment 2 the largest. Segment 4 represented by a small sclerite along anterior margin of segment 3. Segment 5 swollen with one well-developed large aesthetasc. Segment 7 with three-dimensional process as figured in Figure 7B. Segment 8 with triangular distal half. Armature formula: 1-[1pinnate], 2-[7 pinnate], 3-[5 pinnate], 4-[1 pinnate], 5-[1 + 6 pinnate + (1 pinnate + ae)], 6-[0], 7-[2 modified process], 8-[6 + 1 pinnate + trithek]. Apical trithek consisting of a minute aesthetasc and two naked setae.

P5 (Figure 7D) fused medially, defined at base, and whole surface covered with minute spinules. Baseoendopod with a long, cylindrical setophore bearing one plumose outer basal seta. Exopod fused to endopod and represented by three pinnate setae. Minute and coarse spinules on anterior surface.

P6 (Figure 7C) asymmetrical, represented on both sides by a small plate without any additional ornamentation: fused to ventral wall of supporting somite along right side, articulating at base and covering gonopore along left side.

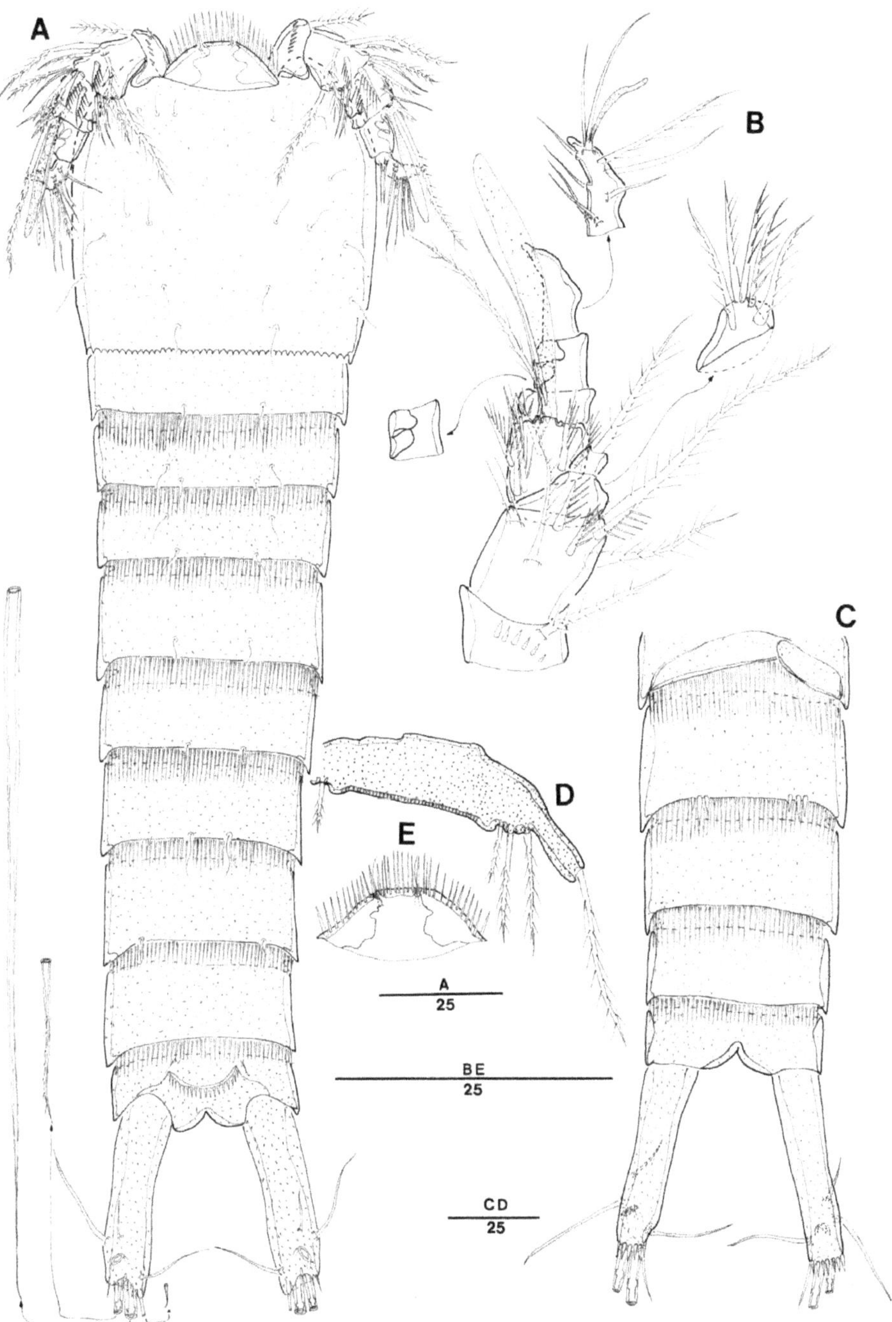

Figure 7. *Doolia ara* gen. et. sp. nov. paratype male. (**A**), habitus, dorsal; (**B**), antennule; (**C**), urosome (excluding P5 bearing somite); (**D**), P5; (**E**), rostrum, dorsal.

4. Discussion

The new species displays interesting characteristics: the basally-defined rostrum, the well-developed mandibular palp with one segmented exopod and endopod, the presence of a seta on the maxilliped syncoxa, three segmented exopods and two segmented endopods in P1–P4, and the separate exopod of P5 in the female. The family Nannopodidae Por, 1986 could harbor the combination of these character sets based on character comparisons (Table 1). Among the seven extant genera of the Nannopodidae, *Nannopus* Brady, 1880 is the only member having the rostrum fused to the cephalothorax, while all other genera have basally-defined rostrum, as in the new species (Table 1). The new species, *Pontopolites* T. Scott, 1894, and *Talpacoxa* Corgosinho, 2012 have a separate exopod on the mandibular palp, but in other genera, these are absent (*Nannopus*, and *Huntemannia* Poppe, 1884) or represented by a single seta (*Rosacletodes* Wells, 1985, *Laophontisochra* George, 2002, and *Acuticoxa* Huys & Kihara, 2010). The maxilliped syncoxa has a pinnate distal seta in the new species, as well as in *Nannopus*, *Huntemannia*, and *Rosacletodes*, while it is absent in *Pontopolites*, *Laophontisochra*, *Acuticoxa*, and *Talpacoxa*. The segmentation of P1–P4 in the new species is the most conservative state within the family, while *Rosacletodes* displays the most derived state with degenerated, or simply absent endopods in P2–P4. Finally, in the new species the P5 exopod of the female is clearly separated from the baseoendopod as in *Huntemannia*, *Rosacletodes*, *Laophontisochra*, *Talpacoxa*, and most *Nannopus* species. These five characters are suggested here as plesiomorphies of the Nannopodidae, and consequently, the new genus forms a basal group with *Nannopus* and *Pontopolites* within the family.

Doolia ara gen. et sp. nov. is the probable sister genus to *Nannopus* due several common plesiomorphic characteristics: the anterior spinular row on the apical margin of the rostrum, the presence of a single seta on the maxilliped syncoxa, the three-segmented exopod in P1—P4, a maximum two-segmented endopod in P1–P4, and the separate P5 exopod from baseoendopod in the female. However *D. ara* also has autapomorphies separating it easily from *Nannopus*: the four-segmented antennule in the female (five-segemented in *Nannopus*), the single abexopodal seta on the antennary allobasis (two abexopodal setae in *Nannopus*), no inner seta on the exp-2 in the exopods of P1–P4 (always present in P2–P3 of *Nannopus*), no sexually-modified P3 endopod in the male (apophysis present in *Nannopus*), and no seta on the P5 endopodal lobe in the male (3–4 setae in *Nannopus*) [7].

Pontopolites forms the basal group in Nannopodidae, with *Doolia* and *Nannopus* having primitive characters: the well-developed and defined triangular rostrum, the one-segmented mandibular exopod, the three-segmented exopods in P1–P4, and the two-segmented P1 endpod. Furthermore, this genus has a six-segmented antennule in the female, and the most primitive two-segmented antennary exopod (Table 1). As pointed out by Karanovic and Cho [9], *Pontopolites* are isolated from two other genera by having one plated P5 in both sexes. *Acuticoxa* also has a similarly-shaped P5 in the female [4]. The other genera, *Huntemannia*, *Rosacletodes*, *Laophontisochra*, *Acuticoxa*, and *Talpacoxa* have reduced segmentation in the legs (Table 1). *Huntemannia*, *Rosacletodes*, and *Talpacoxa* have sexual modifications in the male P3 endopod. *Laophontisochra* is the only genus having no sexual modification in the legs, except for *Doolia*. The male of *Acuticoxa*, and the status of modification in the legs are unknown.

Table 1. Characteristic comparisons among genera of Nannopodidae.

Character\Species	Doolia gen. nov.	Nannopus Brady, 1880	Huntemannia Poppe, 1884	Pontopolites T.Scott, 1894	Rosacletodes Wells, 1985	Laophontisochra George, 2002	Acuticoxa Huys & Kihara, 2010	Talpacoxa Corgosinho, 2012
Rostrum	defined	fused to cephalothorax	defined	defined	defined	defined	defined	defined
Rostral spinules	present	present	present	absent	absent	present	absent	absent
A1 segment no., female	4	5	5	6	5	4	4	5
A2 exopodal segment	4 setae	3–4 setae	4 setae	2-segmented; 1, 120	3 setae	1 segmented or, 1 seta	absent, or 1 minute seta	3 setae
Mandibular exopod	1 segment	absent	absent	1 segment	fused, 1 seta	fused, 1 seta	fused, 1 seta	1 segment
Maxilliped syncoxa	1 seta	1 seta	1 seta	no seta	1 seta	no seta	no seta	no seta
P1 exp:enp segment	3:2	3:1-2	3:1	3:2	2:1	2:2	2:2	1:2
P2 exp:enp segment	3:2	3:2	1–2:1	3:1	2:0	2:1	1–2:1	2:1
P3 exp:enp segment	3:2	3:1-2	1–2:1	3:1	1:0	2:1	1–2:1	2:1
P4 exp:enp segment	3:2	3:1	1–2:1	3:1	1:0	2:1 (1:1 in male)	1:1	2:1
P3 endopod in male	no modification	with apophysis	with additional seta	with apophysis	with apophysis	no modification	unknown	with apophysis
P5 exp:enp seta in female	4:5	4–5:3–4	5:4	fused to 1 plate with 10 setae	5:6	4:1–3	fused to 1 plate with 8 setae	5:3
P5 exp:enp seta in male	3:0	4–5:3–4	4:4	1 plate with 8 setae	5:4	4:2–4	unknown	4:2
P6 seta F:M	2:0	1:2–3	1:3	2:2	0:3	3:3	2: unknown	0:3
No. of valid species	1	11	5	2	1	2	2	1
Reference	present study	[7]	[8]	[9]	[10]	[11,12]	[4]	[13]

In addition, *Doolia* gen. nov. has superficial similarities with *Rhizothrix* Sars G.O., 1909, and they share the following character sets: the presence of pinnate setae on the antennule and its segmentation, the segmentation of exopods and endopods in P1–P4, lack of sexual dimorphism in legs P1–P4, and lack of ornamentation on P6 in the male. *Tryphoema* Monard, 1926 shares similar characters with *Doolia* except for the segmentation of the legs [14]. Based on previous reports [15] and the present study, *Rhizothrix* has clear discrepancies from *Doolia*: clearly separated rostrum with anterior row of long setules (spinules) in *Doolia ara* (fused to cephalosome and lacking the spinules in *Rhizothrix*), the well-developed comb-shaped hyaline frill on the body somites in *D. ara* (all frills serrulate in *Rhizothrix*), the long and cylindrical caudal rami in *D. ara* (short and robust, or ovoid in *Rhizothrix*), absence of the brush setae in the exopod and endopod in P1 in *D. ara* (present in *Rhizothrix*), and presence of setae on the P6 of female in *D. ara* (absent in *Rhizothrix*).

Similarities between *Doolia* and Rizotrichidae are probably the result of parallel evolution; however, this hypothesis should be tested with additional molecular evidence in further studies. There are no related studies of lineages of Nannopodidae based on molecular data, except for an intrageneric discussion of *Nannopus* [16]. There are a few scattered reports of molecular data for presumably related families, Cletodidae [17] and Laophontidae [18]. Unfortunately, I was unable to rescue any DNA markers for the new species. However, the phylogenetic relationships among genera in Nannopodidae could potentially be confirmed by molecular data in a future study.

Key to genera of Nannopodidae Por, 1986 amended from [4]

1. Exopods of P2–P4 three-segmented ... 2

    Exopods of P2–P4 two-segmented ... 4

2. Rostrum bell-shaped, anterior margin with multiple rows of long setules; P1 endopod distinctly shorter than exopod; P2 endopod two-segmented ... 3

    Rostrum triangular, without setular ornamentations; P1 endopod extending beyond distal margin of exp-3; P2 endopod one-segmented with one apical seta ... *Pontopolites* T.Scott, 1894

3. Antennule ♀five-segmented; P2 enp-2 with 2–3 setae; P3 endopod ♂with apophysis ... *Nannopus* Brady, 1880

    Antennule ♀five-segmented; P2 enp-2 with one seta; P3 endopod ♂with no modification ... *Doolia* gen. nov.

4. Antennary exopod one-segmented; P1 exopod one-segmented ... *Talpacoxa* Corgosinho, 2012

    Antennary exopod absent, or represented by a seta; P1 exopod two- or, three-segmented ... 5

5. Antennule ♀four-segmented; P1 endopod prehensile, distinctly longer than exopod ... 6

    Antennule ♀five-segmented; P1 endopod not prehensile, distinctly shorter than exopod ... 7

6. P2–P4 with outer spinous projection on coxa; P5 exopod and baseoendopod fused in ♀, forming a single plate with eight setae/spines ... *Acuticoxa* Huys & Kihara, 2010

    P2–P4 without coxal processes; P5 biramous in ♀ ... *Laophontisochra* George, 2002

7. Rostrum with setula ornamentation around apex: antennule ♀with aesthetasc on segment three; distal exopod segment of P1 with innermost seta much longer than inner distal spine and penicillate apically ... *Huntemannia* Poppe, 1884

    Rostrum without setula ornamentation around apex: antennule ♀with aesthetasc on segment four; distal exopod segment of P1 with innermost seta much shorter than inner distal spine and penicillate at tip ... *Rosacletodes* Wells, 1985

Funding: This research was supported by the Discovery of Korean Indigenous Species Project, NIBR (National Institute of Biological Resources, grant no. 2020), the Marine Biotechnology Program (No. 20170431) of the Korea Institute of Marine Science and Technology Promotion (KIMST) funded by the Ministry of Oceans and Fisheries (MOF), and a grant (2018R1D1A1B07050117) from Korea Research Foundation.

Acknowledgments: The author would like to thank Joon Back Lee and the crew of the R/V Ara (Jeju National University) for their help in collecting samples and support during the study period. I would also like to thank Hee Jin Moon and Jisu Yeom (Hanyang University) for their valuable help in the drawings, and in the preparation of the manuscript, respectively.

Conflicts of Interest: The author declares no conflicts of interest.

References

1. Por, F.D. A Re-evaluation of the family Cletodidae Sars, Lang (Copepoda, Harpacticoida). *Syllogeus* **1986**, *58*, 420–425.
2. Lang, K. Die familie der Cletodidae Sars, 1909. *Zool. Jahrb. Abt. Anat. Ontog. Tiere* **1936**, *68*, 446–480.
3. Huys, R. Unresolved cases of type fixation, synonymy and homonymy in harpacticoid copepod nomenclature (Crustacea: Copepoda). *Zootaxa* **2009**, *2183*, 1–99. [CrossRef]
4. Huys, R.; Kihara, T.C. Systematics and phylogeny of Cristacoxidae (Copepoda, Harpacticoida): A review. *Zootaxa* **2010**, *2568*, 1–38. [CrossRef]
5. Nam, E.-J.; Lee, W. Two new species of the genus Heteropsyllus (Crustacea, Copepoda, Harpacticoida) from Jeju Island, Korea and Devon, England. *J. Nat. Hist.* **2006**, *40*, 1719–1745. [CrossRef]
6. Huys, R.; Gee, J.M.; Moore, C.G.; Hamond, R. Marine and Brackish Water Harpacticoid copepods. Part 1. In *Synopses of the British Fauna (New Series)*; Field Studies Council: Shrewsbury, UK, 1996; pp. 4–19.
7. Vakati, V.; Lee, W. Five new species of the genus Nannopus (Copepoda: Harpacticoida: Nannopodidae) from intertidal mudflats of the Korean West Coast (Yellow Sea). *Zootaxa* **2017**, *4360*, 001–066. [CrossRef] [PubMed]
8. Song, S.J.; Rho, H.S.; Kim, W. A new species of *Huntemannia* (Copepoda: Harpacticoida: Huntemanniidae) from the Yellow Sea, Korea. *Zootaxa* **2007**, *1616*, 37–48. [CrossRef]
9. Karanovic, T.; Cho, J.-L. Second members of the harpacticoid genera *Pontopolites* and *Pseudoleptomesochra* (Crustacea, Copepoda) are new species from Korean marine interstitial. *Mar. Biodivers.* **2018**, *48*, 367–393. [CrossRef]
10. Pallares, R.E. Copépodos harpacticoides marinos de Tierra del Fuego (Argentina). IV. Bahía Tethis. *Contrib. Cient. Centr. Investig. Biol. Mar. Buenos Aires* **1982**, *186*, 3–39.
11. Björnberg, T. Three new species of benthonic Harpacticoida (Copepoda, Crustacea) from São Sebastião Channel. *Nauplius* **2014**, *22*, 75–90. [CrossRef]
12. George, K.H. New phylogenetic aspects of the Cristacoxidae Huys (Copepoda, Harpacticoida), including the description of a new genus from the Magellan region. *Vie Milieu* **2002**, *52*, 31–41.
13. Corgosinho, P.H.C. *Tapacoxa brandini* get. et sp. nov. a new Nannopodidae Brady, 1880 (Copepoda: Harpacticoida) from submersed sands of Pontal do Sul (Paraná, Brazil). *J. Nat. Hist.* **2012**, *46*, 2865–2879. [CrossRef]
14. Alper, A.; Sak, S.; Metin, O. First record of the family Rhizotrichidae (Copepoda, Harpacticoida) from Turkey with description of a new species. *Mar. Biodivers.* **2018**, *48*, 357–365. [CrossRef]
15. Nam, E.J.; Lee, W. A new species of the genus *Rhizothrix* (Copepoda: Harpacticoida: Rhizotrichidae) from Korean waters. *Proc. Biol. Soc. Wash.* **2005**, *118*, 692–705. [CrossRef]
16. Vakati, V.; Eyun, S.-I.; Lee, W. Unraveling the intricate biodiversity of the benthic harpacticoid genus *Nannopus* (Copepoda, Harpacticoida, Nannopodidae) in Korean waters. *Mol. Phylogenet. Evol.* **2019**, *130*, 366–379. [CrossRef] [PubMed]
17. Karanovic, T.; Kim, K.; Lee, W. Concordance between molecular and morphology-based phylogenies of Korean *Enhydrosoma* (Copepoda: Harpacticoida: Cletodidae) highlights important synapomorphies and homoplasies in this genus globally. *Zootaxa* **2015**, *3990*, 451–496. [CrossRef] [PubMed]
18. Yeom, J.; Nikitin, M.A.; Ivanenko, V.N.; Lee, W. A new minute ectosymbiotic harpacticoid copepod living on the sea cucumber *Eupentacta fraudatrix* in the East/Japan Sea. *PeerJ* **2018**, *6*, e4979. [CrossRef] [PubMed]

 *diversity*

Article

A New Species of *Monstrillopsis* Sars, 1921 (Copepoda: Monstrilloida) with an Unusually Reduced Urosome

Donggu Jeon [1], Wonchoel Lee [2], Ho Young Soh [3] and Seong-il Eyun [1,*]

[1]   Department of Life Science, Chung-Ang University, 84 Heukseok-ro, Seoul 06974, Korea; donggu84@gmail.com

[2]   Department of Life Science, Hanyang University, 222 Wangsimni-ro, Seoul 04763, Korea; wlee@hanyang.ac.kr

[3]   Faculty of Marine Technology, Chonnam National University, 50 Daehak-ro, Yeosu, Jeollanam-do 59626, Korea; hysoh@chonnam.ac.kr

*   Correspondence: eyun@cau.ac.kr

http://zoobank.org/urn:lsid:zoobank.org:pub:36893F6D-A5D0-46C1-9D56-9FABBD6BF212

Received: 27 November 2019; Accepted: 18 December 2019; Published: 20 December 2019

Abstract: Male monstrilloid copepods, described herein as *Monstrillopsis paradoxa* sp. nov., were collected from the Chuja Islands, Jeju, Korea, using a light trap. They display many of the common features of *Monstrillopsis*, including large, prominent eyes, an anteriorly positioned oral papilla, and four setae on each caudal ramus. Type-2 modification of the antennules further supports the assignment of the new species to *Monstrillopsis*. However, the present specimens have an unusually low number of urosomal somites, just three in total, compared to five in males of all congeneric species, and from four (in *Cymbasoma*) to five in males of all other monstrilloid genera. Up until now, in the Monstrilloida only females of *Cymbasoma* have been known to have as few as three urosomal somites.

Keywords: Monstrillidae; *Monstrillopsis paradoxa* sp. nov.; *Monstrillopsis planifrons*; morphological taxonomy; tagmosis; male genitalia; pore pattern; male/female matching; marine invertebrate host; semi-parasitic; Korea

---

1. Introduction

Monstrilloida Sars, 1901 is a small group of marine copepods that comprise approximately 170 nominal species in seven genera currently considered as valid: *Monstrilla* Dana, 1849; *Cymbasoma* Thompson, 1888; *Monstrillopsis* Sars, 1921; *Maemonstrilla* Grygier & Ohtsuka, 2008; *Australomonstrillopsis* Suárez-Morales & McKinnon, 2014; *Caromiobenella* Jeon, Lee & Soh, 2018; *Spinomonstrilla* Suárez-Morales, 2019 [1–7]. This group is distinguished from other copepods by their peculiar life cycle and strange morphology. They have a protelean life-cycle consisting of an endoparasitic juvenile phase and a planktonic adult stage. The juveniles are known to infect various marine invertebrates such as polychaetes, gastropods and bivalve mollusks [8–18]. The adults, often regarded as an exclusively reproductive stage, lack all mouth parts as well as the antennae and are non-feeding.

Most of the recent work on monstrilloids has focused on taxonomy and the morphological description of adults, but studies on the parasitic life stages and host-parasite interactions have been on the increase [2,17–19]. As well, research into monstrilloid diversity has been expanding into lesser-studied regions. For example, Suárez-Morales and McKinnon [6,20] recently described 34 species of four genera from Australian waters, with a promised treatment of similar numbers of Australian *Monstrilla* still to come, and such findings imply the possible occurrence of large numbers of undescribed taxa in other such areas as well.

East Asia is clearly a region where the monstrilloid fauna has not been yet fully revealed [16,21]. Serious research on monstrilloids in Korea began less than a decade ago with the first record of *Cymbasoma striifrons* Chang, 2012. So far nine species in four genera have been reported from Korea's eastern and southern coastal waters [3,22–26]. The site of the present study, the Chuja Islands, is a group of 42 small islands, four of which are inhabited. They are located between Jeju Island and the mainland of Korea and are affected by the Tsushima Warm Current. Temperate and subtropical creatures coexist there [27,28], leading to the expectation of high biodiversity in their surrounding waters. Here we report a new type of monstrilloid with a novel and unusual morphological feature for the group, and described it as a new species.

2. Materials and Methods

A plankton sample was collected by Min Ho Seo (Marine Ecology Research Center, Korea) using a hand-made polyvinyl chloride (PVC) light trap containing an light emitting diode (LED) flashlight as a light source. The type materials of the present new species were collected from 19:00 to 7:00 on 11–12 September 2017 alongside a wharf at the type locality (Figure 1). The captured contents were filtered using a 63 μm mesh test sieve, and the retained material was immediately washed with 99.5% ethanol. The sample was initially fixed with 99.5% ethanol on-site and the fixative was exchanged for fresh 99.5% ethanol in the laboratory. Monstrilloids were sorted out from the bulk collection under a SMZ645 stereomicroscope (Nikon, Tokyo, Japan) and kept refrigerated at 4 °C. The osmotic shock of ethanol fixation caused the cephalothorax of many specimens to collapse, so monstrilloid specimens used for morphological analysis were exposed to 0.25% trisodium phosphate dodecahydrate solution ($Na_3PO_4 \cdot 12H_2O$; Daejung Chemicals and Metals, Siheung, Korea) to restore its original shape [3,12,31]. An Eclipse 80i compound microscope with a drawing tube and differential interference optics was used for preparing illustrations. The holotype specimen was dissected into five parts, and each was mounted on a slide separately with lactophenol for detailed examination. Body measurements were obtained by using AxioVision LE64 software (AxioVs40x64 v 4.9.1.0; Carl Zeiss, Oberkochen, Germany) based on the illustrations of the type material. Terminology from Grygier and Ohtsuka [2] and Jeon, et al. [32] was used to describe body segmentation and antennular setation patterns, respectively.

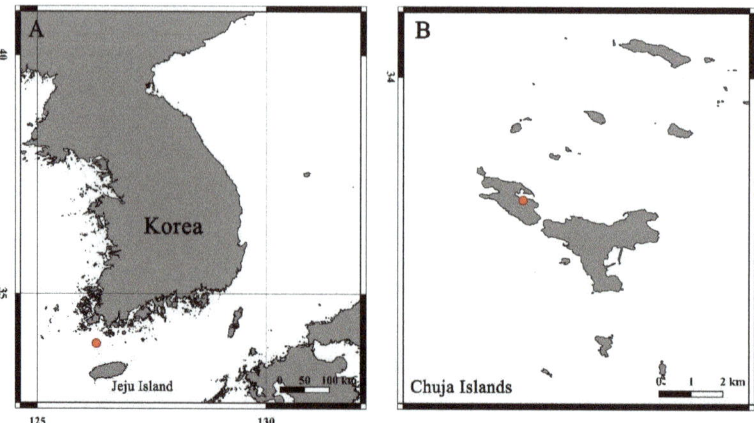

Figure 1. Maps showing the investigated area and the sampling locality. (**A**) Location of Chuja Islands (red dot), Korea; (**B**) sampling site (red dot) in Chuja Islands. The maps were prepared using QGIS [29], a free and open-source geographic information system, with the OpenStreetMap data [30].

3. Results

3.1. Systematics

Order Monstrilloida Sars, 1901
Family Monstrillidae Dana, 1849
Genus *Monstrillopsis* Sars, 1921
Monstrillopsis paradoxa sp. nov.
Figures 2–4

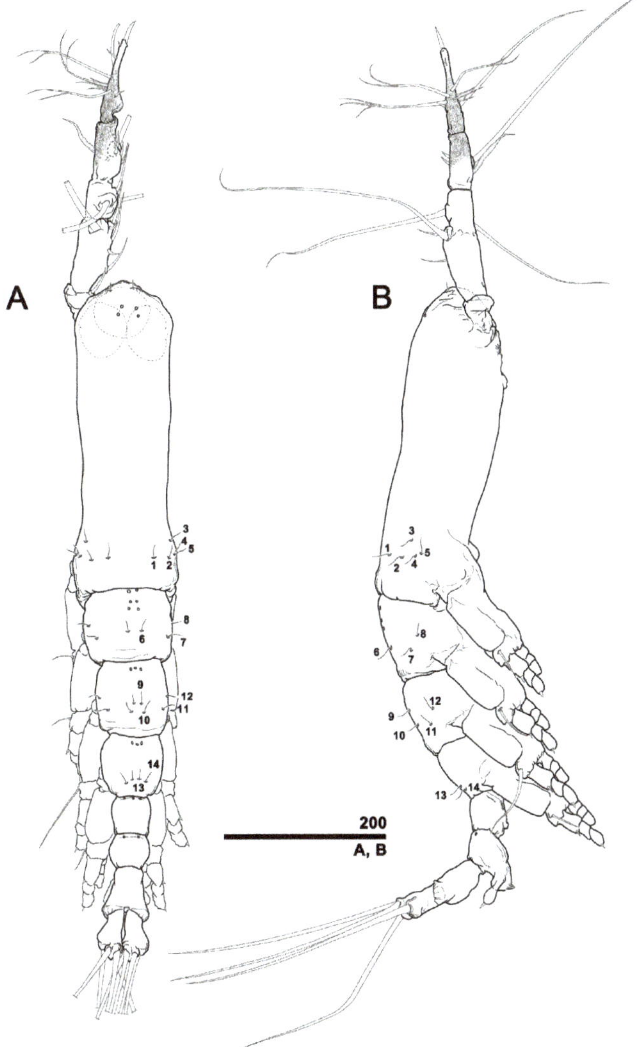

Figure 2. *Monstrillopsis paradoxa* sp. nov., male holotype (MABIK CR00246526) (**A**) Habitus with pit-setae 1–14 of right side indicated, dorsal; (**B**) habitus, lateral. Scale bars are in micrometers.

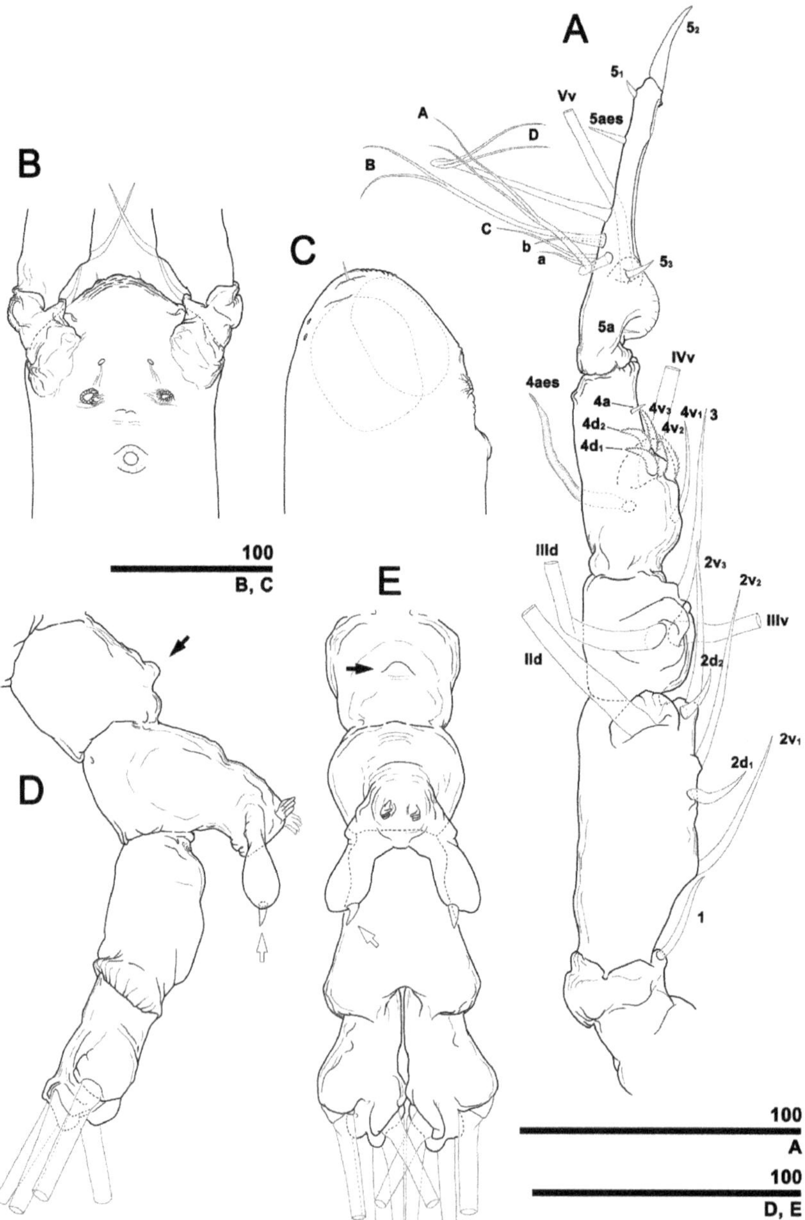

Figure 3. *Monstrillopsis paradoxa* sp. nov., male holotype (MABIK CR00246526). (**A**) Antennule, left, dorsal; (**B**) anterior ventral part of cephalothorax showing crumpled margin; (**C**) anterior part of cephalothorax, lateral; (**D**) urosome showing ventral protuberance (black arrow) on first urosomal somite and apical spinous element (white arrow) on genital lappet, lateral; (**E**) urosome showing ventral protuberance (black arrow) on the first urosomal somite and apical spinous element (white arrow) on right genital lappet. Scale bars are in micrometers.

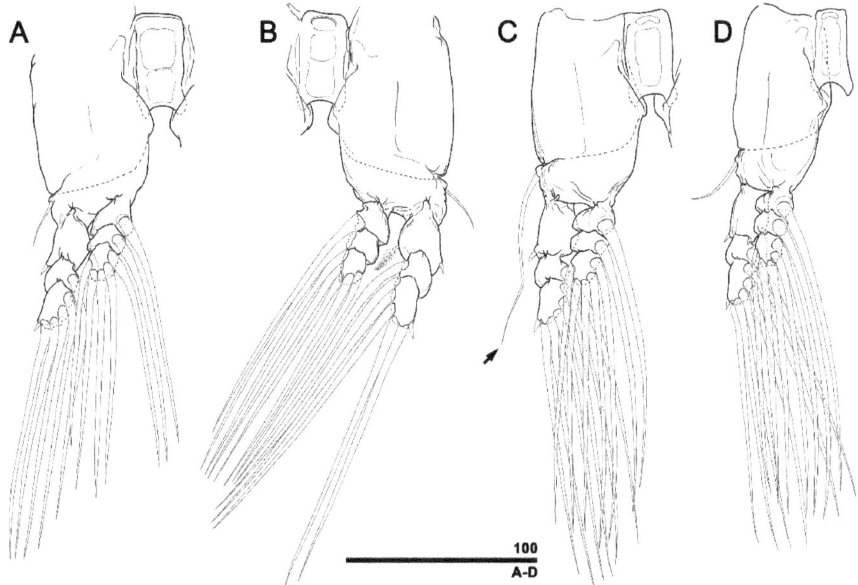

Figure 4. *Monstrillopsis paradoxa* sp. nov., male holotype (MABIK CR00246526). (**A**) Leg 1 with intercoxal sclerite, right, anterior; (**B**) leg 2 with intercoxal sclerite, left, anterior; (**C**) leg 3 showing long basal seta (arrow) and intercoxal sclerite, right, anterior; (**D**) leg 4 with intercoxal sclerite, left, anterior. Scale bars are in micrometers.

3.2. Type Locality

Yeongheung-ri (33°57.59″ N, 126°17.82″ E), Chuja-myeon, Jeju-si, Jeju-do, Republic of Korea. English equivalents of political divisions in Korea: ri = village; myeon = township; si = city; do = province.

3.3. Type Material Examined

The male holotype (MABIK CR00246526; dissected and body parts mounted on five separate slides) and five intact paratypes (MABIK CR00246527; in a 99.5% ethanol vial) are deposited in the National Marine Biodiversity Institute of Korea (MABIK), Seochon, Korea.

3.4. Species Diagnosis. Male

Anterior end of cephalothorax rugose with transverse striations. Antennule 5-segmented with geniculation between fourth and fifth segments. Fifth segment showing type-2 modification [13]: inner proximal one-third with hyaline bump, rest of distal part thin, elongate. Two distal segments dark brown in ethanol-preserved specimens (coloration becoming fainter with time, and immediate partial bleaching occurring upon exposure to Na_3PO_4 solution and/or lactophenol): outer lateral side of the fourth segment most strongly pigmented; the fifth segment with gradual distal weakening of pigmentation, completely lacking pigment near the tip. Outermost natatory seta on the third exopodal segment of all legs smooth, lacking typical serration along its outer side. Outer basal seta on third legs extremely long, well exceeding its exopod. Anterior dorsum of first free pediger with three pairs of pores arrayed in two rows flanking midline. Succeeding three somites, i.e. second and third free pedigers and first urosomal somite, each with a transverse row of three pores near the anterior dorsal margin, with middle pore on midline. Urosome consisting of three somites, viz., limbless fourth free pediger (= first urosomal somite) and genital somite followed directly by anal somite, with no post-genital or penultimate (= preanal) somites. First urosomal somite lacking fifth legs, but with small posterioventral protuberance instead. Genital somite with ventral genital apparatus consisting of a shaft and two lappets. Shaft robust with abrupt notch at midlength in lateral view, more distal part slightly thinner than proximal part and bearing two hand-shaped opercular flaps anteriorly at tip. Genital lappets diverging from posterior distal part of the shaft, clavate in lateral view, rather lamelliform in ventral view. Each lappet armed with a short, robust spinous element at the inner distal corner. Caudal rami twice as long as wide, each with four seate. All caudal setae subequal in length, with inner ventral seta almost always extending diagonally downwards. Measurements: total body length ranged from 0.66–0.88 mm (mean 0.78 mm; $N = 6$ for all measurements reported here). Length ratio (means, with ranges in parentheses) of cephalothorax, metasome, and urosome 49.0 (46.6–53.2):34.1 (30.2–36.6):16.9 (15.9–19.6) in lateral view. Oral papilla located 25.9% (24.3–27.4%) back along the ventral side of cephalothorax. Length of antennules in relation to total body length 45.1% (42.1–51.0%), and to length of cephalothorax 91.9% (86.0–97.0%). Ratio of antennular segment length from proximal to distal 12.3 (11.8–13.8):23.8 (21.7–26.6):14.2 (11.9–16.3):19.9 (19.4–20.4):29.8 (28.9–30.4).

3.5. Description of Male Holotype (MABIK CR00246526)

Total body length 0.76 mm in dorsal view, 0.78 mm in lateral view (Figure 2A,B). Body consisting of cephalothorax incorporating first pediger, free somites 1–3, first urosomal somite, genital somite, and anal somite. Length of somites as percent of total body length 49:12:12:9:6:5:7 in dorsal view, 49:14:12:9:5:4:7 in lateral view. Cephalothorax cylindrical, 0.37 mm long in dorsal view, 0.38 mm long in lateral view, accounting for almost half of total body length. Anterior margin convex, its surface corrugated with strong transversal striations. Anterior part of the forehead with two short, thin sensilla. Width of cephalothorax almost unvarying along its length: narrowest (= waist width; 0.11 mm) at 76.5%, widest (= incorporated first pediger; 0.13 mm) at 91.9% of distance from anterior end. Anterior one-fourth nearly fully occupied by one ventral and two lateral eyes (Figure 2A, Figure 3C). Ventral eye located more anteriorly than lateral eyes, with 1.29 times greater diameter than the latter (70.3 µm versus 54.5 µm). Dorsal surface at the level of eyes with at least four pores in the trapezoidal array (Figure 2A). Ventral surface at the same level with pair of pores and pair of scars (Figure 3B); pores located medially between antennular bases, and scars slightly behind of antennular bases but still medial to them. Moderately developed oral papilla located behind scars, 26% back from anterior margin of cephalothorax. Tergite of incorporated first pediger with five pairs of pit-setae *sensu* Grygier and Ohtsuka [19]: one pair (no. 1) situated dorsally, four pairs laterally (nos. 2–5).

Body somites from first pedigerous somite to genital somite with several pores in various places (Figure 2A,B). First free pediger with three pairs of pit-setae posteriorly (nos. 6–8: one pair dorsally, two pairs laterally) and two longitudinal rows of four pores each, arranged in pairs across midline, with most anterior pore pair covered by posterior cuticular extension of cephalothorax. Second free pediger with four pairs of pit-setae posteriorly (nos. 9–12: two pairs dorsally, two pairs laterally) and three simple pores aligned transversally across anterior midline. Third free pediger with two pairs of pit-setae posteriorly (nos. 13, 14), aligned transversely across dorsum, and transverse row of three simple pores on anterior dorsum. The first urosomal somite also with three simple pores on anterior dorsum and with protuberance near the posterior end of the ventral side. Genital somite with a pair of simple pores in anterior dorsum.

Antennules 5-segmented, with geniculation between fourth and fifth segments (Figure 3A). Length 0.35 mm excluding apical spine 5_2, equal to 44.1% of total body length, 90.2% of cephalothorax length. Length ratio of segments 12:26:12:20:30. First antennular segment with spinous element 1 on inner distal corner, arising slightly dorsally. Second segment armed with five spinous elements ($2v_{1-3}$, $2d_{1,2}$) plus long, strap-like, biplumose seta IId. Ventral spinous elements (2v-series) well developed, slightly longer than dorsal ones (2d-series). Third segment with three elements (3, IIId, IIIv); spinous element 3 located distally on inner side, long IIIv and IIId setae extending ventrally and dorsally, respectively, from midlength of segment. Fourth segment with eight elements ($4v_{1-3}$, $4d_{1,2}$, 4a, IVv, 4aes): naked spinous element $4v_1$ thin, twice as long as other 4d-, 4v-elements; four spinous elements $4v_{2,3}$ and $4d_{1,2}$ all robust, pinnate, curved toward bearing segment; spinous element 4a minute, arising at inner distal one-third of segment; IVv and 4aes arising from ventral surface of segment, 4aes more proximally. Fifth antennular segment modified: inner proximal margin with semicircular expansion, rest of distal part relatively thin, elongate, and twice as long as proximal expansion (Figure 3A). Inner side of distal elongated part with thin, film-like hyaline edge. Segment armed with 12 setal elements. Among these, spinous element 5_1 short, located on outer distal margin; apical spinous element 5_2 elongate, 26% as long as bearing segment; inner spinous element 5_3 arising from dorsal distal part of inner expansion; minute spinous element 5a arising from proximal part of expansion; and long, strap-like seta Vv arising from ventral surface at level of spinous element 5_3. Six outer setae present at midlength of fifth segment: four branched setae (A–D) and two short, simple setae (a, b). Branched seta C rather simple, significantly shorter than other branched setae but still longer than two simple setae. Aesthetasc (5aes) located on the outer side of the segment's distal quarter, longer than spine 5_1. Fourth antennular segment dark brown, especially prominent along the outer lateral side. Proximal half of fifth segment except for hyaline expansion also dark, but pigmentation fading distally, completely absent at tip.

Genital somite rounded in dorsal view, with robust genital shaft constricted at midlength so as to appear notched in lateral view (Figure 2A,Figure 3D,E). Distal part armed with two hand-shaped opercular flaps. Pair of genital lappets diverging from posterior distal part of the shaft, clavate in lateral view, relatively flat and wide in ventral view (Figure 3D,E). Each lappet armed with a short spinous element at the inner distal corner (Figure 3D,E).

Caudal rami (Figure 3D,E) diverging from posterior margin of anal somite, 1.8 times longer than wide. Each ramus armed with four setae: one outer lateral, two dorsal apical, one inner ventral. All caudal setae biplumose, subequal in length.

Incorporated first pediger and three succeeding free pedigers each with pair of well-developed swimming legs (Figure 4A–D). Protopods consisting of a large coxal portion and a relatively small basis. Anterior articulation between coxa and basis poorly defined whereas posterior diagonal articulation clearly expressed. Coxae of each leg pair joined by longitudinally elongate, rectangular intercoxal sclerite 1.9 times longer than wide. Basis of legs 1, 2 and 4 with short seta proximally on outer margin, reaching at most to middle of first exopodal segment; on leg 3 this seta coarsely plumose and distinctly longer, extending well beyond the tip of exopod. Basis with tri-articulate endopod and exopod on distal margin, with endopod positioned more anteriorly than exopod and reaching only to the distal end of the second exopodal segment. First and third exopodal segments 2 times longer than the second segment, but all three endopodal segments subequal in length. Setal patterns of legs 1–4 are shown in Table 1.

Table 1. Setal armature patterns of legs 1–4.

	Coxa	Basis	Exopod	Endopod
Leg 1	0-0	1-0	I-1; 0-1; I, 2, 2	0-1; 0-1; 1, 2, 2
Legs 2–4	0-0	1-0	I-1; 0-1; I, 2, 3	0-1; 0-1; 1, 2, 2

Roman numerals indicate the number of spines, Arabic numerals indicate number of seate.

Outermost setae of third exopodal segments of all legs smooth, with no serration along outer side. Natatory setae subequally long except short inner setae on first exopodal segment not extending beyond tip if endopod. Fifth legs absent.

3.6. Females

Unrecognized.

3.7. Remarks

The present males are assignable to *Monstrillopsis* by virtue of the presence of type-2 male antennular modification with the fifth segment modified into a gradually tapering, elongate distal half with a slightly inner curved spine on the tip and a hyaline expansion on the inner proximal margin [12,21]. This kind of modification is shared with seven congeners, including *M. sarsi* Isaac, 1974, *M. dubioides* Suárez-Morales, 2004 [4] (for male), *M. chilensis* Suárez–Morales, Bello–Smith & Palma, 2006 [33] (for male), *M. boonwurrungorum* Suárez-Morales & McKinnon, 2014 [34] (correct original spelling), *M. hastata* Suárez-Morales & McKinnon, 2014, *M. nanus* Suárez–Morales & McKinnon, 2014, and *M. pontoeuxinensis* Suárez–Morales & Üstün, 2018. Males of six other congeners, *M. reticulata* (Davis, 1949), *M. fosshageni* Suárez-Morales & Dias, 2001, *M. chathamensis* Suárez-Morales & Morales–Ramírez, 2009, *M. cahuitae* Suárez–Morales & Carrillo, 2013, *M. coreensis* Lee, Kim & Chang, 2016 and *M. longilobata* Lee, Kim & Chang 2016 have a much less elongate distal portion of the fifth segment, with a rather long and robust apical spinous element 5_2. Four species of the former group, *M. dubioides*, *M. sarsi*, *M. nanus*, and *M. pontoeuxinensis* can be immediately excluded from further morphological consideration because of significant differences in body size compared to *M. paradoxa* sp. nov., either larger (2.1 mm for *M. dubioides*, 1.2 mm for *M. sarsi*) or smaller (0.5 mm for *M. nanus* and 0.6 mm for *M. pontoeuxinensis*) [4,6,35,36].

Suárez-Morales and McKinnon [6] recognized mainly two types of male genitalia in *Monstrillopsis*: type I with a strong, well-developed median shaft and relatively short, rounded lappets, and type II with a relatively short median shaft and very elongate lappets arising from the posterolateral corners. The latter is characteristic of all three congeners with type-2 male antennules that were not excluded above on account of body size. These have long, slender, cylindrical lappets [6,26,33], however, while the genital lappets of the new species appear lamellar in the ventral view. In addition, the lappets of *M. paradoxa* sp. nov. are armed with a very short but robust inner distal spinous element, something not previously reported in males of any species of *Monstrillopsis*. The most stunning morphological character of the present males is the unusually low number of somites in the urosome. Up to now, both sexes of *Cymbasoma* have been known to have the fewest urosomal somites among the known monstrilloid genera—females with three, including post-genital somite, and males with four, including post-genital and penultimate somites [8,20,21]. Males of the present new species have one fewer somite than males of *Cymbasoma* while matching the number in females of *Cymbasoma*. Despite this disparity compared to other species of *Monstrillopsis*, which have five urosomal somites in general, nevertheless, we have assigned the present males to *Monstrillopsis* in consideration of its possession of several other *Monstrillopsis*-like features: large, prominent eyes, an anteriorly-located oral papilla, and four caudal setae (see Discussion).

3.8. Etymology

The specific epithet is derived from the Latin adjective *paradoxus, -a, -um* (adopted from the Greek *paradoxos*), meaning strange, with a feminine ending to match the gender of the genus name. It pertains to the contradictory morphological characteristics, a type-2 antennular modification supporting placement in *Monstrillopsis*, and an unusually low number of body somites, which argues against such placement.

4. Discussion

The number of body somites has been considered one of the most important, convenient, and evident key features in the monstrilloid taxonomy [4,7,8,37]. The species of *Cymbasoma*, for instance, have been distinguished from the other monstrilloid genera based almost on this criterion alone [38–40]. A review of the most recent 20 years of taxonomic works dealing with over 40 species of *Cymbasoma* (more than half of the total species recorded in this genus) shows that authors continue to use this feature as the primary feature for generic assignment [20,22,41–46]. In light of this history, the even lower number of urosomal somites in the present specimens may provide grounds for proposing a new genus of Monstrilloida. On the other hand, the new species presents a typical type-2 antennular modification, which may be just as important. Full expression of this feature is shared by more than half of the hitherto known males of *Monstrillopsis*, and including partial expression, it has been regarded as a diagnostic feature for the genus [12,47]. Therefore, antennular morphology unambiguously suggests a close morphological affinity of the present specimens to *Monstrillopsis*, and the presence of other *Monstrillopsis*-like characteristics in them further supports their assignment to this genus.

The cuticular pore pattern described above for *Monstrillopsis paradoxa* sp. nov. can be compared with that reported from female *M. planifrons* Delaforge, Suárez-Morales, Walkusz, Campbell & Mundy, 2017. Unlike any other congeneric species, this species has three pairs of dorsal structures, described as "minute papilla-like processes" [48] (p. 4), flanking the midline of the first free pediger. The present new species has six pores arrayed the same way on this somite. Perhaps the structures in *M. planifrons* were also sort of pores. Be this as it may, it is worth noting that Grygier and Ohtsuka [2] have already considered the potential utility of cuticular ornamentation patterns for differentiation and identification. Based on a limited number of examples, they found a stereotypical pore pattern among monstrilloids that distinguishes them from other copepod groups. In addition, based on detailed SEM photographs of four species of *Maemonstrilla*, Grygier and Ohtsuka showed that these species share a similar pore arrangement on the second free pediger [2] (fig. 29). The species of *Caromiobenella* provide another example in that they share a set of four pairs of pores flanking the midline along the posterior medio-dorsal part of the cephalothorax [3,19,23,24]. The currently available evidence thus suggests that pore pattern may be useful in identifying monstrilloids at the genus-level; if so, the present new species still falls into *Monstrillopsis*.

A distinctive pore pattern shared between individuals of different sexes can occasionally reveal their conspecificity. The presence of a similar pore pattern in certain males and females was taken as evidence that both belonged to *Monstrillopsis longilobata*, and this was corroborated by molecular analysis [32]. With this example in mind, the possibility that *M. planifrons* and *M. paradoxa* sp. nov. are the females and males, respectively, of one species must be considered. We think it unlikely, on account of substantial differences in other morphological details besides the dorsal ornamentation of the first free pediger. In particular, unlike males of *M. paradoxa* sp. nov., females of *M. planifrons* are characterized in having (1) a produced, flat, corrugated forehead and (2) a cephalothorax entirely covered in numerous minute papillae, that are considered as species-specific features [48]. None of those features is presented in the present male specimens: the present males have a rounded, convex forehead with transverse striae. The whole cephalothorax is smooth without additional ornamental structures. The outermost seta of the third exopodal segment of legs 1–4 is also different in each species. This seta is, in general, with an inner row of setules and an outer row of coarse spinules as so in *M. planifrons*. The seta of all legs in the present new species is, however, presented without the outer serration. The significant size difference in body length (1.92 mm for the female holotype of *M. planifrons* versus a range of 0.66–0.88 mm for the present males) and considerations of biogeography and habitat–*M. planifrons* being from the Canadian Arctic and *M. paradoxa* sp. nov. from warm-temperate or subtropical waters in East Asia–mitigate against conspecificity [48]. On the other hand, the characteristics mentioned above of the present males may be useful as morphological markers for finding the currently unknown female counterparts. The distinct pigmentation pattern exhibited on the distal part of the antennules in the present males might also be expected to appear in the unknown females.

Female monstrilloids usually have one fewer urosomal somite than the corresponding males [4,8,12,21]. In this rule holds for *M. paradoxa* sp. nov., the unknown females should have a urosome consisting of just two somites: either the first urosomal somite (= fourth free pediger) and the genital (compound) somite, or the genital (compound) somite and anal somite. Loss of the genital somite with retention of the other two would not be supportable, but from a functional point of view, none of these options seems likely. Therefore, we presume that the females of *M. paradoxa* sp. nov. will prove to have at least three somites in their urosome. If so, they will be similar to females of *Cymbasoma* in terms of the body plan, but distinguishable by the number of caudal setae. Most female *Cymbasoma* have three caudal setae [20,21], and the number of caudal setae is the same (four) in both sexes of those species of *Monstrillopsis* [47].

Our discovery of a new body plan in specimens of the greatly modified, but nonetheless highly stereotyped group Monstrilloida is a sign that further collection and detailed examination of these copepods will continue to turn up surprises. The difficulty in classifying the new species also shows that a phylogenetic systematic approach to monstrilloid taxonomy is overdue. We hope to make progress on this in upcoming works.

Author Contributions: Conceptualization, D.J., W.L., H.Y.S. and S.-i.E.; Data curation, D.J., W.L., H.Y.S. and S.-i.E.; formal analysis, D.J., W.L., H.Y.S. and S.-i.E.; funding acquisition, D.J. and S.-i.E.; investigation, D.J.; methodology, D.J.; project administration, D.J., W.L., H.Y.S. and S.-i.E.; resources, D.J., W.L., H.Y.S. and S.-i.E.; supervision, W.L., H.Y.S. and S.-i.E.; validation, D.J., W.L., H.Y.S. and S.-i.E.; visualization, D.J.; writing—original draft, D.J.; writing—review and editing, D.J., W.L., H.Y.S. and S.-i.E. All authors have read and agreed to the published version of the manuscript.

Funding: This study was supported by Basic Science Research Program through the National Research Foundation of Korea (NRF) funded by the Ministry of Education (MOE) (grant no. 2019R1I1A1A01057892) and the Chung-Ang University Research Grants in 2018.

Acknowledgments: We are grateful to Mark J. Grygier (Center for Excellence for the Oceans, National Taiwan Ocean University, Taiwan) for providing his kind English correction and valuable comments to improve the overall quality of the manuscript. We also appreciate the three anonymous reviewers and the editors for providing their helpful comments and advice.

References

1. Dana, J.D. Conspectus Crustaceorum quae in Orbis Terrarum circumnavigatione, Carolo Wilkes e Classe Reipublicae Faederatae Duce, lexit et descripsit Jacobus, D. Dana. Pars II. *Proc. Am. Acad. Arts Sci.* **1849**, *2*, 9–61.

2. Grygier, M.J.; Ohtsuka, S. A new genus of monstrilloid copepods (Crustacea) with anteriorly pointing ovigerous spines and related adaptations for subthoracic brooding. *Zool. J. Linn. Soc.* **2008**, *152*, 459–506. [CrossRef]

3. Jeon, D.; Lee, W.; Soh, H.Y. A new genus and two new species of monstrilloid copepods (Copepoda: Monstrillidae): Integrating morphological, molecular phylogenetic, and ecological evidence. *J. Crustacean Biol.* **2018**, *38*, 45–65. [CrossRef]

4. Sars, G.O. An account of the Crustacea of Norway with short descriptions and figures of all the species. In *Copepoda Monstrilloida & Notodelphyoida*; The Bergen Museum: Bergen, NJ, USA, 1921; Volume III, p. 91.

5. Suárez-Morales, E. A new genus of the Monstrilloida (Copepoda) with large rostral process and metasomal spines, and redescription of *Monstrilla spinosa* Park, 1967. *Crustaceana* **2019**, *92*, 1099–1112. [CrossRef]

6. Suárez-Morales, E.; McKinnon, A.D. The Australian Monstrilloida (Crustacea: Copepoda) I. *Monstrillopsis* Sars, *Maemonstrilla* Grygier & Ohtsuka, and *Australomonstrillopsis* gen. nov. *Zootaxa* **2014**, *3779*, 301–340.

7. Thompson, I.C. Copepoda of Madeira and the Canary Islands, with descriptions of new genera and species. *J. Linn. Soc. Lond. Zool.* **1888**, *20*, 145–156. [CrossRef]

8. Boxshall, G.A.; Halsey, S.H. *An Introduction to Copepod Diversity*; The Ray Society: London, UK, 2004; p. 966.

9. Caullery, M.; Mesnil, F. Sur deux Monstrillides parasites d'Annélides (*Polydora giardi* Mesn. et *Syllis gracilis* Gr.). *Bull. Sci. Fr. Bel.* **1914**, *48*, 15–29.

10. Gallien, L. Description du mâle de *Monstrilla helgolandica* Claus. Synonymie de *Monstrilla serricornis* G. O. Sars et de *Monstrilla helgolandica* Claus. *Bull. Soc. Zool. Fr.* **1934**, *59*, 377–382.

11. Hartman, O. A new monstrilloid copepod parasitic in capitellid polychaetes in Southern California. *Zool. Anz.* **1961**, *167*, 325–334.

12. Huys, R.; Boxshall, G. *Copepod Evolution*; The Ray Society: London, UK, 1991; p. 468.

13. Huys, R.; Llewellyn-Hughes, J.; Conroy-Dalton, S.; Olson, P.D.; Spinks, J.N.; Johnston, D.A. Extraordinary host switching in siphonostomatoid copepods and the demise of the Monstrilloida: Integrating molecular data, ontogeny and antennulary morphology. *Mol. Phylogenet. Evol.* **2007**, *43*, 368–378. [CrossRef]

14. Malaquin, A. Le parasitisme évolutif des Monstrillides (Crustacés Copépodes). *Arch. Zool. Exp. Gén.* **1901**, *9*, 81–232.

15. Pelseneer, P. Éthologie de quelques *Odostomia* et d'un Monstrillide parasite de l'un d'eux. *Bull. Sci. Fr. Bel.* **1914**, *48*, 1–14.

16. Suárez-Morales, E. Monstrilloid copepods: The best of three worlds. *Bull. South. Calif. Acad. Sci.* **2018**, *117*, 92–103. [CrossRef]

17. Suárez-Morales, E.; Harris, L.H.; Ferrari, F.D.; Gasca, R. Late postnaupliar development of *Monstrilla* sp. (Copepoda: Monstrilloida), a protelean endoparasite of benthic polychaetes. *Invertebr. Reprod. Dev.* **2014**, *58*, 60–73. [CrossRef]

18. Suárez-Morales, E.; Paiva Scardua, M.; Da Silva, P.M. Occurrence and histopathological effects of *Monstrilla* sp. (Copepoda: Monstrilloida) and other parasites in the brown mussel *Perna perna* from Brazil. *J. Mar. Biol. Assoc. UK* **2010**, *90*, 953–958. [CrossRef]

19. Grygier, M.J.; Ohtsuka, S. SEM observation of the nauplius of *Monstrilla hamatapex*, new species, from Japan and an example of upgraded descriptive standards for monstrilloid copepods. *J. Crustacean Biol.* **1995**, *15*, 703–719. [CrossRef]

20. Suárez-Morales, E.; McKinnon, A.D. The Australian Monstrilloida (Crustacea: Copepoda) II. *Cymbasoma* Thompson, 1888. *Zootaxa* **2016**, *4102*, 1–129.

21. Suárez-Morales, E. Diversity of the Monstrilloida (Crustacea: Copepoda). *PLoS ONE* **2011**, *6*, e22915. [CrossRef]

22. Chang, C.Y. First record of Monstrilloid copepods in Korea: Description of a new species of the genus *Cymbasoma* (Monstrilloida, Monstrillidae). *Anim. Syst. Evol. Divers.* **2012**, *28*, 126–132. [CrossRef]

23. Chang, C.Y. Two new records of monstrilloid copepods (Crustacea) from Korea. *Anim. Syst. Evol. Divers.* **2014**, *30*, 206–214. [CrossRef]

24. Jeon, D.; Lee, W.; Soh, H.Y. New species of *Caromiobenella* Jeon, Lee & Soh, 2018 (Crustacea, Copepoda, Monstrilloida) from Chuja Island, Korea. *ZooKeys* **2019**, *814*, 33–51.

25. Lee, J.; Chang, C.Y. A new species of *Monstrilla* Dana, 1849 (Copepoda: Monstrilloida: Monstrillidae) from Korea, including a key to species from the north-west Pacific. *Zootaxa* **2016**, *4174*, 396–409. [CrossRef] [PubMed]

26. Lee, J.; Kim, D.; Chang, C.Y. Two new species of the genus *Monstrillopsis* Sars, 1921 (Copepoda: Monstrilloida: Monstrillidae) from South Korea. *Zootaxa* **2016**, *4174*, 410–423. [CrossRef] [PubMed]

27. Cho, I.-Y.; Kang, D.-W.; Kang, J.; Hwang, H.; Won, J.-H.; Paek, W.K.; Seo, S.-Y. A study on the biodiversity of benthic invertebrates in the waters of Seogwipo, Jeju Island, Korea. *J. Asia Pac. Biodivers.* **2014**, *7*, e11–e18. [CrossRef]

28. Yang, H.J.; Seo, J.E.; Gordon, D.P. Sixteen new generic records of Korean Bryozoa from southern coastal waters and Jeju Island, East China Sea: Evidence of tropical affinities. *Zootaxa* **2018**, *4422*, 493–518. [CrossRef]

29. Van Cleave, H.J.; Ross, J.A. A method for reclaiming dried zoological specimens. *Science* **1947**, *105*, 318. [CrossRef]

30. QGIS. A Free and Open Source Geographic Information Systam. Available online: https://qgis.org (accessed on 20 December 2019).

31. Data Derived from OpenStreetMap for Download. Available online: https://osmdata.openstreetmap.de/ (accessed on 20 December 2019).

32. Jeon, D.; Lim, D.; Lee, W.; Soh, H.Y. First use of molecular evidence to match sexes in the Monstrilloida (Crustacea: Copepoda), and taxonomic implications of the newly recognized and described, partly *Maemonstrilla*-like females of *Monstrillopsis longilobata* Lee, Kim & Chang, 2016. *PeerJ* **2018**, *6*, e4938.

33. Suárez-Morales, E.; Ramírez, F.C.; Derisio, C. Monstrilloida (Crustacea: Copepoda) from the Beagle Channel, South America. *Contrib. Zool.* **2008**, *77*, 217–226. [CrossRef]

34. Grygier, M.J.; Suárez-Morales, E. Recognition and partial solution of nomenclatural issues involving copepods of the family Monstrillidae (Crustacea: Copepoda: Monstrilloida). *Zootaxa* **2018**, *4486*, 497–509. [CrossRef]

35. Isaac, M.J. Copepoda Monstrilloida from south-west Britain including six new species. *J. Mar. Biol. Assoc. UK* **1974**, *54*, 127–140. [CrossRef]

36. Suárez-Morales, E.; Üstün, F. Report on some monstrilloids (Crustacea: Copepoda) from Turkey with description of two new species. *Cah. Biol. Mar.* **2018**, *59*, 547–562.

37. Giesbrecht, W. *Systematik und Faunistik der pelagischen Copepoden des Golfes von Neapel und der angrenzenden Meeres-Abschnitte. Fauna und Flora des Golfes von Neapel und der angrenzenden Meeres-Abschnitte herausgegeben von der Zoologischen Station zu Neapel. XIX. Monographie*; Verlag von R. Friedländer & Sohn: Berlin, Germany, 1892; p. 831.

38. Mageed, A.A.A. *Cymbasoma janetae* n. sp., a new monstrilloid (Copepoda, Monstrilloida) from the Gulf of Aqaba (Red Sea, Egypt). *Crustaceana* **2010**, *83*, 513–523. [CrossRef]

39. Suárez-Morales, E. A new species of *Cymbasoma* (Copepoda: Monstrilloida) from the Mediterranean Sea with remarks on the female of *C. tumorifrons* (Isaac). *Mitt. Mus. Nat. Berl. Zool. Reihe* **2002**, *78*, 87–96. [CrossRef]

40. Suárez-Morales, E.; Morales-Ramírez, A. A new species of Cymbasoma (Crustacea: Copepoda: Monstrilloida) from the Pacific coast of Costa Rica, Central America. *Proc. Biol. Soc. Wash.* **2003**, *116*, 206–214.

41. Lian, X.; Tan, Y. A new species of *Cymbasoma* Thompson, 1888 (Copepoda: Monstrilloida) from the Fujian coast, China. *J. Oceanol. Limnol.* **2019**, *37*, 1709–1713. [CrossRef]

42. Suárez-Morales, E.; Carrillo, A.; Morales-Ramírez, A. Report on some monstrilloids (Crustacea: Copepoda) from a reef area off the Caribbean coast of Costa Rica, Central America with description of two new species. *J. Nat. His.* **2013**, *47*, 619–638. [CrossRef]

43. Suárez-Morales, E.; Dias, C.d.O. Taxonomic report of some monstrilloids (Copepoda: Monstrilloida) from Brazil with description of four new species. *Bull. Inst. R. Sci. Nat. Belg. Biol.* **2001**, *7*, 65–81.

44. Suárez-Morales, E.; Goruppi, A.; de Olazabal, A.; Tirelli, V. Monstrilloids (Crustacea: Copepoda) from the Mediterranean Sea (Northern Adriatic Sea), with a description of six new species. *J. Nat. Hist.* **2017**, *51*, 1795–1834. [CrossRef]

45. Suárez-Morales, E.; Morales-Ramírez, A. New species of Monstrilloida (Crustacea: Copepoda) from the Eastern Tropical Pacific. *J. Nat. Hist.* **2009**, *43*, 1257–1271. [CrossRef]

46. Suárez-Morales, E.; Pilz, D. A new species of *Cymbasoma* (Copepoda: Monstrilloida) from Florida with a redecription of *C. quadridens*. *J. Mar. Biol. Assoc. UK* **2008**, *88*, 527–533. [CrossRef]

47. Suárez-Morales, E.; Bello-Smith, A.; Palma, S. A revision of the genus *Monstrillopsis* Sars (Crustacea: Copepoda: Monstrilloida) with description of a new species from Chile. *Zool. Anz.* **2006**, *245*, 95–107. [CrossRef]

48. Delaforge, A.; Suárez-Morales, E.; Walkusz, W.; Campbell, K.; Mundy, C.J. A new species of *Monstrillopsis* (Crustacea, Copepoda, Monstrilloida) from the lower Northwest Passage of the Canadian Arctic. *Zookeys* **2017**, *709*, 1–16. [CrossRef] [PubMed]

Article

Two New Marine Free-Living Nematodes from Jeju Island Together with a Review of the Genus *Gammanema* Cobb 1920 (Nematoda, Chromadorida, Selachinematidae)

Alexei Tchesunov [1],*, Raehyuk Jeong [2] and Wonchoel Lee [2]

[1] Department of Invertebrate Zoology, Faculty of Biology, Moscow State University, 119991 Moscow, Russia
[2] Department of Life Science, Hanyang University, Seoul 04763, Korea; rhjeong88@gmail.com (R.J.); wlee@hanyang.ac.kr (W.L.)
* Correspondence: AVTchesunov@yandex.ru

http://zoobank.org/urn:lsid:zoobank.org:pub:4C6B07ED-52DB-42A1-80B7-2881F0559E56
Received: 19 November 2019; Accepted: 18 December 2019; Published: 1 January 2020

Abstract: In the context of exploration of meiofauna in a sandy intertidal zone of Jeju Island (South Korea), over 70 nematode species are identified, some which have been proven to be new for science. Two new free-living marine nematode species of the family Selachinematidae (Chromadorida, Selachinematidae, Choniolaiminae) are described from the intertidal sandy sediments of Jeju Island (South Korea). *Gammanema okhlopkovi* sp. n. is closest to *Gammanema anthostoma* (Okhlopkov, 2002) and differs by having longer cephalic setae (8.5–19 μm in *G. okhlopkovi* versus 6–7.5 μm in *G. anthostoma*) and by the presence of precloacal supplementary organs. The genus diagnosis of *Gammanema* is updated. The genus includes fourteen valid species, while three species are considered species inquirendae due to incomplete diagnoses and illustrations impeding their correct recognition. An annotated list of valid and invalid *Gammanema* species is provided. A pictorial key for valid *Gammanema* species is constructed, which consists of two components: (1) simplified images of heads, and (2) a table summarizing most of the significant measured and numeric characters between species. *Latronema obscuramphis* sp. n. differs from its related species *Latronema aberrans* (Allgén 1934), *Latronema annulatum* (Gerlach, 1953), and *Latronema spinosum* (Andrássy, 1973) by body size, number of supplementary organs, tail shape, length of spicules, and cuticle ornamentation.

Keywords: free-living marine nematodes; pictorial key; taxonomy

---

1. Introduction

Data on Jeju marine free-living nematodes have been presented by Rho and Kim [1,2], Rho et al. [3], and Lim and Chang [4]. However, the exploration of the rich nematofauna of Jeju is still in the early stages. In connection with our project on Jeju intertidal meiofauna, we have recorded almost 70 nematode species at four points along a transect through the sandy intertidal zone at Shinyang, in the eastern corner of Jeju Island. A taxonomic survey of Jeju nematodes started with the family Thoracostomopsidae by Jeong et al. [5,6]. Now, we present a communication on the family Selachinematidae.

The selachinematid nematodes are usually characterized by a stout cylindrical body with a truncate anterior end. They have a complicated buccal armament of rhabdions, denticles, or heavy mandibles, enabling them to capture nematode prey. The selachinematids have been known for a long time as carnivores in the strict sense, ingesting other smaller nematodes [7,8]. The percentage of selachinematids and other predator taxa in nematode assemblages are usually low in mud and silty sediment, but more abundant in sandy sediment with higher species diversity [9,10]. Particularly, most species of *Gammanema* and *Latronema* are confined to sandy sediments of upper shelf zones.

2. Materials and Methods

Quantitative samples were taken in June 2018, in the intertidal zone at Shinyang, close to the eastern corner of Jeju Island (Figure 1). Sediments were fixed in 5% neutralized formalin solution and brought back to the laboratory. Meiofauna were extracted using the Ludox method [11], and postfixed with 70% ethanol dyed with Rose bengal. Nematodes were counted and specimens were packed into a Syracuse glass filled with mixture of glycerin, 95% ethanol, and distilled water (1:29:70). The glass was placed in a drying oven set to 40 °C for 1–2 days until completely dehydrated, as in the glycerin-ethanol method [12]. Specimens were mounted in a drop of anhydrous glycerin within a wax-paraffin ring. Specimens were identified, measured, photographed, and drawn under a Leica DM5000 light microscope equipped with Leica Application Suite Version 3.8.0 software and a Leica DFC 425 C digital camera. All the measured sizes are given in μm. For scanning electron microscopy, specimens were dehydrated in a series: 40% ethanol, 70% ethanol, 95% ethanol, 95% ethanol + acetone (50:50), acetone I, and acetone II, and then critical point dried. Once dried, specimens were mounted on a stub to be coated with platinum–palladium alloy and examined with Cam Scan S-2.

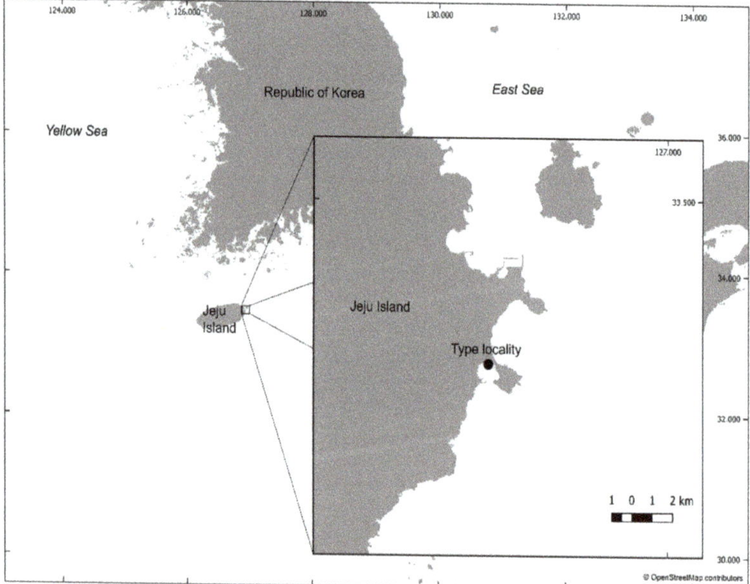

Figure 1. Map of sampling locality. This map is made with QGIS software v.2.18.14, a free and open-source geographic information system (https://qgis.org).

Type specimens were deposited in National Institute of Biological Resources (South Korea).

3. Results—Systematics

3.1. Review of Genus *Gammanema*

Order Chromadorida Chitwood, 1933
Family Selachinematidae Cobb, 1915
Subfamily Choniolaiminae Schuurmans-Stekhoven and Adam, 1931
Genus *Gammanema* Cobb, 1920

3.1.1. Diagnosis

Updated after Leduc 2013 and Tchesunov 2014 [13,14].

Selachinematidae, Choniolaiminae. Cuticle with homogenous annulations, without longitudinal ridges or lateral differentiation. Six inner and six outer labial sensilla, either setose or papillose; four cephalic setae, often slender and longer; outer labial and cephalic sensilla combined in common circle of 10, with dorso- and ventrosublateral sensilla arranged in four pairs with cephalic sensilla. Amphideal fovea spiral or loop-shaped, usually noticeably larger in males than in females. Somatic setae in irregular longitudinal rows; anterior cervical setae may be as long as cephalic setae. Mouth opening surrounded by partly fused lips shaping a circumoral membrane with fine longitudinal striation. There are twelve projections, from small and inconspicuous to prominent and elaborate, at the rim of the mouth opening. Buccal cavity (pharyngostome) consists of two chambers, anterior cup-shaped and posterior cylindrical; walls of each chamber are strengthened with three cuticularized rhabdions; the rhabdions of the anterior chamber terminate posteriorly in minute denticles. Pharynx cylindroids, evenly muscular and devoid of a terminal bulb. Alimentary tract terminates by rectum and anus. Precloacal midventral supplementary organs sucker-like, cup-shaped, tubular, or absent. Tail short, conical, cuticle of its terminal cone levigated or smooth.

Type species: *Gammanema ferox* (Cobb, 1920).

Annotated species list (valid species **in bold**)

1. *Gammanema agglutinans* **Leduc, 2013.** Leduc 2013:21–25, Figures 11–13, Table 2 [13] (males, females); SW Pacific, region to the east of New Zealand, Chatham Rise, depths 350–1238 m.

2. *Gammanema anthostoma* **Okhlopkov, 2002.** Okhlopkov 2002:42–45, Figures 2 and 3 [15] (males and females); White Sea (Northern Russia), Kandalaksha Bay, depth 14 m.

3. *Gammanema cancellatum* **Gerlach, 1955.** Gerlach 1955:271–272, Abbildung. 6 [16] (female); San Salvador, coastal groundwater. **Remark:** the species is retained here as valid, despite the absence of males in diagnosis, since it can be distinguished sound from other congeners by peculiar cuticle ornamentation with rods.

4. *Gammanema conicauda* **Gerlach, 1953.** Gerlach 1953:553–555, Abbildungen 17 a-f [17] (males, females, juvenile); Tyrrhenian Sea, area of Naples, surf zone; North Sea coast.

5. *Gammanema curvata* **Gagarin et Klerman, 2007.** Gagarin and Klerman 2007:780, Figure 2, Table 2 [18] (male, females); Mediterranean coast of Israel, depth 50 m, sandy sediment.

6. *Gammanema fennicum* **(Gerlach, 1953) Gerlach, 1964.** Gerlach 1953:22–23, Abbildung 6 [19] (as *Halichoanolaimus fennicus*) (female, juvenile); east Baltic Sea. Gerlach 1964:37 [20] (as junior synonym of *Gammanema rapax*). Okhlopkov 2002:45–46, Figures 4 and 5 [15] (males, females, juvenile); White Sea (Northern Russia), Kandalaksha Bay, depth 16 m.

7. *Gammanema ferox* Cobb, 1920 **sp. inq.** Cobb 1920:291–293, Figure 74 [21] (male, female); New Hebrides, coral sand. Though the species is described for both genders, the description lacks some necessary measurements, such as anterior setae and spicules; amphids are not depicted and not mentioned in the text (amphideal fovea possibly inconspicuous or reduced); the only illustration presents a head without an indication of gender and position in slide (possibly sublateral). Restitution of valid state is possible after redescription of new specimens from the type area; the species can be recognized through the prominent anterior projections of rugae.

8. *Gammanema kosswigi* **Gerlach, 1964.** Gerlach 1964:38–39, Abbildungen 11 a–d [19] (male, juvenile); Maldive Islands, coastal groundwater, intertidal zone.

9. *Gammanema magnum* **Shi et Xu, 2018.** Shi and Xu 2018, 3–9, Figures 1–4, Table 1 [22] (males, females); East China Sea, intertidal sandy sediment.

10. *Gammanema mediterraneum* **Vitiello, 1970.** Vitiello 1970:491–493, planche XV, 30 a–e [23] (as *Gammanema mediterranea*) (females, juveniles); West Mediterranean, depths 310–650 m.

11. *Gammanema menzelii* (Ditlevsen, 1918) Gerlach, 1964 **sp. inq.** Ditlevsen 1918:172–173, Plate VI, Figure 2; Plate VII, Figures 1 and 8 [24] (as *Halichoanolaimus Menzelii*) (one male); Danish Belt Sea, depth 30 m, clean sand. Gerlach 1964:31 [19] (transfer to *Gammanema*).

12. ***Gammanema okhlopkovi* Tchesunov, Jeong et Lee sp. n.** Present paper.

13. ***Gammanema polydonta* Murphy, 1965.** Murphy 1965:176–179, Figures 3A–C and 4A–H [25] (males, females); Chilean coast, depth 45 m.

14. *Gammanema rapax* (Ssaweljev, 1912) Gerlach 1964 **sp. inq.** Ssaweljev 1912:122 [26] (as *Halichoanolaimus rapax*) (male, female); Barents Sea, Kolafjord, Palafjord, Mogilnojesee (lake with an intermediate layer of sea water on the Kildin Island in the Barents Sea), sand. Murphy (1965: 179 [25]) considered the species; dubious because the original description is not illustrated and type specimens are evidently lacking. Gerlach (1964 [20]) synonymized *Halichoanolaimus menzeli* Ditlevsen 1918, *Gammanema ferox* Cobb, 1920 and *Gammanema fennicum* (Gerlach, 1953) with *Gammanema rapax*.

15. ***Gammanema smithi* Murphy, 1964.** Murphy 1964: 194–198, Figure 3A,B,D,E, Figure 4, Figure 5B [27] (male); Puget Sound (NE Pacific), intertidal sand. **Remark:** there are some discrepancies in the original diagnosis. Caudal glands are described and depicted ([27], Figure 4) as incaudal (i.e., located entirely within tail), while the glands in Figure 5A are definitely excaudal (i.e., gland cell bodies protrude anteriorly and lie along the midgut). Supplementary organs are depicted from a medio-ventral view in Figure 3C, while the hind body in Figure 4 is shown laterally. Taking into account that the diagnosis is based on only one male (holotype), we consider Figures 3C and 5A as wrong and not pertaining to *G. smithi*.

16. ***Gammanema tchesunovi* Gagarin et Klerman, 2007.** Gagarin and Klerman 2007:782-785, Figure 3, Table 3 [18] (males, females); Mediterranean coast of Israel, depths 50–55 m, sand.

17. ***Gammanema uniformis* (Cobb, 1920) Tchesunov et Okhlopkov, 2006.** Cobb 1920:293-294, Figure 75 [21] (as *Trogolaimus uniformis*) (seemingly, one male); Atlantic coast of the United States, New Hampshire. Tchesunov and Okhlopkov 2006:40-43, Figure 12, Table 6 [28] (males, females); Atlantic coast of the United States, Maine.

3.1.2. Remarks on the Species List and Synonymy

Murphy (1965:179 [25]) considered the species *Gammanema rapax* (Ssaweljev, 1912) as dubious because the original description is not depicted, and type specimens are evidently lacking. Gerlach (1964 [20]) synonymized *Gammanema menzelii* (Ditlevsen, 1918), *Gammanema ferox* Cobb, 1920, and *Gammanema fennicum* (Gerlach, 1953) with *G. rapax*. We find this broad synonymization questionable. *G. rapax* is a large species, measuring 5800–6100 μm in body length (Ssaweljev 1912 [26]), while all other species and specimens are at least half the size (except *G. menzelii*) and found far from the type locality. *G. menzelii* corresponds to *G. rapax* in body length (male 5900 μm) but differs in its higher number of preanal midventral supplementary organs (35 versus 20–22). Diagnosis of *G. menzelii* also lacks some important details, such as sizes of anterior sensilla, amphideal fovea, and spicules. Descriptions of *G. ferox* (New Hebrides [21]), *G. rapax* (North Sea [20]), and *G. rapax* (White Sea [15]) differ from the original diagnosis of Ssaweljev (1912 [26]) in their much smaller body size. Hence, we do not accept all these described specimens to be conspecific. Okhlopkov [15] redescribed females of *G. fennicum* and for the first time described males of *G. fennicum*. Though the specimens have not been sampled in the area of the type locality (Gulf of Finland, east Baltic Sea), but instead in the White Sea, the newly found females correspond to the original diagnosis in most details and morphometrics. Hence, *G. fennicum* is restored as a valid species.

3.1.3. Identification of Species of *Gammanema*

Since *Gammanema* species show evident sexual dimorphism in size and shape of amphids and other structures, it is highly desirable to have both sexes (or at least males) for species identification.

Due to deficiency of material, identification based on males is more reliable than identification solely based on females, since the males possess additional significant characteristics, such as supplementary organs and spicules.

For species identification of *Gammanema*, we propose a pictorial key constructed according to the principles of Platt [29], who first introduced pictorial keys in marine nematology. The pictorial key consists of two components, simplified images (caricatures) of heads (Figures 2 and 3), and Tables 1 and 2, with some metric and numeric characters being most important for species differentiation. Because of considerable sexual dimorphism, male and female head images are arranged in two separate plates, where the species are ordered in rows of gradual decrease of amphideal fovea (males) and cephalic setae (females), from top left to bottom right position.

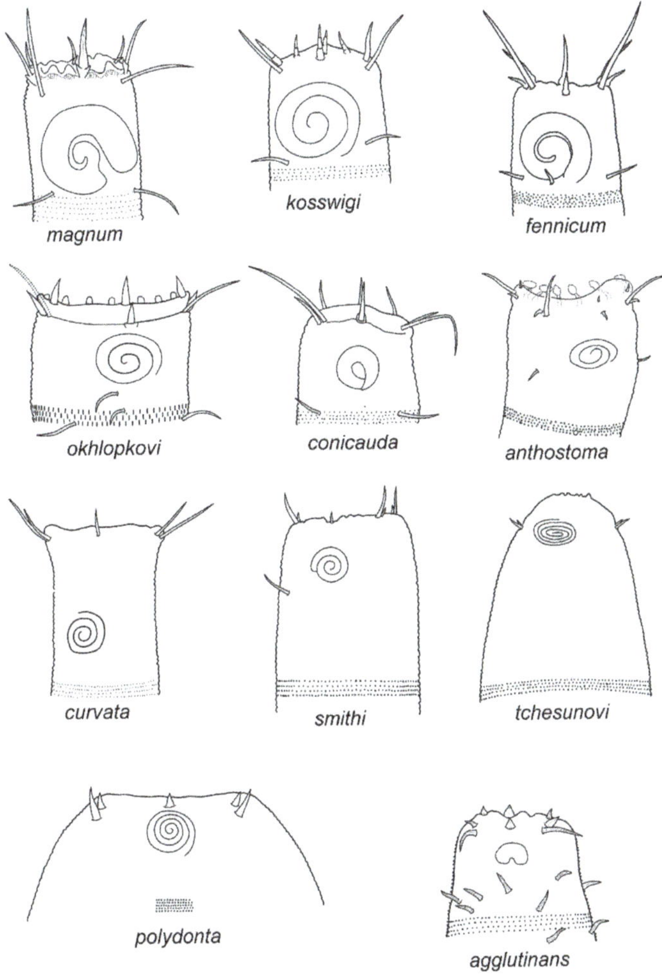

Figure 2. Caricatured male heads of valid *Gammanema* species arranged from top left to bottom right along decrease of amphid size and cephalic setae in relation to head diameter.

Table 1. *Gammanema* valid species, comparison of males.

Species	Characters									
	Body Length	a	Inner Labial Setae Length	Outer Labial Setae Length	Cephalic Setae Length	Amphid Width	Supplements Shape	Number of Supplements	Spicule Sength	Some Particular Bold Features
agglutinans	544–696	10–11	2	2	10–11	10	tubular	≥6	34–36	Cuticle with minute spines, somatic setae leaf-shaped
anthostoma	1002–1950	22–29	5–6	5–8	6–8	14–26	absent	absent	62-87	Prominent circumoral projections each consisted of peduncle and triangular cap
conicauda	1985–2484	41–48	16	17	30–33	15	conical papillae	22	33–40	Prominent circumoral hair-like projections
curvata	3286	76	not indicated	17	13	8–8.5	cup-shaped	16	28	-
fennicum	2165–2600	28–49	17–24	8–11	34–50	28–31	sucker-like	13–21	62–64	-
kosswigi	1200	29	8	10-12	20	30	weak setose	4	39	Outer labial setae jointed
magnum	2900–4085	48–52	15–16	22–25	31–35	42–45	absent	absent	left 51–56, right 62–69	Spicules slightly unequal
okhlopkovi	961–1150	25–31	4–5	3–7	8–19	11–18	sucker-shaped	10–13	32–39	Prominent circumoral Projections peg-like
polyodonta	3200–4200	21–27	not indicated	papillae	16	42–45	cup-shaped	43–52	135	Very massive appearance; Anterior half of body opaque because of densely pigmented intestine
smithi	2210	41	not indicated	5	20	14	cup-shaped	20	78	-
tchesunovi	1603–1889	15–16	not indicated	5	8–8	15–18	cup-shaped	23–24	85–87	Tail rounded and conical, terminal cone not developed
uniformis	2700–2950	45–61	1–2	2–2	9–10	10–11	cup-shaped	19–22	82–87	Cephalic cuticle thickened and internally sclerotized around anterior chamber of pharyngostome

Table 2. *Gammanema* valid species, comparison of females.

Species	Characters						
	Body Length	**a**	**Inner Labial Setae**	**Outer Labial Setae**	**Cephalic Setae**	**Amphid Width**	**Some Particular Bold Features**
agglutinans	763–918	11–15	2	2	11–19	5	Cuticle with minute spines, somatic setae leaf-shaped
anthostoma	1487–1782	17–24	5–6.5	5–7.5	15–19	9.5–10.5	Prominent circumoral projections; each consisted of peduncle and triangular cap
cancellatum	2465	24	no data	40	15	8	Complex ornamentation of cuticle consists of dots and longitudinally arranged short rods
conicauda	2492–3724	30–35	16	12	30–33	8–9	Prominent circumoral hair-like projections
curvata	2392–2639	64–71	no data	15–17	12–13	8–8.5	-
fennicum	1672–2782	13–25	9–13	10–13	16–37	9.5–11	-
magnum	2540–3145	28–33	10–14	20–21	40–45	9–10	-
mediterraneum	660–745	18–20	papillae	papillae	7.5	5–6	-
polyodonta	3730–4050	23–31	papillae	papillae	16	20	Very massive appearance; anterior half of body opaque because of densely pigmented intestine
tchesunovi	1562–1682	11–12	papillae	5	7–8	16–18	Tail rounded and conical, terminal cone not developed
uniformis	2260–2790	40–57	1–2	2–2	9–12	9–10	Cephalic cuticle thickened and internally sclerotized around anterior chamber of pharyngostome

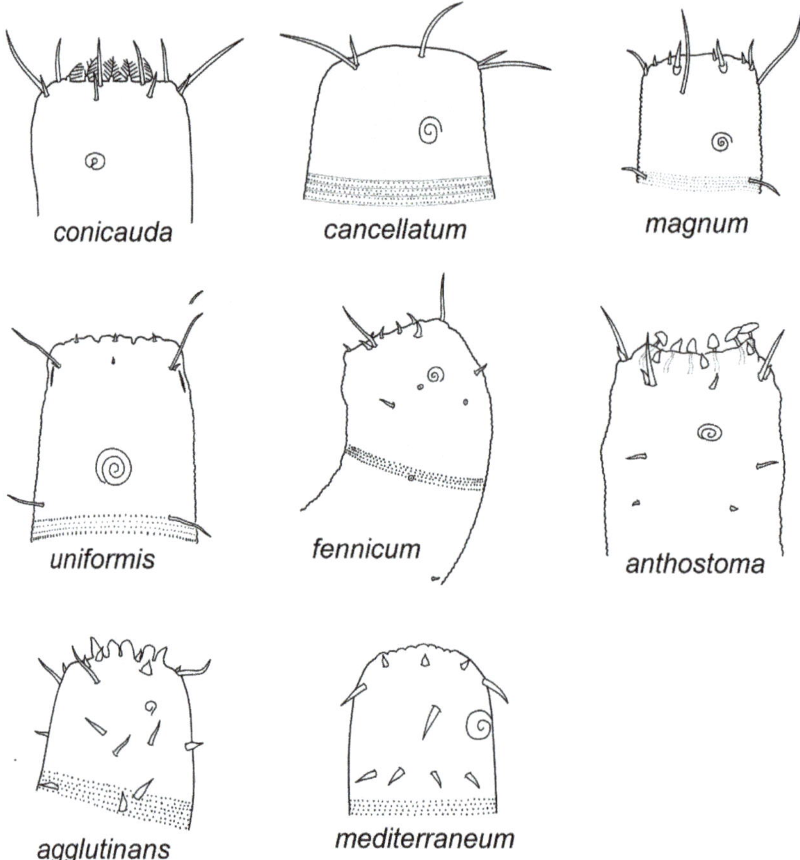

Figure 3. Caricatured female heads of valid *Gammanema* species arranged from top left to bottom right along decrease of cephalic setae in relation to head diameter.

3.2. Description of New Species of Gammanema

Gammanema okhlopkovi sp. n.
Figures 4–7, Table 3.
http://zoobank.org/urn:lsid:zoobank.org:act:3B8336E6-9F74-4861-BCD3-94B27F52697B

3.2.1. Etymology

The species is named in honor of Yuri R. Okhlopkov, a PhD student who published a few papers on Selachinematidae, but unfortunately passed away in 2014 by mischance.

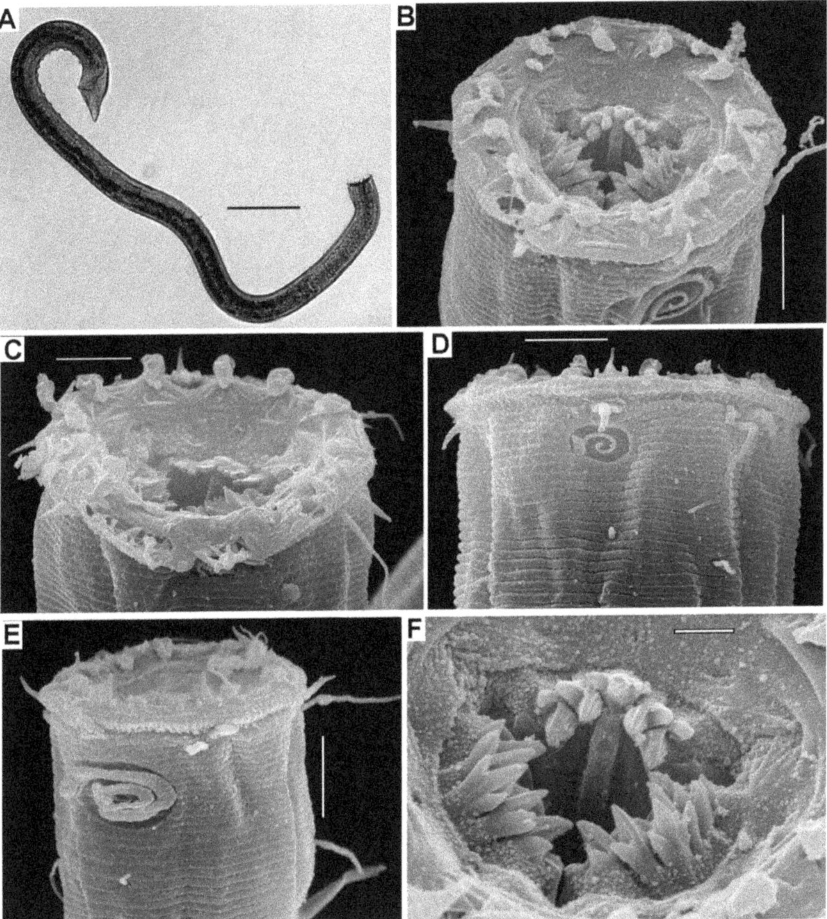

Figure 4. *Gammanema okhlopkovi* sp.n., entire view and anterior body: (**A**) holotype, male, entire view; (**B**) apical view, male, scanning electron micrograph; (**C**) subapical view, female, scanning electron micrograph; (**D**) head, lateral view, female, scanning electron micrograph; (**E**) head, sublateral view, male, scanning electron micrograph; (**F**) mouth opening, male, scanning electron micrograph. Scale bars: (**A**) 100 μm, (**B**) 10 μm, (**C**) 10 μm, (**D**) 10 μm, (**E**) 10 μm, (**F**) 3 μm.

3.2.2. Material Examined

One holotype and seven paratype males were deposited in the National Institute of Biological Resources (South Korea). Inventory numbers of the holotype and paratypes are NIBRIV0000861671 and NIBRIV000086172-NIBRIV0000861677, respectively.

Type locality: Intertidal zone at coast of Jeju Island, South Korea (33°26′05″ N, 126°55′15″ E), sandy beach, June 2018.

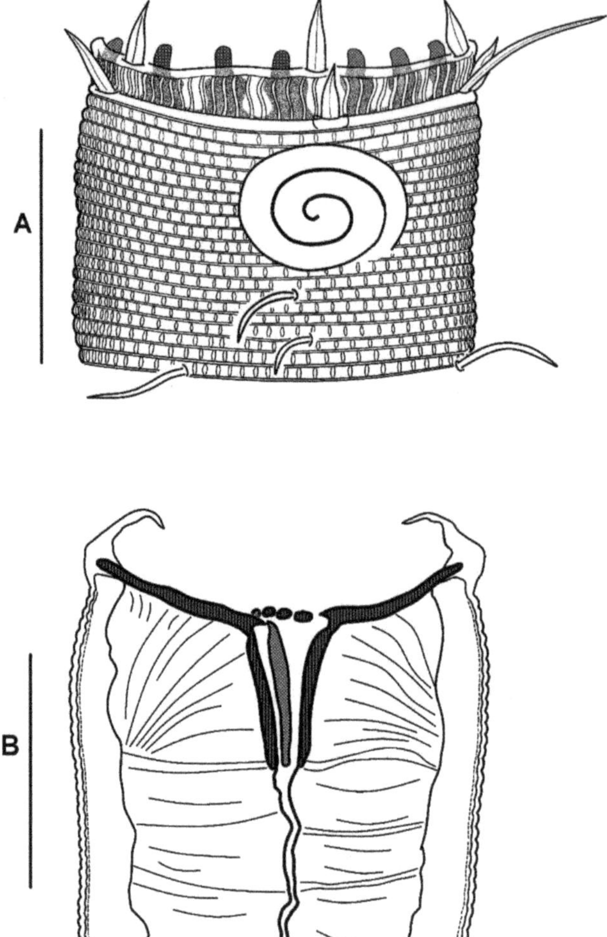

Figure 5. *Gammanema okhlopkovi* sp. n., head of the holotype male: (**A**) surface view; (**B**) optical section. Scale bars: 20 μm.

3.2.3. Description

Body cylindrical, with truncate anterior end and short conical tail. Cuticle annulated and punctuated. There are ca. 13 annules within a 20 μm area, everywhere along the body. Cuticle punctuation homogeneous, without lateral differentiation, the dots arranged in chessboard order. The dots are actually struts in the median zone of the cuticle; the struts are composed of two elements, as seen in sites of damaged cuticles.

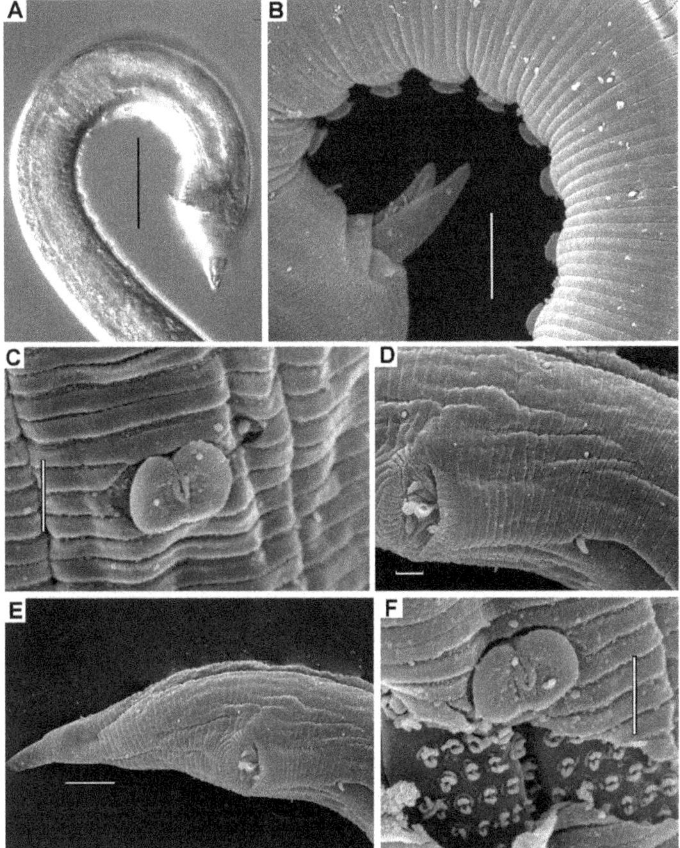

Figure 6. *Gammanema okhlopkovi* sp. n., posterior body, males: (**A**) hind body; (**B**) spicule tips, precloacal seta, and supplementary organs, lateral view, scanning electron micrograph; (**C**) posterior-most supplementary organ, scanning electron micrograph; (**D**) cloacal opening and precloacal seta, subventral view, scanning electron micrograph; (**E**) tail with sensory papillae; (**F**) damaged cuticle near a supplement displays struts in the median zone, scanning electron micrograph. Scale bars: (**A**) 50 µm, (**B**) 10 µm, (**C**) 3 µm, (**D**) 3 µm, (**E**) 10 µm, (**F**) 3 µm.

Truncate anterior end bordered by a membranous fringe with indistinct sectioning in lips. The fringe is marked by longitudinal striations, with about 15 fine, short ribs or rugae between lateral outer labial sensilla and cephalic seta. There are twelve prominent peg-like projections around the rim of the anterior end.

Anterior sensilla arranged in 6 + 10 pattern. Six inner labial sensilla as pointed conical papillae situated just outside or the peg-like projections. Six outer labial sensilla situated at the base of the striated fringe; they are nearly equal in shape and size to the inner labial sensilla. Four cephalic setae at the same level with latero-median outer labial sensilla in tight pairs; within pair, the cephalic seta disposed laterally of the adjacent outer labial sensilla. Somatic sensilla represented by a few irregular postamphideal setae and two subsequent lateral papillae further posteriad. Somatic sensilla on the posterior body and tail region are represented by papillae sparsely arranged in latero-ventral, lateral, and dorso-lateral rows. Amphideal fovea large, transversally oval in outer outline, spirally coiled in three turns.

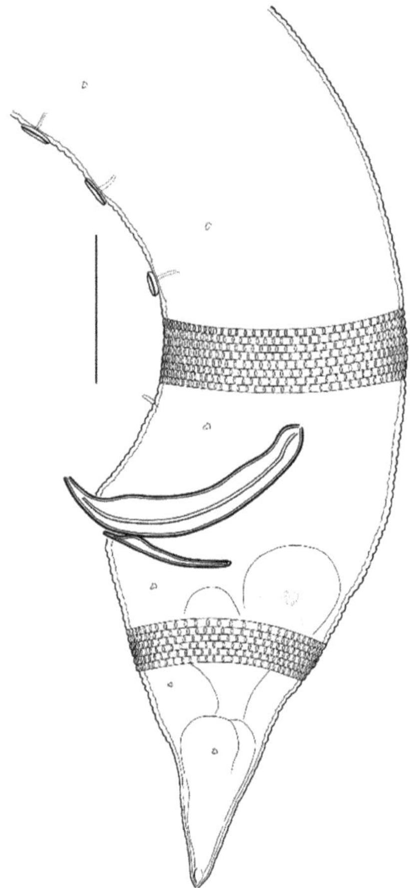

Figure 7. *Gammanema okhlopkovi* sp. n., posterior body of the holotype male. Scale bar: 20 μm.

The mouth opening is situated on the bottom of a funnel-shaped depression on the truncate anterior end. Acute teeth arranged in three radiating groups are visible, even in a semi-open mouth. Cheilostome is narrow and indistinct. Pharyngostome consists of two chambers. Anterior chamber is cup-shaped. Its walls are strengthened by three sclerotized rhabdions terminating posteriorly, with a group or five small cusps or denticles projected freely into the buccal cavity. The anterior rhabdions spread widely, thus making the anterior chamber bowl-shaped. In each group, the median tooth is tridentate, while teeth of two lateral pairs are bidentate. Posterior chamber cylindrical and narrow; its walls are strengthened by three sclerotized rhabdions. Pharynx cylindrical and evenly strongly muscular along its entire length.

Reproductive system is diorchic. Anterior outstretched testis situated ventrally to the intestine; posterior reflexed testis situated to the right of the vas deferens and intestine. Spicules paired, arcuate, distally pointed, and proximally very slightly cephalated. Gubernaculum as paired, nearly straight bars directed dorsally. Midventral supplementary organs, ten to thirteen in number, shaped as a nail-head 5–6 μm in diameter; the posteriormost organ situated at a distance 39 μm from the cloacal opening. A short conical midventral precloacal seta is present.

Tail very short; its conical tip turned left in all the specimens.

Table 3. Morphometrics of males *Gammanema okhlopkovi* sp. n. (holotype and paratypes).

Character	Holotype	Total Number	Min.–Max.	Mean	SD	CV%
Body length	961	8	961–1150	1082	61.8	5.71
pharynx length	178	8	164–207	185	15.1	8.15
tail length	48	7	42–57.5	50.4	5.6	11.1
body diameter at level of cs	37	7	36.9–43.7	39.2	2.40	6.12
body diameter at level of nerve ring	35	8	33.1–40.0	36.8	2.32	6.30
body diameter at level of cardia	35	8	34.1–41.0	36.5	2.45	6.72
body diameter at level of midbody	37	8	37.0–42.0	38.8	2.54	6.65
body diameter at level of cloaca	36	8	36.0–42.5	38.2	2.54	6.65
a	26	8	25.6–31.1	27.8	2.17	7.80
b	5.4	8	5.40–6.56	5.88	0.41	6.97
c	20	7	18.5–24.6	21.6	2.32	10.7
c′	1.33	7	1.01–1.53	1.31	0.16	12.2
inner labial setae length	3.5–4	6	4.0–5.0	4.27	0.39	9.13
lateral outer labial setae length	3.7	6	3.0–7.2	5.22	1.63	31.2
latero-median outer labial setae length	5	5	5.0–7.0	5.88	0.76	12.9
cephalic setae length	12	6	8.5-19.0	14.1	4.01	28.5
anterior cervical setae length	7.6	7	7.6–15.6	10.8	2.56	23.8
amphid width	16	7	11.0–18.0	14.8	2.01	9.13
amphid width/corresponding body diameter	43	8	32.5–47.7	37.7	5.22	13.9
distance amphid—cephalic apex	8.6	8	7.3–11.3	8.95	1.37	15.3
anterior pharyngostome rhabdion length	15	8	13.0–16.9	14.9	1.49	10.0
posterior pharyngostome rhabdion length	14	8	11.2-16.0	13.5	1.40	10.3
spicule length (arc)	34	6	32.0–39.0	35.2	2.86	8.12
spicule ′length (chord)	28	6	29.0–36.0	32.9	2.4	7.29
number of supplements	12	6	10–13	11.5	1.05	9.13
tail terminal cone length	17	4	16.8–23.5	19.6	2.93	15.0

3.2.4. Diagnosis

Gammanema with short and stout body (L 961–1150 μm, a 25–31). Twelve prominent peg-like projections protruded outward from inner side of the circumoral membranous fringe. Set of anterior sensilla composed of six inner labial setae 4–5 μm, six outer labial setae 5–7 μm, and cephalic setae 9–19 μm long. Amphideal fovea spirally coiled in three turns, 32–48 μm wide. Rhabdion of anterior pharyngostome chamber 13–17 μm, rhabdion of posterior pharyngostome chamber 11–16 μm long. Spicules 32–39 μm long. Midventral preanal row of supplementary organs consists of precloacal seta and 10–13 nail-head organs. Tail short and conical, c′ 1–1.6.

3.2.5. Differential Diagnosis

G. okhlopkovi sp. n. is closest to *G. anthostoma*, sharing similar body length and body proportions a, b, and c, and similar size of amphids and spicules. The most obvious differences between the two species are longer cephalic setae (8.5–19 μm in *G. okhlopkovi* versus 6–7.5 μm in *G. anthostoma*), as well as a lack of precloacal supplementary organs in *G. anthostoma* (present in *G. okhlopkovi*).

3.3. Description of New Species of Latronema

Genus *Latronema* (Wieser 1954)

Diagnosis of the genus has been recently updated by Leduc and Zhao [30]. A detailed review of the taxon is intended in forthcoming work on a peculiar diversity of *Latronema* species on an Atlantic seamount.

Latronema obscuramphis sp. n.

http://zoobank.org/urn:lsid:zoobank.org:act:5E195F44-8FDA-48D4-9E71-00855406670D

Figures 8 and 9, Table 4.

3.3.1. Etymology

The species name reflects the indistinctness of the amphids (from Latin "obscurus", inconspicuous).

3.3.2. Material Examined

One holotype male, two paratype females, and four juveniles were deposited in National Institute of Biological Resources (South Korea). Inventory numbers of the holotype and paratypes are NIBRIV0000861678 and NIBRIV0000861679–NIBRIV0000861680, respectively.

3.3.3. Description

Body stout cylindrical, with truncate anterior end and short conical tail. Body cuticle finely cross-striated (16 annules in 20 μm in light microscope, 12 annules in 10 μm, from scanning electron micrographs) in the midbody laterally. The cuticle is not punctated under light microscope but marked with longitudinal wings. The wings number approximately five, visible on the lateral body side. Anteriorly, short and weak extra wings are added between the main wings, close to the head.

Labial region is presented by circular labial membrane with fine longitudinal ribbing (about ten ribs in between two adjacent setae of cephalic crown counted under light microscope, or seven to ten in SEM). Since the edges of the labial membrane are rolled inward, it is impossible to discern whether the membrane is divided into labial lobes. Labial region is encircled by a light cuticular ridge (collar).

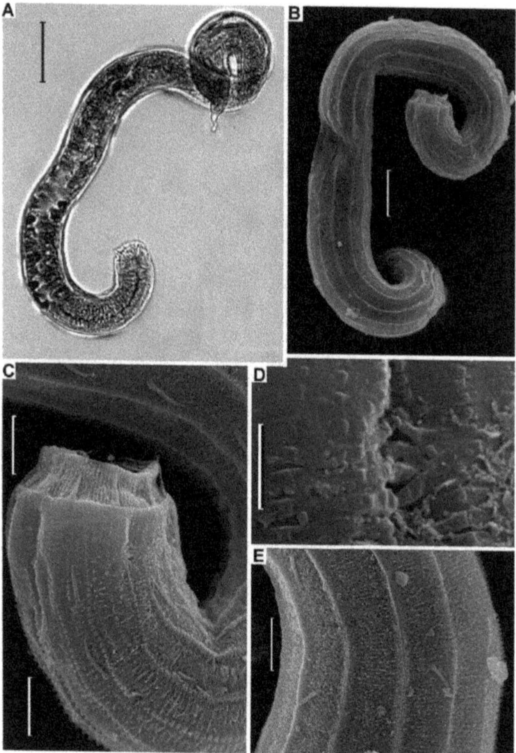

Figure 8. *Latronema obscuramphis* sp. n., entire body of the holotype male and surface cuticular structure: (**A**) holotype male, entire view; (**B**) entire view of a female, scanning electron micrograph; (**C**) head of a female laterally, scanning electron micrograph; (**D**) amphideal fovea of a female, scanning electron micrograph; (**E**) body cuticle rings and alae laterally, scanning electron micrograph. Scale bars: (**A**) 50 μm, (**B**) 30 μm, (**C,E**) 10 μm, (**D**) 3 μm.

Six slender inner labial setae are situated at the edge of the circular labial membrane. Ten equal setae located at the edge of the collar; they are about equal in length to one another and equal to inner labial setae, but wider basally; all the setae of the anterior circle look delicate and nearly transparent, and hence may be hardly visible against the labial cuticle. Fine ribs (striae) on the circumoral membrane may be arranged in fan-like bundles at sites opposite to cheilostomatal lobes. Quite possible the circle of ten equal setae comprises six outer labial and four cephalic setae, but all the setae are located at the same distance between them and do not make close pairs. Somatic setae comparable in length with anterior setae are distributed along the body in irregular longitudinal sublateral rows. Amphids in males are scarcely discernible under light microscope but detected in SEM as a small pit with simplified spiral structure. Amphids in females are not discernible at all under light microscope.

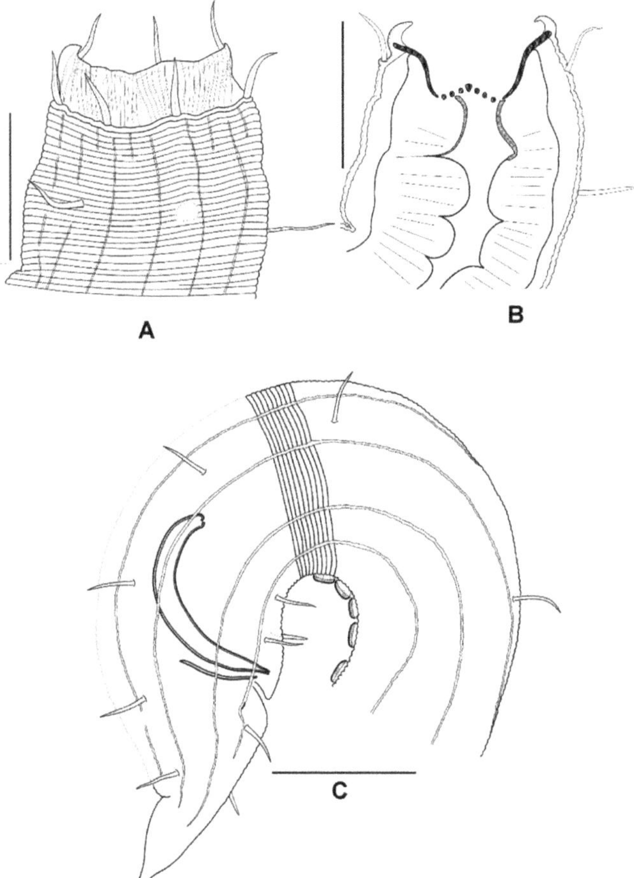

Figure 9. *Latronema obscuramphis* sp. n., male holotype structures: (**A**) surface view of the head; (**B**) sagittal optical section of the head; (**C**) posterior body. Scale bars: 20 μm.

Buccal cavity consists of comparatively narrow cheilostome surrounded by labial membrane and voluminous pharyngostome composed of two compartments, with the anterior being cup-shaped and the posterior cylindrical. Cheilostome walls are differentiated interiorly into twelve rounded lobes that do not continue into anterior free projections, unlike that of *Gammanema*. Walls of the anterior pharyngostome are strengthened by three rhabdions—one dorsal and two lateroventral. The rhabdions terminate posteriorly with about seven acute denticles (one median denticle and two sets of three lateral denticles on both sides of it). Anterior pharyngostome is opened widely, and its rhabdions move apart at obtuse angle. Posterior pharyngostome short and wide, its rhabdions slightly sclerotized, while intermediate cuticle between them thin and soft.

Table 4. *Latronema obscuramphis* sp. n., morphometry.

Character	Holotype Male	Paratype Female 1	Paratype Female 2
Body length	591	666	718
a	15.2	13.9	19.0
b	5.09	4.38	3.78
c	14.4	tail not measurable	tail not measurable
c′	1.60	tail not measurable	tail not measurable
V, %	-	57.4	70.0
body diameter at level of cephalic setae	30.0	38.0	33.1
body diameter at level of nerve ring	30.5	?	30.6
body diameter at level of cardia	35.0	41.5	33.8
body diameter at level of midbody	39.0	48.0	37.0
body diameter at level of cloaca/anus	25.6	30.0	?
inner labial setae length	8.7	7.7	6.0
latero-median outer labial setae length	11.6	not measurable	not measurable
lateral outer labial setae length	10.0	5.5	6.3
cephalic setae length	8.8	not measurable	not measurable
anterior cervical setae length	10.0	12.0	not measurable
anterior pharyngostome rhabdion length	12.0	12.0	12.4
posterior pharyngostome rhabdion length	8.0	7.5	10.7
spicules length (arc)	38.0	-	-
spicules length (chord)	33.0	-	-
gubernaculum length	12.0	-	-
tail terminal cone length	16	not measurable	not measurable

Pharynx cylindrical and strongly muscular. Nerve ring hardly discernible. Intestine is composed of convex enterocytes bulging in the gut lumen. Gut content in the holotype male is a long, fine-granular body, possibly a digested prey nematode in the posterior half of the intestine.

Female reproductive system is didelphic, both ovaries short and small, antidromously reflexed; in one female, anterior ovary to the right and posterior ovary to the left of the intestine, in other female anterior ovary to the left and posterior to the right of the intestine. Both spermathecas and uteri branch are filled with spermatozoa. Posterior ovary contains large oocyte with coarsely granulated cytoplasm and large nucleolated nucleus in the center. Somatic cuticle around vulva is thickened and distinctly set off as a vulvar plate.

Male reproductive system is diorchic. Both anterior outstretched and posterior reflexed testes situated ventrally and to the right of the intestine. Spicules equal, slightly arcuate, distally narrowed, proximally located, with very weakly developed knobs. Gubernaculum as a pair of short bars parallel to posterodistal edge of spicules. Preanal set of midventral supplementary organs composed of a midventral preanal seta and a row of about eight nail-head-shaped organs. The latter are not quite distinctly observed because of their position on the concave body side and covering by foreign particles.

Tail short and conical; its terminal cone covered with smooth cuticle and curved to the left.

3.3.4. Diagnosis

Latronema with small and stout body (L 591–718 μm, a 13.9–19). All the anterior setae (i.e., inner labial setae, outer labial setae, and cephalic setae) are similar in length, ca 9-12 μm in males, 6–6.5 μm in females. Amphideal fovea small and obscure. Anterior pharyngostome rhabdion 12–12.5 μm long, posterior rhabdions 7.5–11 μm. Spicules arcuate, 38 μm long, gubernaculum 12 μm. Preanal midventral row of supplementary organs made up of preanal conical seta and twelve sucker-like organs. Tail very short (c 14.4, c′ 1.6), with terminal cone ca 16 μm long usually turned left.

3.3.5. Differential Diagnosis

L. obscuramphis sp. n. differs considerably from most *Latronema* species by combination of body size, anterior setae length, size and shape of amphideal fovea, size of spicules, number of preanal supplementary organs. *L. aberrans* (Allgén, 1934), *L. annulatum* (Gerlach, 1953), and *L. spinosum* (Andrássy, 1973) are closest to *L. obscuramphis* in characters cited above. *L. obscuramphis* differs from *L. aberrans* (the species known from the west Baltic Sea and White Sea) by smaller body length (male, L 591 versus 960 µm; females, L 666–718 versus 1140 µm) and lesser number of preanal supplementary organs (8 versus 13); from *L. annulatum* (Madagascar, littoral) by tail shape (in male c′ 1.64 versus 2.4–2.8), longer spicules (38 µm versus 27 µm), and lesser number of supplements (8 versus 12); and from *L. spinosum* (Cuba, only female known) by no spinose body cuticular rings, smaller body (666–718 versus 850–1062 µm), and lesser index b (3.8–4.4 versus 4.5–5).

4. Conclusions

The nematofauna of the intertidal sandy beach of Jeju Island is very rich, including over 70 species representing nearly all the families of marine free-living nematodes. Even described species of Thoracostomopsidae [5,6] and Selachinematidae (present paper) constitute only about one-third of the actual diversity of those families from the studied Jeju beach. In forthcoming works, we will try to identify at least the most common nematode species, aiming in the future to reveal fauna composition and species distribution from upper to lower intertidal horizons, as well as vertical distribution in sediment columns.

Author Contributions: Conceptualization, A.T. and W.L.; investigation, A.T. and R.J.; data curation, R.J. and W.L.; writing—original draft preparation, A.T.; writing—review and editing, R.J. and W.L.; funding acquisition, W.L. and A.T. All authors have read and agreed to the published version of the manuscript.

Funding: The study was supported by a grant 18-504-51026 Russian Fund of Basic Research, a grant NRF-2017K2A9A1A06051528 from the Korea Research Foundation, a grant NIBR202000000 entitled "Study of Undiscovered Invertebrate Species from Korea" from the National Institute of Biological Resources (NIBR) funded by the Ministry of Environment (MOE) of the Republic of Korea, as well as by the Marine Biotechnology Program of the Korea Institute of Marine Science and Technology Promotion (KIMST) funded by the Ministry of Oceans and Fisheries (MOF) (No. 20170431).

Acknowledgments: The authors thank Vadim Mokievsky (P.P. Shirshov Institute of Oceanology, Moscow) and Jungho Hong (Biodiversity Institute, Marine Act Co., Seoul) for their help in the field work sampling in June 2018. The authors are grateful to both anonymous reviewers for their remarks and advice, which helped us amend the manuscript.

Conflicts of Interest: The authors declare no conflict of interest.

Abbreviations

a	body length divided by body diameter at midbody
b	body length divided by pharynx length
c	body length divided by tail length
c′	tail length divided by anal body diameter
CV	coefficient of variation (SD divided by mean, in %)
SD	standard deviation
V	distance from anterior end to vulva in % of entire body length

References

1. Rho, H.S.; Kim, W. *Paradraconema jejuense*, a new species of genus *Paradraconema* (Nematoda: Draconematidae) from Korea. *Korean J. Syst. Zool.* **2004**, *21*, 81–91.
2. Rho, H.S.; Kim, W. A new free-living marine nematode species of the genus *Dracogalerus* Allen and Noffsinger (Nematoda: Draconematidae) from a shallow subtidal zone of Jeju Island, Korea. *Integr. Biosci.* **2005**, *9*, 113–122. [CrossRef]

3. Rho, H.S.; Decraemer, W.; Sørensen, M.V.; Min, W.G.; Jung, J.; Kim, W. *Megadraconema cornutum*, a new genus and species from Korea, with a discussion of its classification and relationships within the family Draconematidae (Nematoda, Desmodorida) based on morphological and molecular characters. *Zool. Sci.* **2011**, *28*, 68–84. [CrossRef]

4. Lim, H.W.; Chang, C.Y. First record of *Desmoscolex* Nematoda (Desmoscolecida: Desmoscolecidae) from Korea. *Integr. Biosci.* **2006**, *10*, 219–225. [CrossRef]

5. Jeong, R.; Tchesunov, A.V.; Lee, W. Bibliographic revision of *Mesacanthion* Filipjev, 1927 (Nematoda: Thoracostomopsidae) with description of a new species from Jeju Island, South Korea. *PeerJ* **2019**, *7*, e8023. [CrossRef] [PubMed]

6. Jeong, R.; Tchesunov, A.V.; Lee, W. Two species of Thoracostomopsidae (Nematoda: Enoplida) from Jeju Island, South Korea. *PeerJ*, submitted, under review.

7. Okhlopkov, J.R. Feeding of free-living nematodes of the families Selachinematidae and Richtersiidae in the White Sea. *Proc. Pertsov White Sea Biol. Stn.* **2003**, *9*, 127–139. (In Russian)

8. Moens, T.; Braeckman, U.; Derycke, S.; Fonseca, F.; Galucci, F.; Gingold, R.; Guilini, K.; Ingels, J.; Leduc, D.; Vanaverbeke, J.; et al. Ecology of free-living marine nematodes. In *Handbook of Zoology*; Schmidt-Rhaesa, A., Ed.; De Gruyter: Berlin, Germany, 2014; Volume 2, pp. 109–152.

9. Warwick, R.M. Nematode associations in the Exe estuary. *J. Mar. Biol. Assoc. UK* **1971**, *51*, 439–454. [CrossRef]

10. Gallucci, F.; Steyaert, M.; Moens, T. Can field distributions of marine predacious nematodes be explained by sediment constraints on their foraging success? *Mar. Ecol. Prog. Ser.* **2005**, *304*, 167–178. [CrossRef]

11. Burgess, R. An improved protocol for separating meiofauna from sediments using colloidal silica soils. *Mar. Ecol. Prog. Ser.* **2001**, *214*, 161–165. [CrossRef]

12. Seinhorst, J. A rapid method for the transfer of nematodes from fixative to anhydrous glycerin. *Nematologica* **1959**, *4*, 67–69. [CrossRef]

13. Leduc, D. Two new genera and five new species of Selachinematidae (Nematoda, Chromadorida) from the continental slope of New Zealand. *Eur. J. Taxon.* **2013**, *63*, 1–32. [CrossRef]

14. Tchesunov, A.V. Order Chromadorida Chitwood, 1933. In *Handbook of Zoology*; Schmidt-Rhaesa, A., Ed.; Gastrotricha, Cycloneuralia and Gnathifera, Volume 2: Nematoda; De Gruyter: Berlin, Germany, 2014; pp. 373–398.

15. Okhlopkov, J.R. Free-living nematodes of the families Selachinematidae and Richtersiidae in the White Sea (Nematoda, Chromadoria). *Zoosyst. Ross.* **2002**, *11*, 41–55.

16. Gerlach, S.A. Zur Kenntnis der freilebenden marinen Nematoden von San Salvador. *Z. für Wiss. Zool.* **1955**, *158*, 249–303.

17. Gerlach, S.A. Die Nematodenbesiedlung des Sandstrandes und des Küstengrundwassers an der italienischen Küste. I. Systematischer Teil. *Arch. Zool. Ital.* **1953**, *37*, 517–640.

18. Gagarin, V.G.; Klerman, A.K. New species of predatory chromadorids (Nematoda, Chromadorida) from the Mediterranean Sea. *Zool. Zhurnal* **2007**, *86*, 778–786, (In Russian with English Abstract).

19. Gerlach, S.A. Die Nematodenfauna der Uferzonen und des Küstengrundwassers am finnischen Meeresbusen. *Acta Zool. Fenn.* **1953**, *73*, 1–32.

20. Gerlach, S.A. *Revision der Choniolaimidae und Selachinematidae (Freilebende Meeres-Nematoden)*; Kosswig-Festschrift; Mitteilungen aus dem Hamburgischen Zoologischen Museum und Institut: Hamburg, Germany, 1964; pp. 23–50.

21. Cobb, N.A. One hundred new nemas (type species of 100 new genera). *Contrib. Sci. Nematol.* **1920**, *9*, 217–343.

22. Shi, B.; Xu, K. Two new rapacious nematodes from intertidal sediments, *Gammanema magnum* sp. nov. and *Synonchium caudatubatum* sp. nov. (Nematoda, Selachinematidae). *Eur. J. Taxon.* **2018**, *405*, 1–17. [CrossRef]

23. Vitiello, P. Nématodes libres marins des vases profondes du Golfe du Lion. II. Chromadorida. *Tethys* **1970**, *2*, 449–500.

24. Ditlevsen, H. Marine freeliving nematodes from Danish waters. *Vidensk. Meddelelser Dansk Naturhistorisk Foren. I Kjobenhavn* **1918**, *70*, 147–214.

25. Murphy, D.G. Chilean marine nematodes. *Veröff. Inst. für Meeresforsch. Bremerhav.* **1965**, *9*, 173–203.

26. Ssaweljev, S. Zur Kenntnis der freilebenden Nematoden des Kolafjords und des Relictensee Mogilnoje. *Trav. Soc. Imp. Nat. St. Pétersbg.* **1912**, *43*, 108–126.

27. Murphy, D.G. Free-living marine nematodes, I. Southerniella youngi, Dagda phinneyi, and Gammanema smithi, new species. *Proc. Helminthol. Soc. Wash.* **1964**, *31*, 190–198.

28. Tchesunov, A.V.; Okhlopkov, Y.R. On some selachinematid nematodes (Chromadorida: Selachinematidae) deposited in the collection of the Smithsonian National Museum of Natural History. *Nematology* **2006**, *8*, 21–44.

29. Platt, H.M. Pictorial taxonomic keys: Their construction and use for the identification of free-living marine nematodes. *Cah. Biol. Mar.* **1984**, *25*, 83–91.

30. Leduc, D.; Zhao, Z. *Latronema whataitai* sp. n. (Nematoda: Selachinematidae) from intertidal sediments of New Zealand, with notes on relationships within the family based on preliminary 18S and D2-D3 phylogenetic analyses. *Nematology* **2015**, *17*, 941–952. [CrossRef]

Article

An Introduction to the Study of Gastrotricha, with a Taxonomic Key to Families and Genera of the Group

M. Antonio Todaro [1,*], Jeffrey Alejandro Sibaja-Cordero [2,3], Oscar A. Segura-Bermúdez [2], Génesis Coto-Delgado [2], Nathalie Goebel-Otárola [2], Juan D. Barquero [2], Mariana Cullell-Delgado [2] and Matteo Dal Zotto [1,4]

1   Department of Life Sciences, Università di Modena e Reggio Emilia, via G. Campi, 213/D, 41125 Modena, Italy
2   Centro de Investigación en Ciencias del Mar y Limnología (CIMAR), Universidad de Costa Rica, San Pedro, San José 11501- 2060, Costa Rica
3   Escuela de Biología, Universidad de Costa Rica, San Pedro, San José 11501-2060, Costa Rica
4   Consortium for the Interuniversity Center of Marine Biology and Applied Ecology, viale N. Sauro, 4, 57128 Livorno, Italy
*   Correspondence: antonio.todaro@unimore.it

Received: 27 June 2019; Accepted: 17 July 2019; Published: 23 July 2019

Abstract: Gastrotricha is a group of meiofaunal-sized, free-living invertebrates present in all aquatic ecosystems. The phylum includes over 860 species globally, of which 505 nominal species have been recorded in marine sandy sediments; another 355 taxa inhabit the freshwater environments, where they are recurrent members of the periphyton and epibenthos, and, to a lesser degree, of the plankton and interstitial fauna. Gastrotrichs are part of the permanent meiofauna and, in general, they rank among the top five groups for abundance within meiobenthic assemblages. The diversity, abundance, and ubiquity of Gastrotricha allow us to suppose an important role for these animals in aquatic ecosystems; however, ecological studies to prove this idea have been comparatively very few. This is mainly because the small size and transparency of their bodies make gastrotrichs difficult to discover in benthic samples; moreover, their contractility and fragility make their handling and morphological survey of the specimens rather difficult. Here we offer an overview, describe the basic techniques used to study these animals, and provide a key to known genera in an attempt to promote easy identification and to increase the number of researchers who may be interested in conducting studies on this understudied ecological group of microscopic organisms.

Keywords: benthos; biodiversity; key; meiofauna; taxonomy

---

1. Introduction

Gastrotrichs are minute (from 60 μm to 3.5 mm in total length) vermiform, acoelomate invertebrates; they inhabit the aquatic ecosystems of the world as part of the meiofaunal communities. In freshwater habitats, gastrotrichs are members of the benthos and periphyton and, to a limited degree, also of the plankton and psammon. In marine settings, these tiny animals inhabit (mostly) the interstice of the sandy habitats and are usually the third group in density among the interstitial meiofaunal taxa, behind the nematodes and the harpacticoid copepods (e.g., abundance up to 364 ind./10 cm^2) [1]; however, several studies have found them to be the second or the first most abundant meiobenthic group [2–5]. In inland waters, the group usually figures among the top five most abundant taxa, and populations may attain a density of 2600 ind./10 cm^2 [6]. In marine as well as in freshwater systems, the ecological role of Gastrotricha is accomplished within the detritivorous, microphagous benthic assemblage. Gastrotrichs feed on bacteria, microscopic algae, and small protists; food is ingested by aspiration thanks to the powerful, triradiated, myoepithelial pharynx. In turn, they represent prey for small macrofauna, carnivorous ciliates, and free-living flat worms. The gastrotrichs' ecological disparity is coupled with an ample morphological diversity which may seem amazing when comparing the large and vermiform marine representatives with the tiny and tenpin-shaped freshwater forms. Despite their variety, gastrotrichs are considered to constitute a monophyletic group (phylum) based on the following synapomorphies: (1) cuticle made up of two layers, with the external layer (epicuticle) consisting of one or more plasma-membrane-like sheets (lamellar layer); (2) epicuticle covering the entire body, including the locomotor and sensorial cilia; (3) a "duo-gland system" adhesive apparatus lacking an anchor cell; and 4) peculiar helicoidal muscles enwrapping the anterior portion of the alimentary canal [7,8]. The phylum has a worldwide distribution with some 860 nominal species (as of July 2019) distributed into two orders: Chaetonotida, including 483 tenpin- or bottle-shaped species, two-thirds of which are found in inland ecosystems, and Macrodasyida, grouping 377 vermiform species, the vast majority of which are marine or, more rarely, estuarine. Only four macrodasyidan species, belonging to the genera *Marinellina* (1 sp.) and *Redudasys* (3 spp.), have been reported from freshwater habitats to date. The current classification sees the order Chaetonotida divided into 8 families and 32 genera, whereas the order Macrodasyida counts 10 families and 36 genera. The continuous description of new taxa (species, genera) and the ongoing process of re-systematization suggest that we should consider the statistics reported above as being highly conservative.

Phylogenetic relationships of the Gastrotricha have been questioned for a long time. By virtue of their morphological traits, many researchers have considered Gastrotricha to be close relatives of Nematoda, Rotifera, Gnathostomulida, or Kinorhyncha, within large assemblages such as the Aschelminthes, Pseudocoelomates, etc. However, phylogenetic analyses of the "Aschelminthes", grounded on genetic traits (e.g., 18S rRNA gene) showed such groupings to be polyphyletic and Gastrotricha as part of the Lophotrochozoa but with unstable alliances within the clade [9]. Recent phylogenomic studies have also dismissed the Platyzoa clade, within which Gastrotricha has been allocated for some time, and have convincingly shown Gastrotricha together with the Platyhelminthes allied in a clade named Rouphozoa as a subset within the protostomian Spiralia [10,11]. Parts of the in-group phylogenetic relationships remain unclear, e.g., the evolutionary relationships between the representatives of the two orders or within the clearly paraphyletic family Chaetonotidae. Fortunately, relationships among taxa belonging to several families, especially of the Macrodasyida, are becoming less obscure [9,12–19].

Recent overviews of the gastrotrichs' biology and morphology have been offered by several authors [20–22]. Updated information regarding, e.g., classification, distribution, literature, etc., can be found at the dedicated Gastrotricha World Portal [23] and through the World Register of Marine Species (WoRMS) [24].

2. Materials and Methods

2.1. Sampling

Sampling procedures in freshwater environments and marine ecosystems are usually analogous; qualitative studies implicate the gathering of sediment by mean of a scoop, spoon, plastic jar, or a hand-held planktonic net, while quantitative research typically uses corers of clear plastic or Plexiglas (2–5 cm inner diameter, 10–20 cm long). The sea-dweller taxa are typically interstitial, inhabiting preferentially clean, fine to medium sands, with some occurring in muddy substrata (e.g., *Musellifer* spp.) and a few that are tolerant of high sulphide or organic loads [25–30].

Qualitative intertidal sampling is typically carried out at low tide; pits are dug in the beach, and the sand from the bottom and the wall of the pits is then removed with a spoon or scoop and transferred to plastic jars (Supplementary Material Figure S1); subtidal material for qualitative studies can be taken directly by scooping up the upper 10 cm sediment surface with a plastic container (e.g., a 500 mL jar), which is immediately closed off underwater (Supplementary Material Figure S2). Jars filled with sand are then transported to the laboratory and allowed to rest for some time (1 h to overnight) at a suitable temperature. Over the hours, the fauna move upward and will enrich the top layers of the sand, facilitating the following extraction process (see below). The horizontal distribution of Gastrotricha is patchy; consequently, the collection of several small samples is more illustrative of the taxonomic assemblage of a location than a sole big sample. Interstitial forms of freshwater habitats may be collected using similar techniques. Freshwater gastrotrichs that live on the surfaces of rooted aquatic plants, along with benthic, periphytic, and semiplanktonic taxa, are qualitatively sampled by gathering bunches of vegetation together with the bottom deposits and filtering the water through a net or a sieve with mesh of appropriate size (e.g., 25–30 µm) (Supplementary Material Figure S3). The gastrotrich-enriched sample is placed in buckets and rapidly transported to the laboratory where it is subsequently moved to small aquaria, kept at a suitable temperature, and moderately oxygenated with an air stone (Supplementary Material Figure S4).

2.2. Extraction

Freshwater and marine samples should both be processed within 5–6 days to obtain the living specimens, which are normally better suited than preserved animals for taxonomic purposes (e.g., identification) since fixation generally causes artifacts that alter and/or obscure the diagnostic characteristics. For freshwater samples only, additional checks 2–4 weeks after sampling are advisable, since taxa initially absent may be found later due to the hatching of resting eggs.

Interstitial fauna can efficiently be separated (extracted) from the sand by narcotization and decantation, using a solution of $MgCl_2$ (7% marine sample or 1% freshwater sample) as a narcotic. For this purpose, 1–2 spoons of the fauna-enriched top layer of sand (see above) are placed into a small vessel with a sufficient amount of added narcotic solution to cover the sand. The material is then swirled and allowed to sit for 5 minutes, after which it is gently swirled again and the liquid decanted into small Petri dishes (5.5 cm). At this stage, a small amount of either seawater (marine samples) or freshwater (freshwater sample) is added to each Petri dish, which is then scanned for gastrotrichs using a binocular microscope at 40–50× magnification, preferably in transmitted light (Supplementary Material Figures S5,S6).

The freshwater, non-sandy samples placed in the small aquaria, as reported above, may be processed for gastrotrichs by sucking up with a large pipet a small amount of the detritus and the overlying water and by transferring the sucked material to a large Petri dish (9.5–12 cm); the dish is then scanned for active (motile) gastrotrichs under a dissecting microscope as described above (Supplementary Material Figure S4). Alternatively, material collected with the large pipet may be transferred to a glass flask, with an equal quantity of 2% $MgCl_2$ solution added, aliquoted into Petri dishes of suitable diameter, and thence analyzed for narcotized gastrotrichs under a dissecting microscope.

For obvious reasons, quantitative studies should be based on fixed material. To reduce artifacts that may hamper species identification, treatment of the freshly collected material (samples) with a solution of magnesium chloride (7% for marine or 1% for freshwater samples) for 5–10 min is very much suggested before the material is fixed. Fixation may be done using a solution of 10% borax-buffered formalin; later, some rose bengal (1%) may be added to ease sorting. Gastrotrich specimens in the quantitative samples may be extracted from the sandy substrata using the same techniques used for other meiofaunal taxa, e.g., by elutriation and multiple decantations. Extraction from samples containing fine sediment and rich in detritus can be carried out by centrifugation using the silica gel LUDOX AM (d = 1.210) to create a gradient [31]. The supernatant should be filtered using a 20–30 µm mesh sieve to concentrate the gastrotrichs.

2.3. Morphological Analysis

Morphometric data should be acquired on living, relaxed specimens mounted on a microscope slide and covered with a square coverslip (15–18 mm). As the mounting of a gastrotrich may be tricky, the following practice carried out routinely at the first author laboratory may facilitate the task. To mount the specimen of interest, a drop of the same medium the specimen is extracted from is put on a clean microscope slide and a single gastrotrich is transferred to it by using a micropipette (mouth or hand held). In the case of freshwater medium, to relax/anesthetize the specimen, a small amount of 1% magnesium chloride solution can be added to the liquid containing the gastrotrich; alternatively, small crystals of alkaloids such as novocaine or procaine are put at the edge of the water so they dissolve gradually in the water, anesthetizing the animal. Thereafter, a clean coverslip (cover glass) is carefully put on the water. To avoid excessive animal compression, the coverslip should not be used as it is; instead, small modelling clay posts are attached beneath its corners before it is put in place (Supplementary Material Figures S7,S8). As deep morphological survey requires the use of oil immersion optics (e.g., 60×, 100×) it is important the specimen be positioned far from the sides of the coverslip. Proper positioning of the specimen at (or near) the center of the slide and its dorso-ventral orientation may be attained by adding a tiny amount of the liquid medium to the cover glass sides or by absorbing the liquid with a piece of blotting paper. Animals gently compressed between the slide and the coverslip are then observed under an upright biological microscope, preferably using DIC (differential interference contrast) lenses (Supplementary Material Figure S9). Fine anatomical traits may necessitate SEM observation; for this purpose, specimens are opportunely prepared (e.g., by hesamethildysilazane or the CPD (critical point drying) technique) [32,33].

Identification of formalin-fixed gastrotrichs from a location may be facilitated by a preliminary identification of the local fauna based on living specimens. Regardless, identification of preserved material can be executed on animals included in watery solutions, or better, established on (semi)permanent mounts. The latter can be set by including the gastrotrichs in a solution of glycerol–formalin (1:3), and the coverslip is then sealed with nail polish or Glyceel. Alternatively, specimens may be mounted in absolute glycerol on an H-S slide after immersion in a 10% glycerol–ethanol solution, which is allowed to evaporate in an oven at 40 °C for 2–4 days [34]. However, in many cases, permanent mounts—even in the case of uncontracted, well-oriented specimens—do not permit a full identification as many of the diagnostic traits deteriorate over time. Consequently, for taxonomic purposes, photos or high-resolution video sequences of living, relaxed specimens may deliver superior long-lasting records of the anatomical characteristics of a species compared with specimens mounted on microscope slides.

2.4. Taxonomic Key

The following key, modified from [35], encompasses the valid families and genera of Gastrotricha (Figures 1–20) described to date [24]. Two families (Redudasyidae and Hummondasyidae, belonging to Macrodasyida) and five genera (*Bifidochaetus* and *Cephalionotus* belonging to Chaetonotida and *Anandrodasys*, *Hummondasys*, *Thaidasys*, and *Kryptodasys* belonging to Macrodasyida) included herein have been established since the publication of the previous keys [12–15,36,37]. For the inclusion of *Megadasys* among the Planodasyidae (order Macrodasyida), see [18].

The key is designed to be used by researchers and students who have a general knowledge on how to identify animals but may not have much expertise on Gastrotricha; it is practical in style and is grounded on relevant discriminatory traits as they appear in relaxed mature animals. In most cases, anatomical traits are those which are easily visible using differential interference contrast optics and which are countable. However, to facilitate the assignment, it is imperative that the mounted specimen to be identified is oriented in a dorso-ventral fashion. The following abbreviations are used in the key: PhIJ, pharyngo-intestinal junction; TbA, adhesive tubes of the anterior series; TbD, adhesive tubes of the dorsal series; TbP, adhesive tubes of the posterior series; TbV, adhesive tubes of the ventral series.

3. Results

Key to Families and Genera of Gastrotricha

1a   Body flask-, bottle- or tenpin-shaped; posterior body region usually furcate (furca), less often rounded off or bifurcate; cuticle usually forming ornamentations such as scales and spines; TbA, TbD, and TbL absent; TbP present, numbering 2 (rarely 4 or 0) at the distal end of the furcal rami; mouth narrow; pharynx lacking pores. Freshwater, marine, and brackish: periphytic, epibenthic, and interstitial, occasionally semiplanktonic. Order CHAETONOTIDA, Suborder PAUCITUBULATINA (Figure 1A–C) **38**

1b   Different from the above. **2**

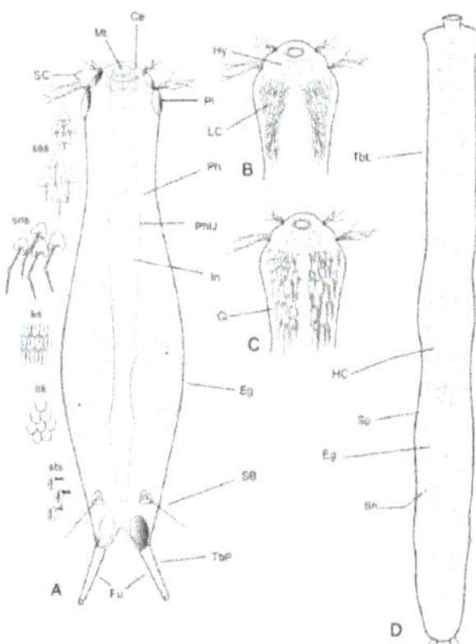

Figure 1. Drawings of hypothetical Chaetonotida: (**A–C**) Paucitubulatina and (**D**) Multitubulatina (*Neodasys*). (**A**) Habitus, dorsal view; (**B,C**) anterior region, ventral view, showing different arrangement of the locomotor ciliations. Abbreviations: **Ce**, cephalion; **Ci**, locomotor cirri; **eg**, egg; **Fu**, furca; **HC**, hemoglobin-containing cell; **Hy**, hypostomion; **In**, intestine; **LC**, locomotor cilia; **Mt**, mouth; **Ph**, pharynx; **PhIJ**, pharyngo-intestinal junction; **Pl**, pleuria; **SB**, sensorial bristle; **SC**, sensorial cilia; **sk**, scales with a keel; **sns**, scales with notched spines; **Sh**, spermatophores; **Sp**, spermatozoa; **ss**, scales smooth; **sss**, scales with simple spines; **sts**, scale with a stalk; **TbL**, lateral adhesive tubes; **TbP**, posterior adhesive tubes. (**A–C**), original; scales redrawn and modified from [38]; (**D**), redrawn and modified from [39].

2a (1b)   Body vermiform; cuticle naked, not forming scales and/or spines; TbA and TbD absent; TbL present in the form of numerous papillae along each side; TbP, some per side, fused at their bases forming two adhesive structures; mouth narrow, pharynx lacking pores. Uncommon; marine: interstitial. Order CHAETONOTIDA, Suborder MULTITUBULATINA, NEODASYIDAE . . . . *Neodasys* (Figure 1D)

2b (1b)   Body usually vermiform, occasionally tenpin-shaped; cuticle naked or forming ornamentations such as plates and multi-pointed hooks; TbA, TbL, and TbP present, usually numerous; TbD present in several taxa; mouth opening narrow to broad; pharyngeal pores usually present. Marine and brackish, rarely fresh water: interstitial. Order MACRODASYIDA (Figure 2) **3**

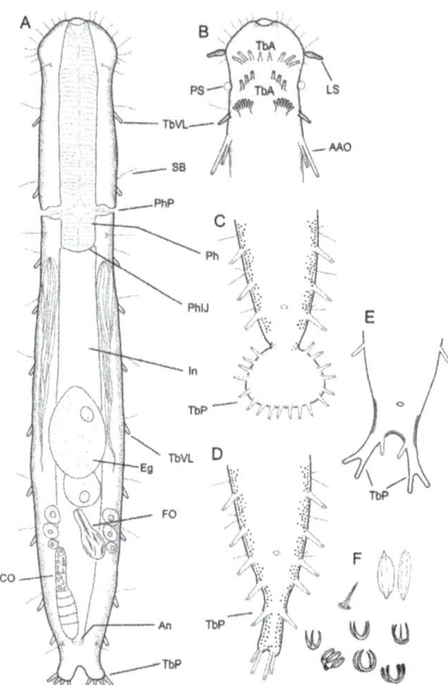

Figure 2. Drawing of a hypothetical Macrodasyida. (**A**) Habitus showing the internal organs, dorsal view; (**B**) anterior region, ventral view, showing some of the possible arrangements of adhesive tubes of the anterior series, i.e., borne, singly, directly from the body surface (the two most anterior ones), and borne in a group on a fleshy extensible base (the posterior one); (**C–E**), some of the possible configurations of the posterior region, ventral view; (**F**) some of the possible spines and scales found among Macrodasyida. Abbreviations: **AAO**, accessory adhesive organs; **An**, anus; **CO**, caudal organ; **Eg**, egg; **Fo**, frontal organ; **In**, intestine; **LS**, leaf-like sensorial organ; **Ph**, pharynx; **PhIJ**, pharyngo-intestinal junction; **PhP**, pharyngeal pores; **PS**, piston pit sensorial organ; **TbA**, anterior adhesive tubes; borne singly, directly from the body; **TbP**, posterior adhesive tubes; **TbVL**, ventrolateral adhesive tubes; **Te**, testicle. (**A–C**) original; (**D**) modified from [13]; (**E**) modified [15].

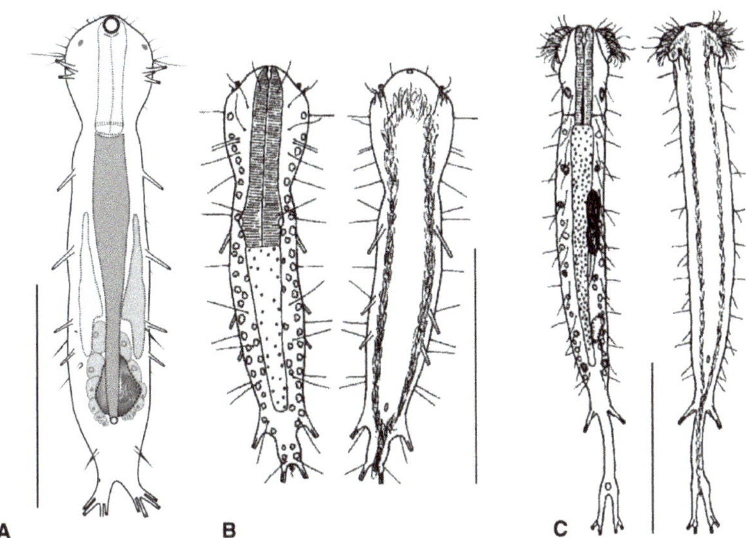

Figure 3. Macrodasyida, Dactylopodolidae: models of (**A**) *Dactylopodola*, (**B**) *Dendropodola*, and (**C**) *Dendrodasys*. Scale bars = 100 μm. (**A**) from [40] with modifications, (**B**) from [41] with modifications, and (**C**) from [42] with modifications.

6a (5a)   Head simple or bearing two sensorial tentacles; cuticular covering bare; posterior body region bilobed; TbL present. Regionally common; marine: interstitial. .. .. *Dactylopodola* (Figure 3A)

6b (5a)   Head simple or with crenulated lateral lobes; cuticular covering bare; posterior body region bifurcate; TbL absent. **7**

7a (6b)   Head simple, cuticle naked. Rare; marine: interstitial. ... ... *Dendropodola* (Figure 3B)

7b (6b)   Head with elongate crenulated lateral lobes. Uncommon; marine: interstitial. ... ... *Dendrodasys* (Figure 3C)

8a (5b)   Trunk region without tentacles, but presenting dented lateral sides; posterior body region furcate; distal rami, each showing a small TbP. Rare; marine: interstitial. .... *Xenodasys* (Figure 4A)

8b (5b)   Trunk region bearing numerous tentacles; lateral sides of the trunk region parallel, lacking indentations; posterior region furcate; each ramus showing an adhesive pad at the end. Rare; marine: interstitial. ... ... *Chordodasiopsis* (Figure 4B)

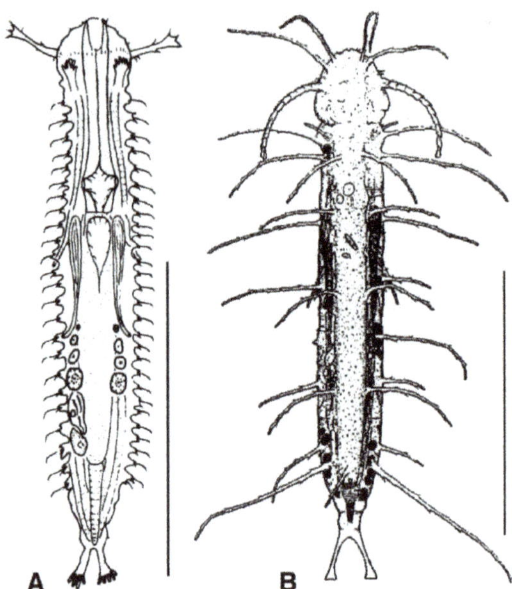

Figure 4. Macrodasyida, Xenodasyidae: models of (**A**) *Xenodasys* and (**B**) *Chordodasiopsis*. Scale bars = 200 μm. (**A**) from [39] with modifications, (**B**) from [43] with modifications.

9a (4b)   TbA, usually 4 or more per side, occasionally 2 or 3, at the end of extensible fleshy base (Figure 2B); pharynx with pores located at the base. **10**

9b (4b)   TbA, generally 1 to 3 per side, occasionally 4 or more, arising singly and directly from the body surface; pharynx with pores at the base or in the middle. **16**

10a (9a)   Head generally well demarcated posteriorly by a furrow; posterior body region tapered into a medial process, truncated, rounded, or broadly expanded, but never two-lobed. CEPHALODASYIDAE (partim) (Figure 5) **11**

10b (9a)   Head normally not clearly delimited; posterior body region two-lobed. TURBANELLIDAE (partim) (Figure 6) **12**

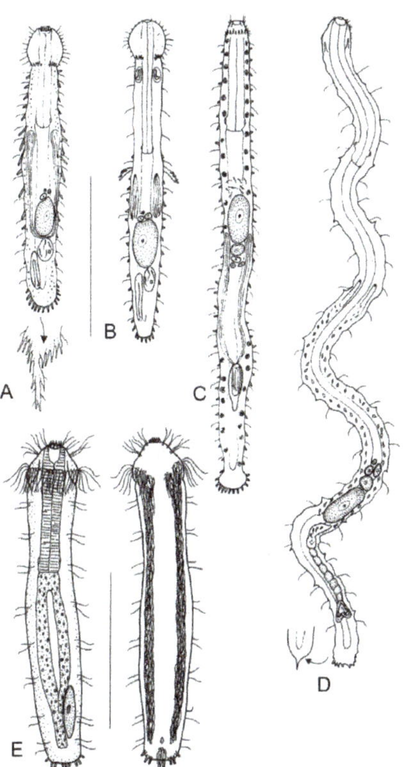

Figure 5. Macrodasyida, Cephalodasyidae: models of (**A**) *Cephalodasys*, (**B**) *Pleurodasys*, (**C**) *Mesodasys*, (**D**) *Dolichodasys*, and (**E**) *Paradasys*. Scale bars = 200 μm. (**A–D**) from [39] with modifications, (**E**) from [35] with modifications.

11a (10a) Head surrounded by very thick and dense sensory cilia; a couple of accessory adhesive organs present near the PhIJ, laterally directed; each organ comprising 3–4 tubes of unequal length; a couple of club-shaped gravireceptor organs on the dorsal side of the posterior cephalic region may be present. Rare; marine: interstitial. ... ... *Pleurodasys* (Figure 5B)

11b (10a) Cephalic sensory cilia and accessory adhesive organs described above are absent. Regionally common; marine and brackish: interstitial. ... ... *Cephalodasys* (Figure 5A)

12a (10b) Head showing elongate lateral tentacles. 13

12b (10b) Head without tentacles, occasionally with conical lobes. 14

13a (12a) TbL numerous. Uncommon; marine: interstitial. ... ... *Dinodasys* (Figure 6A)

13b (12a) TbL lacking, paired TbV inserted just past the PhIJ. Rare; marine: interstitial. ... ... *Pseudoturbanella* (Figure 6B)

14a (12b) Paired accessory adhesive organs in the anterior pharyngeal region; organs are posteriorly directed, and each is made up of two tubes of unequal lengths. Common; marine and brackish: interstitial. ... ... *Paraturbanella* (Figure 6C)

14b (12b) Accessory adhesive tubes described above are either absent or present in different body regions. 15

15a (14b) Accessory adhesive tubes not present. Common; marine and brackish: interstitial. ... ... *Turbanella* (Figure 6E)

15b (14b) Accessory adhesive tubes present, close to the PhIJ. Rare; marine: interstitial. ... ...
... ... ***Prostobuccantia*** (Figure 6D)

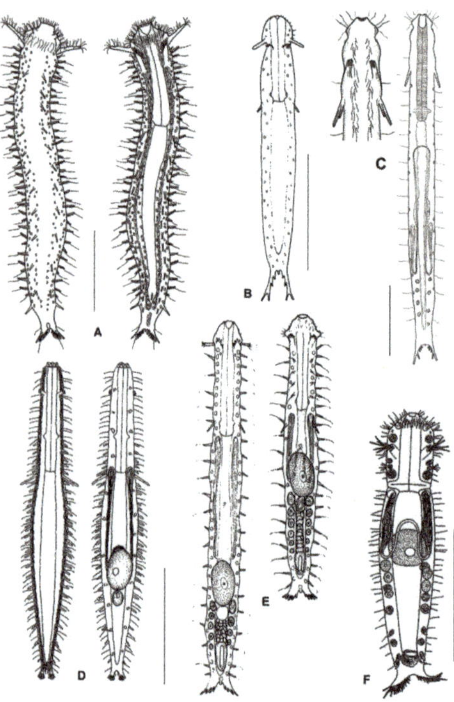

Figure 6. Macrodasyida, Turbanellidae: models of (**A**) *Dinodasys*, (**B**) *Pseudoturbanella*, (**C**) *Paraturbanella*, (**D**) *Prostobuccantia*, (**E**) *Turbanella*, and (**F**) *Desmodasys*. Scale bars = 200 μm. (**A**) from [44] with modifications, (**B**) from [39] with modifications, (**C**) from [45] with modifications, (**D**) from [46] with modifications, and (**F**) from [47] with modifications.

16a (9b) Pharynx with pores far from the base; posterior body region unilobed, ovoidal in shape, or tapering off. MACRODASYIDAE (Figure 7) **17**

16b (9b) Pharynx with pores at the base; posterior end of body not tapering off **20**

17a (16a) Head bearing a lateral leaf-like sensorial organ; posterior body region unilobed, ovoidal in shape. Rare; marine: interstitial. ... ... ***Thaidasys*** (Figure 7B,C)

17b (16a) Head bearing lateral piston pit sensorial organs; posterior body region tapering into a medial process. **18**

18a (17b) Posterior process in the form of a long tail. Regionally common; marine: interstitial and epibenthic. ... ... ***Urodasys*** (Figure 7E)

18b (17b) Posterior process short or in the form of a short tail. **19**

19a (18b) Frontal organ posterior to the largest egg; spermatozoa filiform. Common; marine: interstitial. ... ... ***Macrodasys*** (Figure 7D)

19b (18b) Frontal organ anterior to the largest egg; spermatozoa stout. Uncommon; marine: interstitial. ... ... ***Kryptodasys*** (Figure 7A)

20a (16b) Cuticle forming ornamentations such as hooks, papillae, scales, or thickenings. **21**

20b (16b) Cuticle smooth, without ornamentation such those reported above. **27**

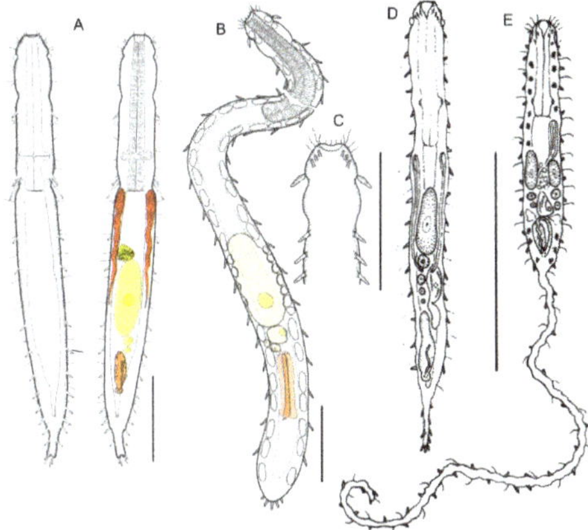

Figure 7. Macrodasyida, Macrodasyidae: models of (**A**) *Kryptodasys*, (**B**, **C**) *Thaidays*, (**D**) *Macrodasys*, and (**E**) *Urodasys*. Scale bars: (**A–C**) = 100 μm, (**D**, **E**) = 200 μm. (**A**) from [15] with modifications, (**B**,**C**) from [14] with modifications, and (**D**, **E**) from [39] with modifications.

21a (20a) Presence of elongate scales; mouth narrow; pharynx without pores. Uncommon; marine: interstitial. LEPIDODASYIDAE .... *Lepidodasys* (Figure 8)

21b (20a) Presence of variously spined hooks, large scales, or papillae; mouth opening generally broad; pharyngeal pores present. THAUMASTODERMATIDAE (partim) (Figure 9) **22**

22a (21b) Presence of papillae or large scales. **23**

22b (21b) Presence of uni- or multi-spined hooks. **24**

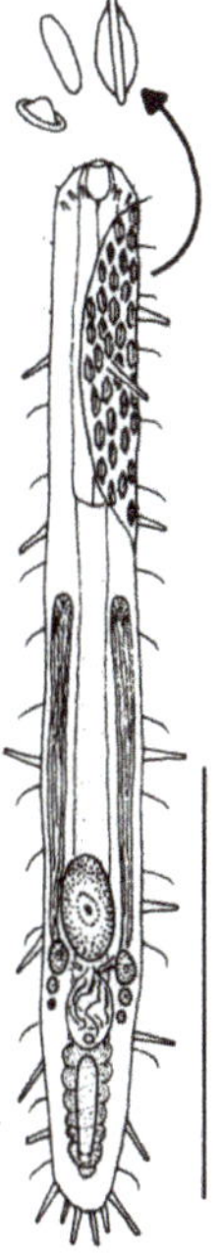

Figure 8. Macrodasyida, Lepidodasyidae: model of *Lepidodasys*. Scale bar = 200 μm. From [39] with modifications.

23a (22a) Cuticle with large scales, but not papillae; on either side of the body a single row of wide spines present. Regionally common; marine: interstitial. .... *Diplodasys* (Figure 9A)

23b (22a) Cuticle with papillae, but not scales or spines. Uncommon; marine: interstitial. ... ... *Oregodasys* (Figure 9B)

24a (22b) Cuticle with hooks showing a single spine; right and left testicles present; Common; marine: interstitial. ... ... *Acanthodasys* (Figure 9C)

24b (22b) Cuticle with hooks showing more than one spine; a single testicle on the right-hand body side. **25**

25a (24b) Anterior body region showing conspicuous, grasping structures on either side of the mouth funnel (buccal palps); hooks bearing 5, 4, or 3 spines (penta-, tetra-, or triancres). Common; marine: interstitial. ... ... *Pseudostomella* (Figure 9D)

25b (24b) Anterior body region without buccal palps; hooks showing 5, 4, 3, or 2 spines (penta-, tetra-, tri-, or biancres). **26**

26a (25b) Head bearing two pairs of sensoria tentacles on the lateral sides; mouth narrow, hooks with four spines. Common; marine: interstitial. ... ... *Thaumastoderma* (Figure 9E)

26b (25b) Head bearing no or one pair of sensorial tentacles on the lateral sides; hooks with 5, 4, 3, or 2 spines. Very common; marine: interstitial. ... ... *Tetranchyroderma* (Figure 9F)

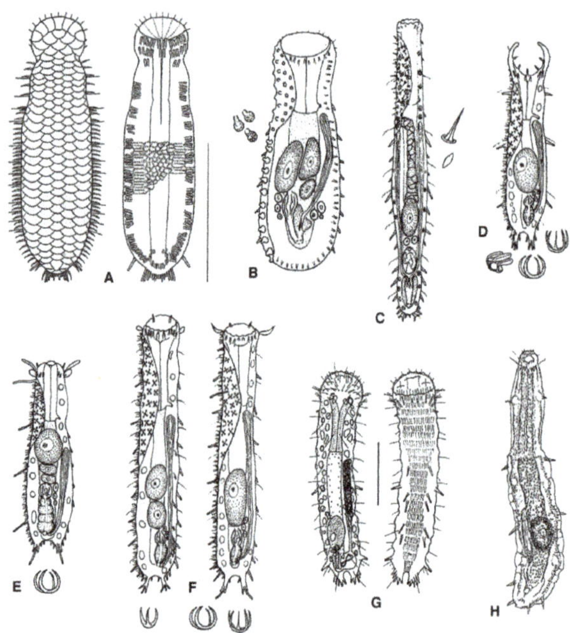

Figure 9. Macrodasyida, Thaumastodermatidae: models of (**A**) *Diplodasys*, (**B**) *Oregodasys*, (**C**) *Acanthodasys*, (**D**) *Pseudostomella*, (**E**) *Thaumastoderma*, (**F**) *Tetranchyroderma*, (**G**) *Ptychostomella*, and (**H**) *Hemidasys*. Scale bars: (**A–F**, **H**) = 200 μm, (**G**) = 50 μm. (**A**) from [48] with modifications, (**B–F**) from [39] with modifications, (**G**) from [41] with modifications, and (**H**) from [49] with modifications.

27a (20b) Male apparatus absent (i.e., parthenogenetic); TbA, two groups of three tubes per side; TbL, four or five per side, TbP up to five per side. Rare; marine: interstitial. REDUDASYIDAE (partim) ... ... *Anandrodasys* (Figure 10A)

27b (20b) These characteristics not combined. **28**

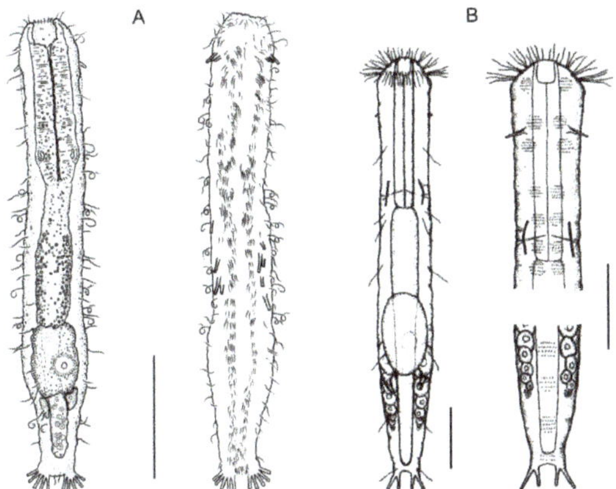

Figure 10. Macrodasyida, Redudasyidae: models of (**A**) *Anandrodasys* and (**B**) *Redudasys*. Scale bars = 50 μm. (**A**) from [50] with modifications, (**B**) from [51] with modifications.

28a (27b) TbA, several to many, arranged in two tufts; TbL absent. Rare; marine: interstitial. TURBANELLIDAE (partim) .... *Desmodasys* (Figure 6F)

28b (27b) TbA, few to many, but not arranged in tufts; TbL normally present or, if absent, then TbA few in number. **29**

29a (28b) TbA, few; TbL few; body elongate (to about 1 mm in length) and narrow; posterior end in the form of two distinct pedicles. HUMMONDASYIDAE. .... *Hummondasys* (Figure 11)

29b (28b) These characteristics not combined. **30**

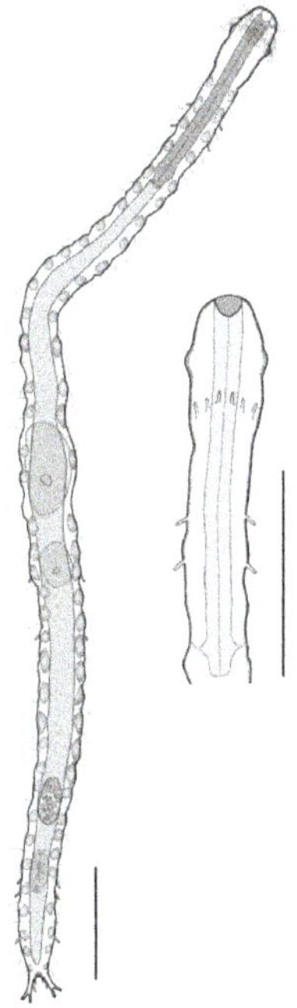

Figure 11. Macrodasyida, Hummondasyidae: model of *Hummondasys*. Scale bars = 100 μm. From [13] with modifications.

30a (29b) TbA, few to many; TbL and TbP, numerous (more than 10 per side); mouth narrow (< 0.4 ×
head width); posterior body region in the form of a large round lobe or clearly two-lobed.
PLANODASYIDAE (Figure 12) **31**

30b (29b) TbA or TbL numbering fewer than six tubes per side; oral opening narrow to broad, and if
narrow, then posterior body region not clearly two-lobed. **33**

31a (30a) TbA, present in low numbers; body very long (up to 3.5 mm in length) and rather narrow;
posterior body region ending as a large lobe. Uncommon; marine: interstitial. ... ...
... ... *Megadasys* (Figure 12A)

31b (30a) Posterior body region distinctly two-lobed. **32**

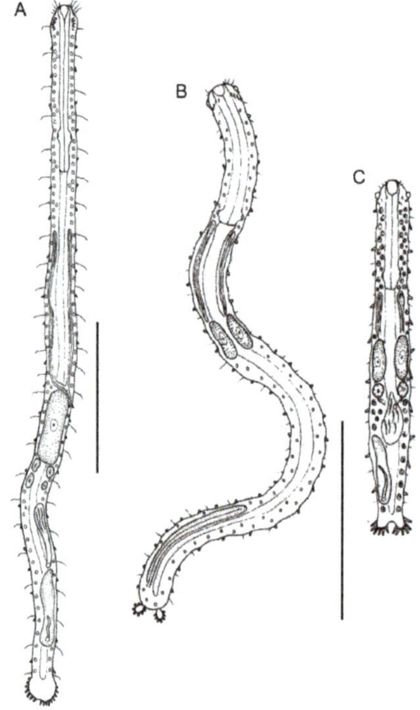

Figure 12. Macrodasyida, Planodasyidae: models of (**A**) *Megadasys*, (**B**) *Planodasys*, and (**C**) *Crasiella*.
Scale bars = 200 µm. From [39] with modifications.

32a (31b) Posterior lobes in the form of oval appendages; most anterior TbA arranged transversely;
caudal organ elongate. Rare; marine: interstitial. ... ... *Planodasys* (Figure 12B)

32b (31b) Posterior lobes in the form of furcate extensions; most anterior TbA arranged longitudinally;
caudal organ ovate. Uncommon; marine: interstitial. ... ... *Crasiella* (Figure 12C)

33a (30b) Oral opening, narrow (< 0.4 × head width); right and left testicles present.
CEPHALODASYIDAE (partim) (Figure 5)... **34**

33b (30b) Oral opening broad (> 0.6 × head width) or, if narrow, leading to a large buccal cavity surrounded
by an oral hood; a single testicle, on the right-hand side. THAUMASTODERMATIDAE (partim)
(Figure 9) **36**

34a (33a) Total body length > 1 mm; TbA, one per side; TbL in form of numerous papillae along the
body sides. Uncommon; marine: interstitial. ... ... *Dolichodasys* (Figure 5D)

34b (33a) Total body length < 1 mm. **35**

35a (34b) TbA, 1–4 tubes per side, arranged in two groups; TbL, 0–6 tubes per side. Uncommon; marine: interstitial. .... *Paradasys* (Figure 5E)

35b (34b) TbA, few to several per side; TbL, several to many. Common; marine: interstitial. ... ... *Mesodasys* (Figure 5C)

36a (33b) Oral opening, broad; locomotor cilia extending over the entire ventral surface; male genital pore not surrounded by cuticular plates. Common; marine: interstitial. ... ... *Ptychostomella*(Figure 9G)

36b (33b) Oral opening, narrow, leading to a large buccal cavity covered by an oral hood; ventral locomotor cilia restricted to the pharyngeal region; male genital pore surrounded by cuticular plates. Very rare (possibly extinct); marine: interstitial ... ... *Hemidasys* (Figure 9H)

37a (3b) Total body length 300–400 μm; TbA, 1–2 per side; pharyngeal pores present. Rare; interstitial. REDUDASYIDAE (partim) .... *Redudasys* (Figure 10B)

37b (3b) Body length up to 220 μm; TbA, one per side; pharyngeal pores absent. Rare; interstitial. INCERTAE SEDIS. ... ... *Marinellina* (Figure 13)

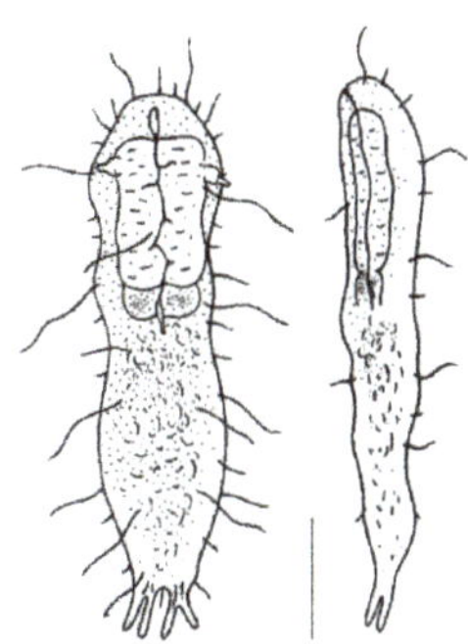

Figure 13. Macrodasyida, INCERTAE SEDIS: model of *Marinellina*. Scale bar = 50 μm. From [52] with modifications.

38a (1a) Ventral locomotor ciliation made up of cirri (Figure 1C). Marine and brackish. XENOTRICHULIDAE (Figure 14) **39**

38b (1a) Ventral locomotor ciliation formed by single cilia, occurring in longitudinal bands or tufts, never composed of cirri (Figure 1B). Freshwater, marine, and brackish. **41**

39a (38a) Cirri of the head and pharyngeal regions of two different sizes, with 1–2 transverse rows of small and short cirri anteriorly followed by transverse rows of big and longer cirri; frontal portion of pharynx with a swelling (bulb). Common; marine and brackish: interstitial. .. .. *Heteroxenotrichula* (Figure 14A)

39b (38a) Cirri, all of similar size, pharynx without anterior swelling (bulb). **40**

40a (39b) Ovary and eggs present, testicles and spermatozoa absent; head clearly distinct; scales on the dorsal side, flat; scales of the lateral mid-trunk, pedunculated; a pair of lateral spines at the base of the furcal branches. Common; marine: interstitial. .... *Draculiciteria* (Figure 14B)

40b (39b) Testicles and spermatozoa present; head in general not clearly defined; scales of the lateral mid-trunk bearing a stalk or flat; if stalked, similar to the dorsal scales. Common; marine and brackish: interstitial. ... ... *Xenotrichula* (Figure 14C)

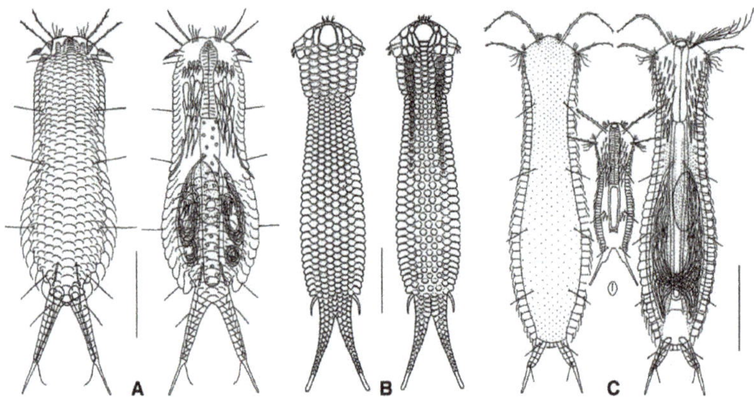

Figure 14. Chetonotida, Xenotrichulidae: models of (**A**) *Heteroxenotrichula,* (**B**) *Draculiciteria,* and (**C**) *Xenotrichula.* Scale bars = 50 μm. (**A, C**) modified from [35], (**B**) modified from [53].

41a (38b) Posterior body region furcate or bifurcate; caudal rami with or without TbP. **42**

41b (38b) Posterior body region rounded or truncated; perhaps showing two caudal protuberances or spines. **60**

42a (41a) Posterior body region bifurcate, bearing four TbP or two TbP and two spiniform cuticular processes; elsewhere, cuticle smooth, not forming scales or spines. Rare; freshwater: interstitial or periphytic/epibenthic. DICHAETURIDAE ... ... *Dichaetura* (Figure 15A)

42b (41a) Posterior body region furcate; cuticle smooth or forming spines and/or scales. **43**

43a (42b) Body cuticle smooth; caudal rami with TbP, sickle-shaped; cilia of the head not arranged into tufts. Very rare; freshwater: semiplanktonic or hyperbenthic. PROICHTHYDIIDAE (Figure 15B,C) **44**

43b (42b) Body cuticle generally forming spines and scales; caudal rami with or without TbP; if present, caudal rami and TbP generally straight, short to very long; cilia of the head emerging as tufts or forming a continuous band around the elongate, muzzle-like frontal end. **45**

44a (43a) Cilia of the head arranged as a transverse row of small elements on the dorsal side; locomotor cilia limited to head and neck, emerging as separate tufts. Freshwater: hyperbenthic ... ... *Proichthydium* (Figure 15B)

44b (43a) Cilia of the head emerging mostly from the lateral sides as single, short to very long elements; locomotor cilia distributed in two bands that run from under the head to the posterior trunk region. Freshwater: semiplanktonic. ....*Proichthydioides* (Figure 15C)

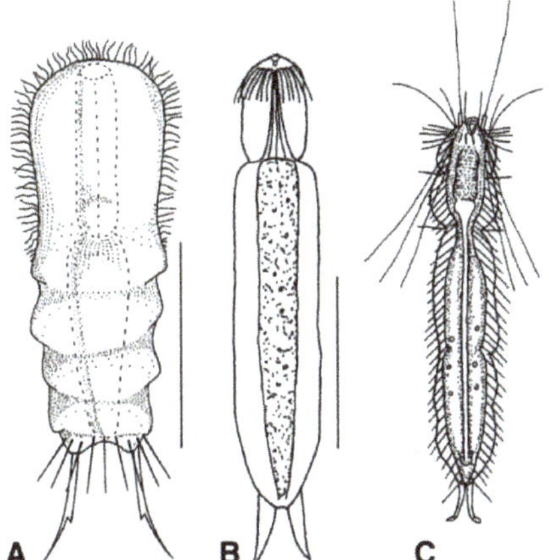

Figure 15. Chetonotida: (**A**) Dichaeturidae, a model of *Dichaetura*. (**B, C**) Proichthydiidae, models of (**B**) *Proichthydium* and (**C**) *Proichthydioides*. Scale bars = 50 μm. (**A**) from [54] with modification, (**B**) from [55] with modification, and (**C**) re-sketched from [56].

45a (43b) Cilia of the head organized as one or two pairs of dorsal or ventrodorsal tufts. Common (with the exception of *Arenotus* and *Undula*); freshwater, marine, and brackish: periphytic, epibenthic, and interstitial CHAETONOTIDAE (Figures 16 and 17). **46**

45b (43b) Cilia of the head organized in a band, encircling a muzzle-like frontal end; TbP numbering two or four. Uncommon to rare, marine: epibenthic or interstitial. MUSELLIFERIDAE (Figure 20) **67**

46a (45a) TbP at the end of the furcal rami absent. Rare; freshwater: epibenthic. ... ... *Undula* (Figure 16A)

46b (45a) TbP at the end of the furcal rami present. **47**

47b (46b) Furcal base narrow (pedunculated); caudal rami segmented; cephalion and hypostomion extremely large; scales without a keel, notch, or spine. Rare; freshwater: epibenthic. ... ... *Cephalionotus* (Figure 16B)

47b (46b) These characteristics not combined. **48**

48a (47b) Furcal rami very long (up to one-third of the total body length), multi-segmented, bare or with tiny spines or scales. Common, freshwater: periphytic and epibenthic. ... ... *Polymerurus* (Figure 16C)

48b (47b) Furcal rami from very short to mid length, not segmented, scales or spines limited to the proximal portion or lacking altogether. **49**

49a (48b) Cuticular covering bare (or mostly bare) or made up of scales lacking spines; seldomly, some spines may be present at the base of the furca. **50**

49b (48b) Cuticular covering including scales bearing spines (spined scales) and/or a keel (spined, keeled scales and keeled scales, respectively); spines from long to very short, bearing 1–2 indentations laterally (notched spines), or simple. **57**

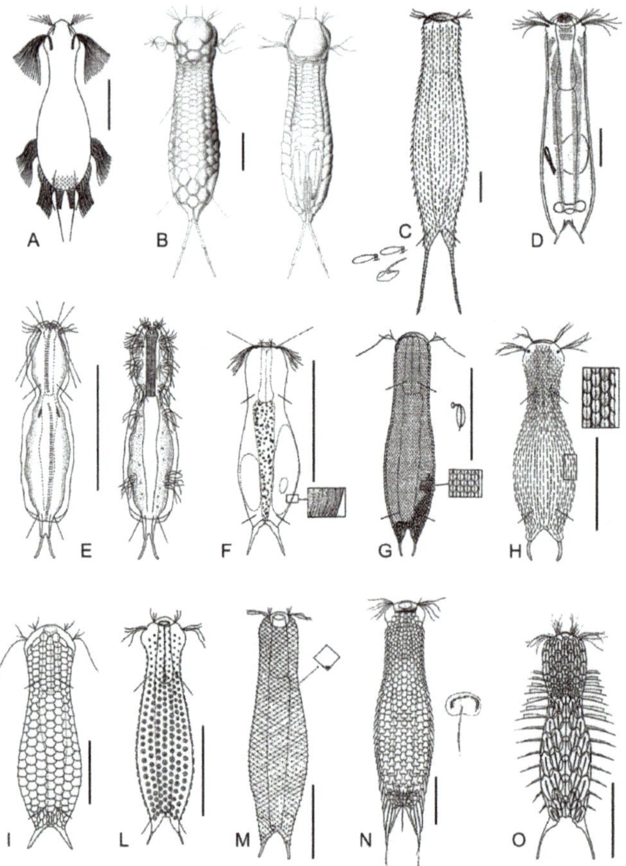

Figure 16. Chetonotida, Chaetonotidae: models of (**A**) Undula, (**B**) Cephalionotus, (**C**) Polymerurus, (**D**) Arenotus, (**E**) Caudichthydium, (**F**) Ichthydium, (**G**) Aspidiophorus, (**H**) Heterolepidoderma, (**I**) Lepidodermella, (**L**) Fluxiderma, (**M**) Rhomballichthys, (**N**) Lepidochaetus, and (**O**) Halichaetonotus. Scale bars (**A**, **C–D**) = 50 μm, (**B**) = 25 μm. (**A**) from [57] with modifications, (**B**) from [37] with modifications, (**C**) from [51] with modifications, (**D**) from [58] with modifications, (**E–H**, **M**) from [54] with modifications, (**I**, **L**) from [59] with modifications, and (**O**) from [60] with modifications.

50a (49a) Cuticular covering bare, rarely a few scales and/or spines at base of the furca may be present. **51**

50b (49a) Cuticular covering wholly or prevalently made of spineless scales. **53**

51a (50a) Cuticle completely bare, very thick, obviously distinguishable from the underlying epidermal layer. Rare; freshwater: interstitial. .... *Arenotus* (Figure 16D)

51b (50a) Cuticle thin, mostly bare, except for perhaps two terminal scales at the end ventral interciliary field; occasionally, weak striations along the body or few spines and/or scales at the furcal base may be present. Common; freshwater, rarely marine or brackish water: periphytic, epibenthic, and interstitial. **52**

52a (51b) Furcal base pedunculated; locomotor cilia distributed in separated tufts. Uncommon; marine: interstitial. .... *Caudichthydium* (Figure 16E)

52b (51b) Furcal base not pedunculated; locomotor cilia mostly forming two longitudinal bands. Common; freshwater, rarely brackish or marine: epibenthic, periphytic, and interstitial. . . . . ***Ichthydium*** (Figure 16F)

53a (50b) Scales, small, keeled, or stalked. **54**

53b (50b) Scales, large and bare, round, rhomboidal, or polygonal in shape. **55**

54a (53a) Most scales with a stalk; occasionally, few scales may lack a stalk and bear a keel or a spine instead. Common; freshwater, brackish, and marine: epibenthic, periphytic, and interstitial. . . . . ***Aspidiophorus*** (Figure 16G)

54b (53a) Numerous keeled scales; occasionally, few scales may bear a spine. Common; freshwater, brackish, and marine: periphytic, epibenthic, and interstitial. . . . . ***Heterolepidoderma*** (Figure 16H)

55a (53b) Scales polygonal in shape. Common; freshwater, rarely brackish or marine: interstitial, epibenthic, and periphytic. . . . . ***Lepidodermella*** (Figure 16I)

55b (53b) Scales rhomboidal or circular in shape. **56**

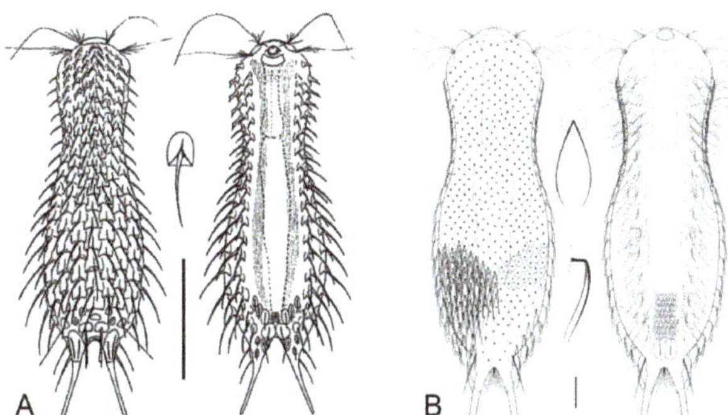

Figure 17. Chaetonotida, Chaetonotidae: models of (**A**) *Chaetonotus* and (**B**) *Bifidochaetus*. Scale bars (**A**) = 50 μm, (**B**) = 10 μm. (**A**) from [61] with modifications, (**B**) from [62] with modifications.

56a (55b) Scales circular. Rare; freshwater: periphytic. . . . . ***Fluxiderma*** (Figure 16L)

56b (55b) Scales rhomboidal. Rare; freshwater: periphytic. . . . . ***Rhomballichthys*** (Figure 16M)

57a (49b) Scales of the ventral interciliary field similar in shape to the scales of the dorsal side; scales of the dorsal side possessing a double edge anteriorly, with or without a spine but always deprived of a keel; several pairs of thin spines of increasing length at the lateral sides of the furcal base. Rather common; freshwater: periphytic and epibenthic. . . . . ***Lepidochaetus*** (Figure 16N)

57b (49b) Scales of the ventral interciliary field dissimilar in shape from scales of the dorsal side; scales of the dorsal side with a single edge anteriorly, keeled or keeled and spined **58**

58a (57b) Scales lateral to the ventral locomotor cilia with spines bearing lamellae (hydrofoil scales); scales of the dorsal side bearing a keel; seldom presence of 1–5 scales bearing spines. Common; marine and brackish: interstitial. . . . . ***Halichaetonotus*** (Figure 16O)

58b (57b) Scales bearing spines with lamellae normally absent; if present, dorsal scales spined. **59**

59a (58b) Dorsal scales round to suboval, without keels and/or notches but carrying distally bifurcating hairlike spines. Rare; freshwater: epibenthic . . . . ***Bifidochaetus*** (Figure 17B)

59b (58b) These characteristics not combined. Very common; freshwater, marine, and brackish: epibenthic, periphytic, and interstitial. . . . . ***Chaetonotus*** (Figure 17A)

60a (41b) Posterior body region rounded-off or truncated with paired lateral projections; head bearing a pair of rod- or club-shaped tentacles; trunk bearing small, spined scales; rarely, trunk scales restricted to a small patch on the ventral side. Uncommon to rare; freshwater: hyperbenthic and semiplanktonic. NEOGOSSEIDAE (Figure 18) **61**

60b (41b) Posterior body region rounded or truncated, occasionally with a very short caudal lobe or paired postero-lateral protuberance; head without tentacles; body scales reduced or absent; trunk bearing very long and movable spines arranged into groups. Uncommon to rare; freshwater: hyperbenthic, epibenthic, and semiplanktonic. DASYDYTIDAE (Figure 19)... **62**

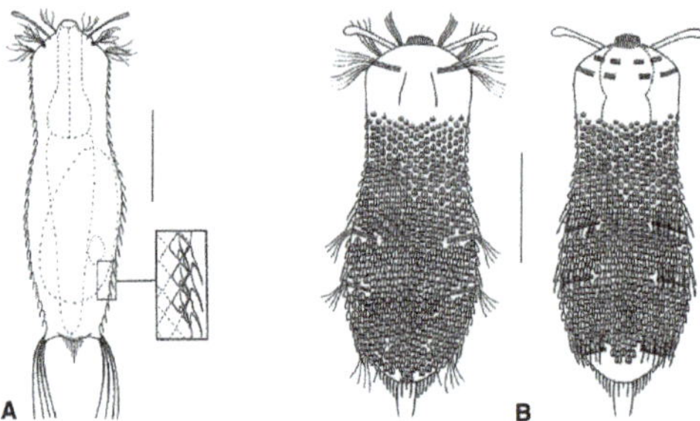

Figure 18. Chaetonotida, Neogosseidae: models of (**A**) *Neogossea* and (**B**) *Kijanebalola*. Scale bars = 50 μm. (**A**) from [54] with modifications, (**B**) from [57] with modifications.

61a (60a) Posterior body region truncated, showing two lateral projections bearing a tuft of long spines; trunk with fine spined scales. Uncommon: epibenthic and semiplanktonic. .... *Neogossea* (Figure 18A)

61b (60a) Posterior body region rounded, with a central group of spines and no lateral projections; trunk with keeled scales, seldom reduced to a small group on the ventral side. Rare: epibenthic and semiplanktonic. ... ... *Kijanebalola* (Figure 18B)

62a (60b) Trunk region bearing long, scattered spines on the dorsal side or two caudal spines only; body scales absent; locomotor cilia arranged in two longitudinal bands; pharynx bearing two robust swellings (bulbs). Rare: epibenthic and semiplanktonic. ... ... *Anacanthoderma* (Figure 19A)

62b (60b) Trunk region bearing long, lateral spines arranged into columns or groups; dorsal spines present or absent; locomotor cilia arranged in tufts; pharynx bearing a single swelling or cylindrical. **63**

63a (62b) Lateral spines, simple or with notches; if present, scales large, elliptical in shape, and few in number; pharynx cylindrical (i.e., without bulbs). **64**

63b (62b) Lateral spines with a single lateral notch and bifurcate apex, or with 2–3 lateral notches and pointed apex; if present, numerous, small, keeled scales; pharynx bearing a swelling at the posterior end. **65**

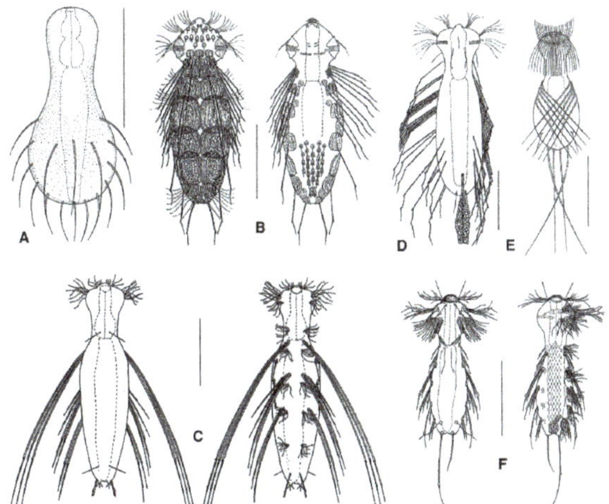

Figure 19. Chaetonotida, Dasydytidae: models of (**A**) *Anacanthoderma*, (**B**) *Ornamentula*, (**C**) *Stylochaeta*, (**D**) *Dasydytes*, (**E**) *Haltidytes*, and (**F**) *Setopus*. Scale bars = 50 μm. (**A**, **F**) from [54] with modifications, (**B**–**E**) from [57] with modifications.

64a (63a) Trunk showing dorsal spines; two pairs of caudal spines; all spines show a noticeable lateral notch; dorsal scales, rather large and of peculiar lace-like appearance. Rare: epibenthic, periphytic, and semiplanktonic. ... ... *Ornamentula* (Figure 19B)

64b (63a) Trunk lacking dorsal spines; a single pair of caudal spines or none; if very long, the lateral spines are thick and bent basally, becoming thinner and thinner distally; lateral notch present or absent; where present, body scales are small and feebly keeled. **66**

65a (65b) Lateral spines, robust, showing pointed tip and 2–3 lateral notches; body scales lacking; posterior body region showing two bristled protuberances on the sides. Uncommon: semiplanktonic and epibenthic. ... ... *Stylochaeta* (Figure 19C)

65b (63b) Lateral spines, almost straight, showing a bifurcate tip and a single lateral notch; body scales present; posterior body region rounded. Uncommon: semiplanktonic, periphytic, and epibenthic ... ... *Dasydytes* (Figure 19D)

66a (63b) Caudal spines absent or present; if present, in general of different length; lateral spines, straight, of medium length; ventral, S-shaped, jumping spines lacking. Rare: semiplanktonic and epibenthic. .... *Setopus* (Figure 19F)

66b (64b) Caudal spines absent; lateral spines very long, strongly bent crossing over the dorsal side; ventral S-shaped jumping spines present. Rare: semiplanktonic and epibenthic. ... ... *Haltidytes* (Figure 19E)

67a (45b) Furcal rami each with a single TbP; body scales bearing fine spines but lacking keels. Rare; marine: interstitial or infaunal. Uncommon: epibenthic and interstitial. ... ... *Musellifer* (Figure 20A)

67b (45b) Furcal rami each with two TbP; body scales, keeled. Rare: interstitial. .... *Diuronotus* (Figure 20B)

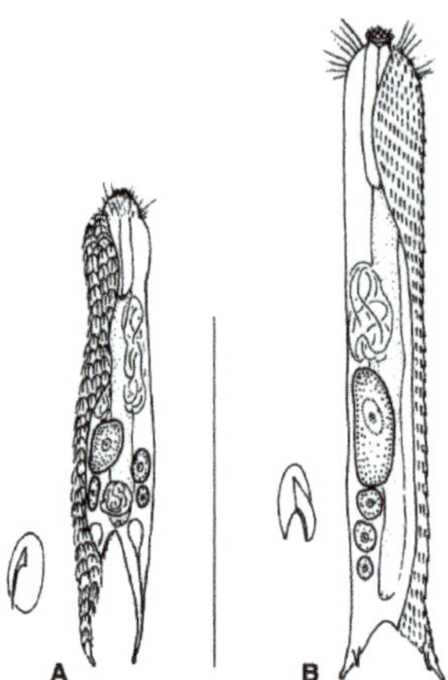

Figure 20. Chetonotida, Muselliferidae: models of (**A**) *Musellifer* and (**B**) *Diuronotus*. Scale bar = 200 μm. From [39] with modifications.

4. Discussion

This paper was derived from a workshop on marine meiofaunal organisms of Costa Rica with a focus on Gastrotricha (and Kinorhyncha) held in January 2019 at the CIMAR (University of Costa Rica) (Supplementary Material Figure S10). During the event, intended for undergraduate and graduate students, general information on the phylum was provided in the classroom and the main techniques regarding sampling, extraction, and observation were illustrated. Techniques were put in practice by some of the students who participated in subsequent 15-day field work. The effectiveness of an early versions of the taxonomic keys was determined by the pupils based on direct observation of living, relaxed specimens (some of the students) and/or based on photographs of freshly sampled specimens (all the students). The proposed version of the keys benefited from the insightful comments which emerged during the testing. At the end of the training period, all the students were able to correctly identify at the genus level the gastrotrich involved in the testing. Based on this outcome, we are confident in the work's usefulness to many others.

Supplementary Materials: The following are available online at http://www.mdpi.com/1424-2818/11/7/117/s1, Figure S1: Sampling of marine gastrotrichs of the littoral zone, Figure S2: Sampling of marine gastrotrichs of the sublittoral zone, Figure S3: Sampling freshwater gastrotrichs, Figure S4: Processing of freshwater sample for *in vivo* studies, Figure S5: Extraction of interstitial fauna, Figure S6: Extraction of interstitial fauna, Figure S7: Mounting of the specimens, Figure S8: Mounting of the specimens, Figure S9: Morphological analysis and documentation, Figure S10: Instructors (background) and students (foreground) participating at the work-shop on marine meiofaunal organisms of Costa Rica with focus on Gastrotricha.

Author Contributions: Conceptualization, M.A.T.; Data curation, M.D.; Funding acquisition, M.A.T. and J.A.S.-C.; Investigation, J.A.S.-C., O.A.S.-B., J.D.B. and M.D.; Supervision, M.A.T.; Writing—original draft, M.A.T.; Writing—review and editing, J.A.S.-C., O.A.S.-B., G.C.-D., N.G.-O., J.D.B., M.C.-D. and M.D.Z.

Funding: The research was funded in part by a Universidad de Costa Rica (project 808-B8-184 Marine meiofauna of Costa Rica) grant to J.A.S.-C. The publishing cost was met by a MIUR grant "Finanziamento annuale individuale delle attività base di ricerca" to M.A.T.

Acknowledgments: The first draft of this paper was prepared during the workshop on marine meiofaunal organisms of Costa Rica with a focus on Gastrotricha and Kinorhyncha, held in January 2019 at the CIMAR (University of Costa Rica). We thank the director of CIMAR for supporting the event.

Conflicts of Interest: The authors declare no conflict of interest.

References

1. Todaro, M.A. Meiofauna from the Meloria Shoals: Gastrotricha, biodiversity and seasonal dynamics. *Biol. Mar. Medit.* **1998**, *5*, 587–590.
2. Gray, J.S. The effects of pollution on sand meiofauna communities. *Thalass. Jugosl.* **1971**, *7*, 79–86.
3. Coull, B.S. Long-term variability of estuarine meiobenthos: An 11 year study. *Mar. Ecol. Prog. Ser.* **1985**, *24*, 205–218. [CrossRef]
4. Todaro, M.A.; Fleeger, J.W.; Hummon, W.D. Marine gastrotrichs from the sand beaches of the northern Gulf of Mexico: Species list and distribution. *Hydrobiologia* **1995**, *310*, 107–117. [CrossRef]
5. Hochberg, R. Spatiotemporal size-class distribution of *Turbanella mustela* (Gastrotricha: Macrodasyida) on a northern California beach and its effect on tidal suspension. *Pacific Sci.* **1999**, *53*, 50–60.
6. Nesteruk, T. Density and biomass of Gastrotricha in sediments of different types of standing waters. *Hydrobiologia* **1996**, *24*, 205–208. [CrossRef]
7. Ruppert, E.E. Gastrotricha. In *Microscopic Anatomy of Invertebrates, Aschelminthes*; Harrison, F.W., Ruppert, E.E., Eds.; Wiley-Liss: New York, NY, USA, 1991; Volume 4, pp. 41–109.
8. Hochberg, R.; Litvaitis, M.K. A muscular double helix in gastrotricha. *Zool. Anz.* **2001**, *240*, 61–68. [CrossRef]
9. Todaro, M.A.; Telford, M.J.; Lockyer, A.E.; Littlewood, D.T.J. Interrelationships of the Gastrotricha and their place among the Metazoa inferred from 18S rRNA genes. *Zool. Scr.* **2006**, *35*, 251–259. [CrossRef]
10. Struck, T.H.; Wey-Fabrizius, A.R.; Golombek, A.; Hering, L.; Weigert, A.; Bleidorn, C.; Klebow, S.; Iakovenko, N.; Hausdorf, B.; Petersen, M.; et al. Platyzoan paraphyly based on phylogenomic data supports a noncoelomate ancestry of Spiralia. *Mol. Biol. Evol.* **2014**, *31*, 1833–1849. [CrossRef]
11. Egger, B.; Lapraz, F.; Müller, S.; Dessimoz, C.; Girstmair, J.; Skunca, N.; Rawlinson, K.A.; Cameron, C.B.; Beli, E.; Todaro, M.A.; et al. A Transcriptomic-phylogenomic analysis of the evolutionary relationships of flatworms. *Curr. Biol.* **2015**, *25*, 1–7. [CrossRef]
12. Todaro, M.A.; Dal Zotto, M.; Jondelius, U.; Hochberg, R.; Hummon, W.D.; Kånneby, T.; Rocha, C.E.F. Gastrotricha: a marine sister for a freshwater puzzle. *PLoS ONE* **2012**, *7*, e31740. [CrossRef] [PubMed]
13. Todaro, M.A.; Leasi, F.; Hochberg, R. A new species, genus and family of marine Gastrotricha from Jamaica, with a phylogenetic analysis of Macrodasyida based on molecular data. *Syst. Biodiv.* **2014**, *12*, 473–488. [CrossRef]
14. Todaro, M.A.; Dal Zotto, M.; Leasi, F. An integrated morphological and molecular approach to the description and systematisation of a novel genus and species of Macrodasyida (Gastrotricha). *PLoS ONE* **2015**, *10*, e0130278. [CrossRef] [PubMed]
15. Todaro, M.A.; Dal Zotto, M.; Kånneby, T.; Hochberg, R. Integrated data analysis allows the establishment of a new, cosmopolitan genus of marine Macrodasyida (Gastrotricha). *Sci. Rep.* **2019**, *9*, 7989. [CrossRef] [PubMed]
16. Leasi, F.; Todaro, M.A. The muscular system of *Musellifer delamarei* (Renaud-Mornant, 1968) and other chaetonotidans with implications for the phylogeny and systematisation of the Paucitubulatina (Gastrotricha). *Biol. J. Linn. Soc.* **2008**, *94*, 379–398. [CrossRef]
17. Kånneby, T.; Todaro, M.A.; Jondelius, U. Phylogeny of Chaetonotidae and other Paucitubulatina (Gastrotricha: Chaetonotida) and the colonization of aquatic ecosystems. *Zool. Scr.* **2013**, *42*, 88–105. [CrossRef]
18. Guidi, L.; Todaro, M.A.; Ferraguti, M.; Balsamo, M. Reproductive system and spermatozoa ultrastructure support the phylogenetic proximity of *Megadasys* and *Crasiella* (Gastrotricha, Macrodasyida). *Contr. Zool.* **2014**, *83*, 119–131. [CrossRef]
19. Kånneby, T.; Todaro, M.A. The phylogenetic position of Neogosseidae (Gastrotricha: Chaetonotida) and the origin of planktonic Gastrotricha. *Org. Divers. Evol.* **2015**, *6*, 1–12.
20. Balsamo, M.; Grilli, P.; Guidi, L.; d'Hondt, J.L. Gastrotricha: Biology, ecology and systematics. Families Dasydytidae, Dichaeturidae, Neogosseidae, Proichthydiidae. In *Identification Guides to the Plankton and Benthos of Inland Waters*; Dumont, H.J.F., Ed.; Backhuys Publisher: Leiden, The Netherlands, 2014; Volume 24, pp. 1–187.
21. Kånneby, T.; Hochberg, R. Phylum Gastrotricha. In *Thorp and Covich's Freshwater Invertebrates: Ecology and General Biology*; Thorp, J., Rogers, D.C., Eds.; Elsevie Academic Press: Amsterdam, The Netherlands, 2015; Volume 1, pp. 211–223.

22. Kieneke, A.; Schmidt-Rhaesa, A. Gastrotricha and Gnathifera. In *Handbook of Zoology*; Schmidt-Rhaesa, A., Ed.; De Gruyter: Berlin, Germany, 2015; Volume 3, pp. 1–134.

23. Gastrotricha Web Portal. Available online: http://www.gastrotricha.unimore.it/ (accessed on 25 June 2019).

24. World Register of Marine Species (WoRMS). Available online: http://www.marinespecies.org/ (accessed on 25 June 2019).

25. Hummon, W.D.; Todaro, M.A.; Balsamo, M.; Tongiorgi, P. Effects of pollution on marine Gastrotricha in the northwestern Adriatic Sea. *Mar. Pollut. Bull.* **1990**, *21*, 241–243. [CrossRef]

26. Todaro, M.A.; Rocha, C.E.F. Diversity and distribution of marine Gastrotricha along the northern beaches of the state of Sao Paulo (Brazil), with description of a new species of *Macrodasys* (Macrodasyida, Macrodasyidae). *J. Nat. Hist.* **2004**, *38*, 1605–1634. [CrossRef]

27. Todaro, M.A.; Rocha, C.E.F. Further data on marine gastrotrichs from the State of São Paulo and the first records from the State of Rio de Janeiro (Brazil). *Meiofauna Mar.* **2005**, *14*, 27–31.

28. Hummon, W.D. Gastrotricha. In *The Light and Smith Manual: Intertidal Invertebrates from Central California to Oregon*; Carlton, J.T., Ed.; University of California Press: Berkeley, CA, USA, 2007; pp. 267–268.

29. Todaro, M.A.; Leasi, F.; Bizzarri, N.; Tongiorgi, P. Meiofauna densities and gastrotrich community composition in a Mediterranean sea cave. *Mar. Biol.* **2006**, *149*, 1079–1091. [CrossRef]

30. Sergeeva, N.G.; Ürkmez, D.; Todaro, M.A. Significant occurrence of *Musellifer profundus* Vivier, 1974 (Gastrotricha, Chaetonotida) in the Black Sea. *Check List* **2019**, *15*, 219–224. [CrossRef]

31. Pfannkuche, O.; Thiel, H. Sampling processing. In *Introduction to the Study of Meiofauna*; Higgins, R.P., Thiel, H., Eds.; Smithsonian Institution Press: Washington, DC, USA, 1988; pp. 134–145.

32. Todaro, M.A. Contribution to the study of the Mediterranean meiofauna: Gastrotricha from the Island of Ponza, Italy. *Boll. Zool.* **1992**, *59*, 321–333. [CrossRef]

33. Hochberg, R.; Litvaitis, M.K. Hexamethyldisilazane for scanning electron microscopy of Gastrotricha. *Biotech. Histochem.* **2000**, *75*, 41–44. [CrossRef] [PubMed]

34. Lee, J.M.; Chang, C.Y. Two new marine gastrotrichs of the genus *Ptychostomella* (Macrodasyida, Thaumastodermatidae) from South Korea. *Zool. Sci.* **2003**, *20*, 481–489. [CrossRef] [PubMed]

35. Todaro, M.A.; Hummon, W.D. An overview and a dichotomous key to genera of the phylum Gastrotricha. *Meiofauna Mar.* **2008**, *16*, 3–20.

36. Kolicka, M.; Dabert, M.; Dabert, J.; Kånneby, T.; Kisielewski, J. *Bifidochaetus*, a new Arctic genus of freshwater Chaetonotida (Gastrotricha) from Spitsbergen revealed by an integrative taxonomic approach. *Invert. Syst.* **2016**, *30*, 398–419. [CrossRef]

37. Garraffoni, A.R.S.; Araujo, T.Q.; Lourenço, A.P.; Guidi, L.; Balsamo, M. A new genus and new species of freshwater Chaetonotidae (Gastrotricha: Chaetonotida) from Brazil with phylogenetic position inferred from nuclear and mitochondrial DNA sequences. *Syst. Biodiv.* **2017**, *15*, 49–62. [CrossRef]

38. Balsamo, M.; Todaro, M.A. Gastrotricha. In *Freshwater Meiofauna Biology and Ecology*; Rundle, S.D., Robertson, A.I., Schmid-Araya, J.M., Eds.; Backhuys Publisher: Leiden, The Netherlands, 2002; pp. 45–61.

39. Pfannkuche, O. Gastrotricha. In *Introduction to the Study of Meiofauna*; Higgins, R.P., Thiel, H., Eds.; Smithsonian Institution Press: Washington, DC, USA, 1988; pp. 302–311.

40. Todaro, M.A.; Perissinotto, R.; Bownes, S.J. Two new marine Gastrotricha from the Indian Ocean coast of South Africa. *Zootaxa* **2015**, *3905*, 193–208. [CrossRef]

41. Hummon, W.D.; Todaro, M.A.; Tongiorgi, P. Italian Marine Gastrotricha: II. One new genus and ten new species of Macrodasyida. *Boll. Zool.* **1993**, *60*, 109–127. [CrossRef]

42. Hummon, W.D.; Todaro, M.A.; Tongiorgi, P.; Balsamo, M. Italian marine Gastrotricha: V. Four new and one redescribed species of Macrodasyida in the Dactylopodolidae and Thaumastodermatidae. *Ital. J. Zool.* **1998**, *65*, 109–119. [CrossRef]

43. Rieger, R.M.; Ruppert, E.E.; Rieger, G.E.; Schoepfer-Sterrer, C. On the fine structure of gastrotrichs, with description of *Chordodasys antennatus* sp. n. *Zool. Scr.* **1974**, *3*, 219–237. [CrossRef]

44. Hummon, W.D. Gastrotricha of the North Atlantic Ocean: 1. Twenty four new and two redescribed species of Macrodasyida. *Meiofauna Mar.* **2008**, *16*, 117–174.

45. Todaro, M.A.; Dal Zotto, M.; Bownes, S.J.; Perissinotto, R. Two new interesting species of Macrodasyida (Gastrotricha) from KwaZulu-Natal (South Africa). *Proc. Biol. Soc. Wash.* **2017**, *130*, 139–154. [CrossRef]

46. Evans, W.A.; Hummon, W.D. A new genus and species of Gastrotricha from the Atlantic coast of Florida, U.S.A. *Trans. Am. Microsc. Soc.* **1991**, *110*, 321–327. [CrossRef]

47. Clausen, C. Gastrotricha Macrodasyida from the Tromsø region, northern Norway. *Sarsia* **2000**, *85*, 357–384. [CrossRef]

48. Luporini, P.; Magagnini, G.; Tongiorgi, P. Contribution a la connaissance des gastrotriches des cotes de Toscane. *Cah. Biol. Mar.* **1971**, *12*, 433–455.

49. Claparède, E. Miscellaneous zoologiques. III. Type d'un nouveau genere de gastrotriches. *Ann. Sci. Nat. Zool.* **1867**, *8*, 16–23.

50. Kieneke, A.; Rothe, B.H.; Schmidt-Rhaesa, A. Record and description of *Anandrodasys agadasys* (Gastrotricha: Redudasyidae) from Lee Stocking Island (Bahamas), with remarks on populations from different geographic areas. *Meiofauna Mar.* **2013**, *20*, 39–48.

51. Kisielewski, J. Two new interesting genera of Gastrotricha (Macrodasyida and Chaetonotida) from the Brazilian freshwater psammon. *Hydrobiologia* **1987**, *153*, 23–30. [CrossRef]

52. Ruttner-Kolisko, A. *Rheomorpha neiswestnovae* und *Marinellina flagellata*, zwei phylogeneticsh interessante Wurmtypen aus dem Susswasserpsammon. *Österr. Zool. Z.* **1955**, *6*, 55–69.

53. Luporini, P.; Magagnini, G.; Tongiorgi, P. Chaetonotoid gastrotrichs of the Tuscan Coast. *Boll. Zool.* **1973**, *40*, 31–40. [CrossRef]

54. Balsamo, M. Gastrotrichi. In *Guide C.N.R. per IL Riconoscimento Delle Specie Animali Delle Acque Interne Italiane*; Consiglio Nazionale delle Ricerche AQ/1/199; CNR (Centro Nazionale Ricerche): Roma, Italy, 1983; Volume 7, pp. 547–571.

55. Cordero, E.H. Notes sur les Gastrotriches. *Physis* **1918**, *4*, 241–244.

56. Sudzuki, M. The Gastrotricha of Japan which live in the capillary water of the interstitial system: II. *Bull. Biogeogr. Soc. Japan* **1971**, *27*, 37–41.

57. Kisielewski, J. Inland-water Gastrotricha from Brazil. *Ann. Zool. (Warsaw)* **1991**, *43*, 1–168.

58. Mock, H. Chaetonotoidea (Gastrotricha) from the North Sea Island of Sylt. *Mikrofauna. Meeres.* **1980**, *78*, 1–107.

59. Schwank, P. Gastrotricha und Nemertini. In *Süsswasserfauna von Mittleuropas*; Brauer, A., Ed.; G. Fisher Verlag: Stuttgart, Germany, 1990; Volume 3, pp. 1–252.

60. Schrom, H. Nordadriatische Gastrotrichen. *Helgoländer Wiss. Meeresunters.* **1972**, *23*, 286–351. [CrossRef]

61. Hummon, W.D.; Balsamo, M.; Todaro, M.A. Italian marine Gastrotricha: I. Six new and one redescribed species of Chaetonotida. *Boll. Zool.* **1992**, *59*, 499–516. [CrossRef]

62. Kånneby, T. A redescription of *Chaetonotus* (*Primochaetus*) *veronicae* Kånneby, 2013 (Gastrotricha: Chaetonotidae). *Zootaxa* **2015**, *4027*, 442–446. [CrossRef]

Article

A New Species of *Paraturbanella* Remane, 1927 (Gastrotricha, Macrodasyida) from the Brazilian Coast, and the Molecular Phylogeny of Turbanellidae Remane, 1926

Ariane Campos [1,*], M. Antonio Todaro [2] and André Rinaldo Senna Garraffoni [1]

[1] Department of Animal Biology, Institute of Biology, University of Campinas, Campinas, SP 13083970, Brazil; arsg@unicamp.br
[2] Department of Life Sciences, University of Modena and Reggio Emilia, via G. Campi, 213/D, I-41125 Modena, Italy; antonio.todaro@unimore.it
[*] Correspondence: arianeecampos@gmail.com

http://zoobank.org/urn:lsid:zoobank.org:pub:EFA8F92E-7298-4982-B832-54F71E4BB995
Received: 1 December 2019; Accepted: 15 January 2020; Published: 21 January 2020

Abstract: The family Turbanellidae includes *Paraturbanella* and five other genera. Despite the fact that the monophyly of these genera were not satisfactorily tested, species belonging to the genus *Paraturbanella* are distinguished from turbanellids by sharing a peculiar group of tubes on the ventrolateral side of the anterior pharyngeal region known as *"dohrni"* tubes. In this study, *Paraturbanella tricaudata* species nova (sp. nov.) from the intertidal zone of a sandy beach in Trindade (Rio de Janeiro State) and the sublittoral sand of Prumirim Island (São Paulo State), Brazil, is described. The new species can be distinguished from all other *Paraturbanella* species by the presence of three caudal cones (one medial and two laterals to it) and peculiar arrangement of the male system. This is the first description of a *Paraturbanella* species from Brazil and the third registered from the Southern Hemisphere (as opposed to 19 species in the Northern Hemisphere); thus, knowledge of marine gastrotrichs biodiversity in this region is far from satisfactory.

Keywords: gastrotricha; meiofauna; biodiversity; taxonomy; South America; South Hemisphere; nuclear genes

1. Introduction

Gastrotricha are meiofauna representatives (encompassing animals ranging in size from about 0.042 to 0.500 mm) living in freshwater and marine ecosystems around the world [1,2]. The *phylum* is divided into two orders—Chaetonotida Remane, 1925 [3] and Macrodasyida Remane, 1925 [3]. The latter group is composed in its majority of hermaphroditic and marine species, which live interstitially in sandy bottoms [2,4]. The order Macrodasyida includes 10 families, 36 genera and approximately 380 described species [2,5,6]. They have a strap-shaped body, a pharynx with pharyngeal pores, and, usually, numerous adhesive tubes in the anterior, lateral and posterior regions.

The family Turbanellidae Remane, 1927 [7] includes six genera: the monospecific *Prostobuccantia* Evans & Hummon, 1991 [8] and Pseudoturbanella d'Hondt, 1968 [9]; *Dinodasys* Remane, 1927 [7] (two species); *Desmodasys* Clausen, 1965 [10] (three species); *Paraturbanella* Remane, 1927 [7] (23 species); and *Turbanella* Schultze, 1853 [11] (32 species).

Although knowledge of phylogenetic relationships within Turbanellidae is still limited, *Turbanella* and *Paraturbanella* species have several common characteristics, but the presence of extraordinary adhesive tubes easily distinguishes the genus *Paraturbanella* from *Turbanella* [2,6,12–16].

The type species of the genus *Paraturbanella* was collected in the Gulf of Naples (Italy) and described as *Paraturbanella dohrni*, based on the presence of an extraordinary pair of accessory adhesive tubes (known also as *"dohrni"*, or *"Seitenfüsschen"* in German). Remane could distinguish these animals from the *Turbanella* species [7].

Nowadays, twenty-three species are currently known: *P. africana* Todaro, Dal Zotto, Bownes & Perissinotto, 2017 [17]; *P. aggregotubulata* Evans, 1992 [18]; *P. armoricana* (Swedmark, 1954) [19]; *P. boadeni* Rao & Ganapati, 1968 [20]; *P. brevicaudata* Rao, 1991 [21]; *P. cuanensis* Maguire, 1976 [22]; *P. dohrni* Remane, 1927 [7]; *P. dolichodema* Hummon, 2010 [23]; *P. eireanna* Maguire, 1976 [22]; *P. intermedia* Wieser, 1957 [24]; *P. levantia* Hummon, 2011 [25]; *P. manxensis* Hummon, 2008 [26]; *P. mesoptera* Rao, 1970 [27]; *P. pacifica* Schmidt, 1974 [28]; *P. pallida* Luporini, Magagnini & Tongiorgi, 1971 [29]; *P. palpibara* Rao & Ganapati, 1968 [20]; *P. pediballetor* Hummon, 2008 [26]; *P. sanjuanensis* Hummon, 2010 [23]; *P. scanica* Clausen, 1996 [30]; *P. solitaria* Todaro, 1995 [31]; *P. stradbroki* Hochberg, 2002 [32]; *P. teissieri* Swedmark, 1954 [19]; and *P. xaymacana* Dal Zotto, Leasi & Todaro, 2018 [33].

Herein, we describe a new species of *Paraturbanella* from Brazil, previously reported as *Paraturbanella* sp. 2 [34,35].

2. Material and Methods

2.1. Sampling and Sample Processing

Information about the description of the new species is mainly taken from specimens found in samples collected by hand from the shallow intertidal area of Praia do Cachadaço, Trindade–Paraty (23°21′15.8″ S, 44°43′41.5″ W), Rio de Janeiro State, Brazil, in October 2017. The sandy sediment was filtered (0.40 μm mesh), and extraction of animals from the sediment was carried out by anaesthetization with $MgCl_2$. The supernatant was poured into a Petri dish, and with a micro-pipette it was possible to separate the gastrotrichs into embryo dishes. Specimens were observed alive with a stereomicroscope ZEISS Stemi 2000. Additional information comes from specimens collected and documented by one of us (M. Antonio Todaro) in 2002 and 2003 [34,35]; the sampling sites and distribution of *Paraturbanella* species along the Brazilian coast were plotted on the map, elaborated using ArcGis [36] (Figure 1), and made available at https://www.arcgis.com/home/webmap/viewer.html?webmap=fafb1d0942b4483a99ddf828fc24039a.

2.2. Microscopical Study—Differential Interference Contrast (DIC)

The gastrotrichs of interest were picked out with a micropipette, mounted on glass slides, and studied using Zeiss Axio Imager M2 plus Differential Interference Contrast optics, with Zen Lite Zeiss software (Zen Blue) and an Axiocam 105 color camera. The observation was taken using a slow-moving living specimen and the measures followed convention [37], according to the position of morphological characteristics along the body. The software used to measure the structures of specimens was Image J.

2.3. Scanning Electron Microscopy (SEM)

The specimens were fixed in glutaraldehyde 2.5%, rinsed with cacodylate buffer 0.1 M, dehydrated in a graded ethanol series, and critically point dried with CO_2. Stubs were coated with gold and analyzed using a scanning electron microscope JSM 5800 LV, with 10 kV voltage.

2.4. Granulometric Analysis

The particle size analyses were made following the sediment screening method using different opening meshes [38]. A fraction of the sample was separated and oven dried at 60 °C for 24 h. The dried material was sieved in different Granutest sieves with progressively smaller openings, and the fraction retained in each sieve was weighed and noted. Through these values, sediment fractions were weighed and granulometric characters (median, standard deviation, skewness and kurtosis) were calculated using GRANPLOTS with line segments [39].

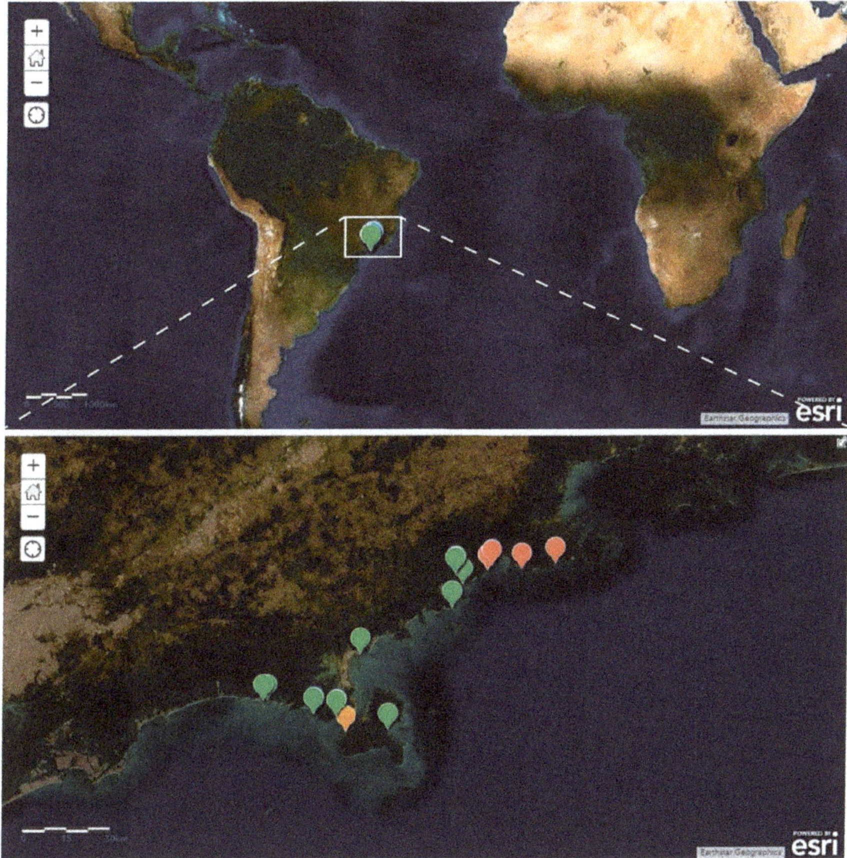

Figure 1. Map of Brazil showing the distribution of *Paraturbanella* species. Red: *Paraturbanella tricaudata* species nova (sp. nov.). Green: *Paraturbanella* sp. 1 [34,35]. Orange: *Paraturbanella* sp. 3 [34].

2.5. DNA Extraction, Amplification and Sequences

Genomic DNA was extracted from entire individuals of *Paraturbanella tricaudata* species nova (sp. nov.) using a QIAmp DNA Micro Kit (Quiagen, Hilden, Germany), following the manufacturer's instructions. PCR amplification was performed in a 20 μL reaction mixture containing 3 μL of genomic DNA, 13.5 μL of water, 2 μL of 10x buffer, 0.5 μL of dNTP, 0.2 μL of Taq Platinum (Quiagen) and 0.4 μL (4 pmol) of specific primers [4]. The DNA fragments were sequenced using BigDye Terminator reactions in a 3500xL Genetic Analyzer (Life Technologies, Carlsbad, CA, USA) at the CBMEG laboratory (Campinas, Brazil). The 18S rDNA and 28S rDNA sequences of *Paraturbanella tricaudata* sp. nov. were deposited in GenBank (Table 1).

2.6. Phylogenetic Analysis

All the formal species of the family Turbanellidae from which sequences of 18S rDNA and/or 28S rDNA were available in GenBank were included as the ingroup in the present analysis (*Paraturbanella*: 5 species; *Turbanella*: 5 species). The species and respective GenBank accession numbers are listed in Table 1. The outgroup was composed of two *Megadasys* species and four *Mesodasys* species due to the close phylogenetic position of these taxa with the taxon Turbanellidae [4,40]. Each gene dataset was aligned separately using BioEdit Sequence Alignment Editor and Mafft v.7.402 (L-INS-I approach) [41].

Parsimony analysis was performed using a heuristic search with equally weighted characters that was available using TNT software [42]. Most parsimonious trees were searched by the heuristic method, with 1000 replications, holding 1,000,000 trees per search (command line: mult = replications 1000 hold 100,000) and collapsing the tree after the search.

For maximum likelihood analysis, GTRCAT model was chosen and RAxML [43] was run with a with 1000 bootstrap replicates.

Table 1. Taxa of Gastrotricha included in this study, with respective GenBank accession numbers of 18S and 28S rDNA sequences.

Species	18S	28S	References
Megadasys sp. 1	JF357656	JF357704	Todaro et al. [44]
Megadasys minor Kisielewski, 1987	AY228131	-	Todaro et al. [45]
Mesodasys littoralis Remane, 1951	JF357658	JF357706	Todaro et al. [44]
Mesodasys laticaudatus Remane, 1951	JF357657	JF357705	Todaro et al. [44]
Mesodasys sp.	AY963690	KF921011	Petrov et al. [46]
Mesodasys adenotubulatus Hummon, Todaro & Tongiorgi, 1993	AM231780	-	Todaro et al. [45]
Paraturbanella sp.	KF921017	-	Petrov et al. [46]
Paraturbanella teissieri Swedmark, 1954	JF357661	JF357709	Todaro et al. [44,45]
Paraturbanella pallida Luporini, Magagnini & Tongiorgi, 1973	JF357660	JF357708	Todaro et al. [44]
Paraturbanella dohrni Remane, 1927	JF357659	JF357707	Todaro et al. [44]
Paraturbanella tricaudata sp. nov.	SUB6819988	SUB6819996	Present study
Turbanella sp.	JF970238	-	Paps & Riutort [47]
Turbanella cornuta Remane, 1925	AF157007	JF357711	Todaro et al. [44]
Turbanella pilosum Kolicka, 2018	MF325920	MF325905	Kolicka et al. [13]
Turbanella lutheri Remane, 1952	JF357669	-	Todaro et al. [44]
Turbanella bocqueti Kaplan, 1958	JF357662	JF357710	Todaro et al. [44]

3. Results

3.1. Taxonomic Account

Phylum Gastrotricha Metschnikoff, 1865.
Order Macrodasyida Remane, 1925 [3] (Rao & Clausen, 1970) [48].
Family Turbanellidae Remane, 1926 [49]
Genus *Paraturbanella* Remane, 1927 [7]
Paraturbanella tricaudata sp. nov. (Table 2 and Figures 2–6).
Synonym *Paraturbanella* sp. 2—Todaro & Rocha [34,35].

Table 2. Morphometric features of *Paraturbanella tricaudata* sp. nov. Min: minimum value. Max: maximum value. N: total number of measured adult specimens. SD: standard deviation.

Character	Min–Max (μm)	Average (μm)	SD (μm)	N
Lt: total body length	417–480	449.00	31.51	5
Lbc: length of buccal cavity	18.6–22.3	19.9	1.2	10
Diameter of mouth opening	13.3–17.4	15.0	1.3	10
Head width at cephalic swellings	35.8–51.5	40.4	5.1	8
Neck constriction width	30.1–34.1	32.3	1.7	8
Maximum trunk width	40.9–49.1	43.2	2.8	8
Lph: length of pharynx	127.1–141.9	136.9	4.9	6
Distance of the pharyngeal pores from PhJIn	20.2–27.1	23.5	2.8	6
"*Dohrni*" longer tube length	22.6–33.8	27.8	4.9	6
"*Dohrni*" shorter tube length	15.6–19.4	18.5	2.5	5
Maximum length of TbA	8.7–9.6	9.2	0.4	4
Length of TbP	6.2–7.9	7.0	0.9	5

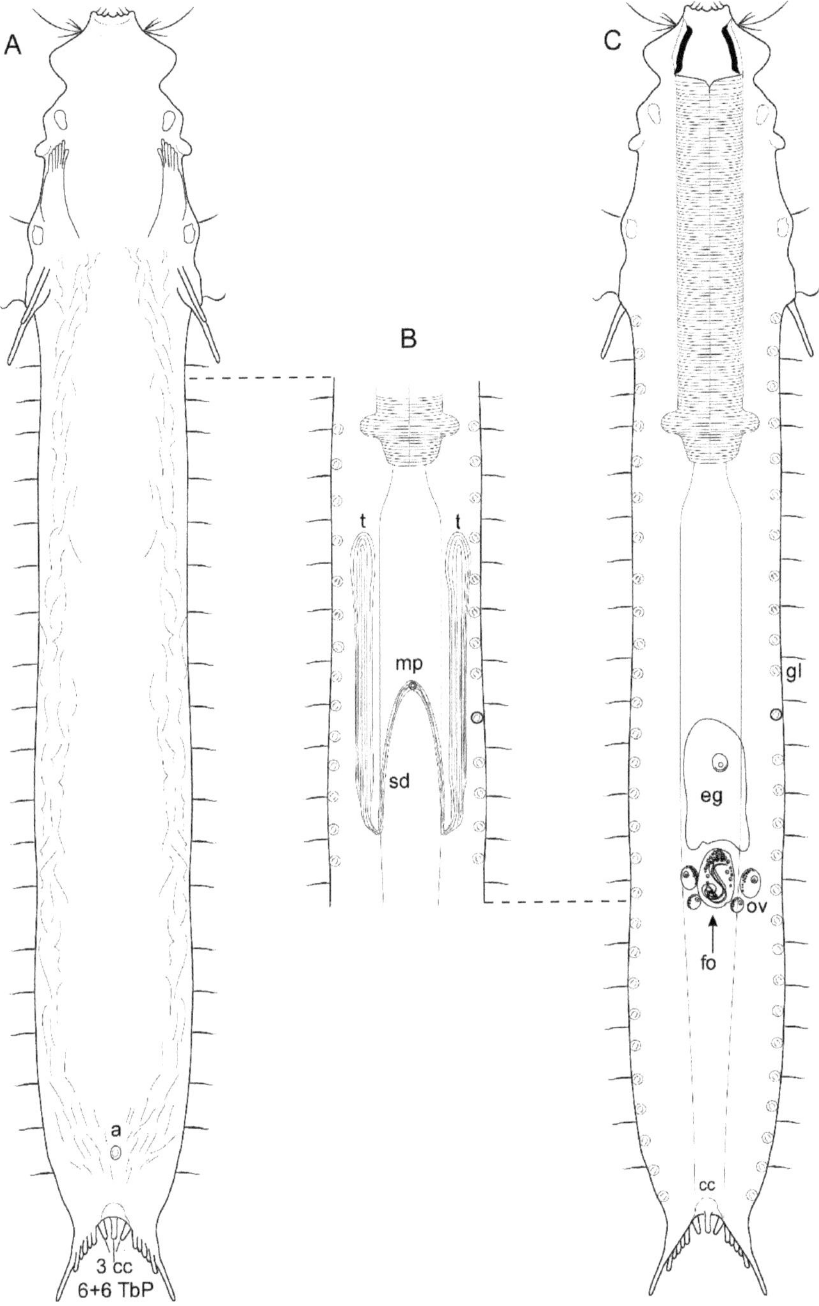

Figure 2. *Paraturbanella tricaudata* sp. nov. schematic drawing. (**A**) Ventral view of the *Habitus*. (**B**) Detail of the mid-trunk region, showing a ventral view of the male system. (**C**) Dorsal view of the *Habitus*, showing a mature egg, the frontal organ and the ovaries. Abbreviations: a = anus, fo = frontal organ, gl = gland, mp = male pore, ov = ovary, sd = sperm duct, and t = testes.

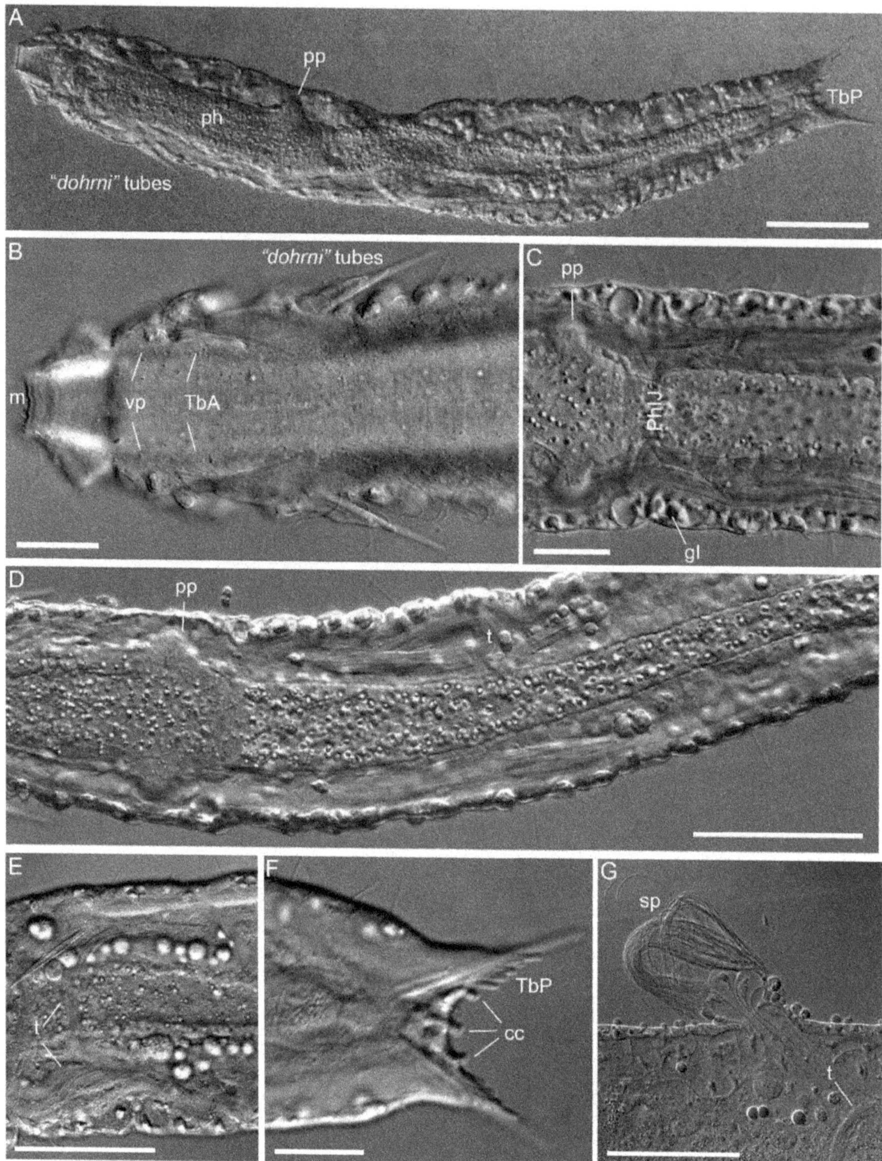

Figure 3. *Paraturbanella tricaudata* sp. nov. DIC photomicrographs. (**A**) Dorsal view of the *Habitus*. (**B**) Anterior region showing the mouth (m), peribuccal swellings, anterior adhesive tubes (TbA), *"dohrni"* tubes and ventral papillae (vp). (**C**) Pharyngeal pores (pp), pharyngo-intestinal junction (PhIJ), and epidermal glands (eg). (**D**) Paired testes (t)—they are lateral and extend posteriorly, with sperm ducts. (**E**) Sperm duct recurving to the front of the mid-intestine. (**F**) Caudal lobes, posterior adhesive tubes (TbP), and caudal cones (cc). (**G**) Sperm (sp) released from the side of the body due to excessive compression. Scale bar: **A** = 50; **B**, **C**, **E**, **F** = 20; **D** = 40; **G** = 10 μm.

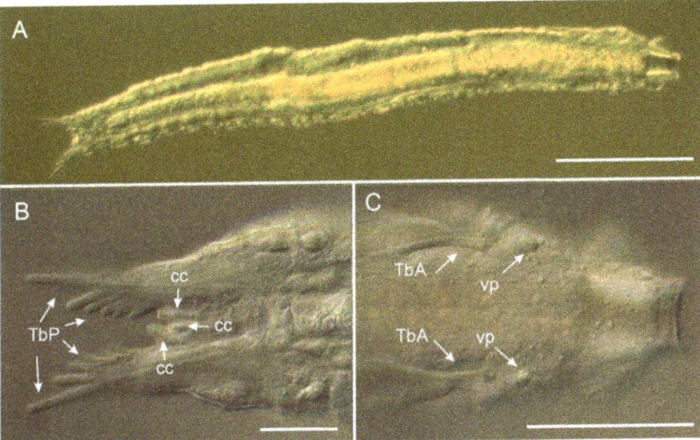

Figure 4. Adult specimen of *Paraturbanella tricaudata* sp. nov. (2002–2003 sampling). (**A**) *Habitus*; (**B**) ventral view of the posterior body end, showing posterior adhesive tubes (TbP) and tree caudal cones (cc); (**C**) ventral view of the anterior region, showing the anterior adhesive tubes (TbA) and ventral papillae (vp). Scale bars: **A** = 100; **B, C** = 50 μm.

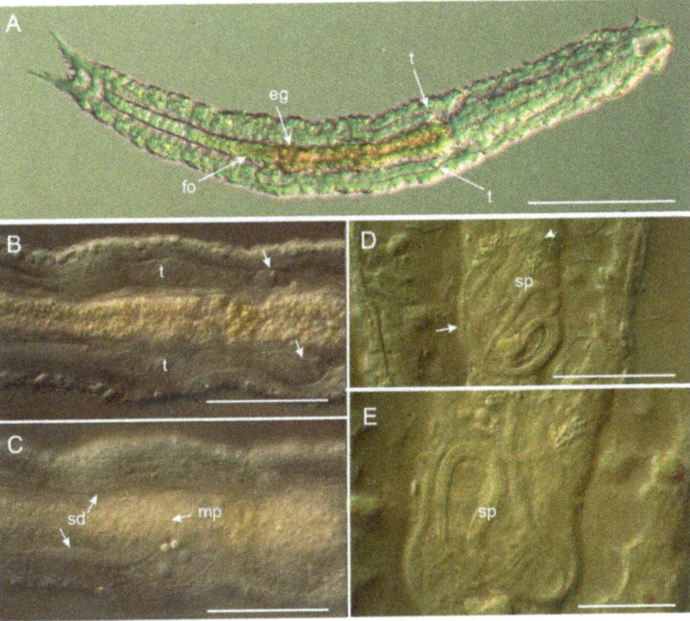

Figure 5. Adult specimen of *Paraturbanella tricaudata* sp. nov. (2002–2003 sampling). (**A**) *Habitus*, indicating the anterior most portion of the testes (t), frontal organ (fo) and the largest egg (egg). (**B**) Trunk region showing the testes (t), including the anterior most portion (arrows). (**C**) Trunk region from a slightly different focal plane showing the anastomosis of the sperm ducts (sd) and the ventral, ephemeral male pore (mp). (**D**) Dorsal view of the trunk region showing the frontal organ with a bundle of allosperm inside; the arrowhead indicates the location of the dorsal external pore through which sperm (sp) have been injected. (**E**) Close-up of the spermatozoa bundle inside the frontal organ. Scale bars: **A** = 100; **B, C** = 50; **D, E** = 20 μm.

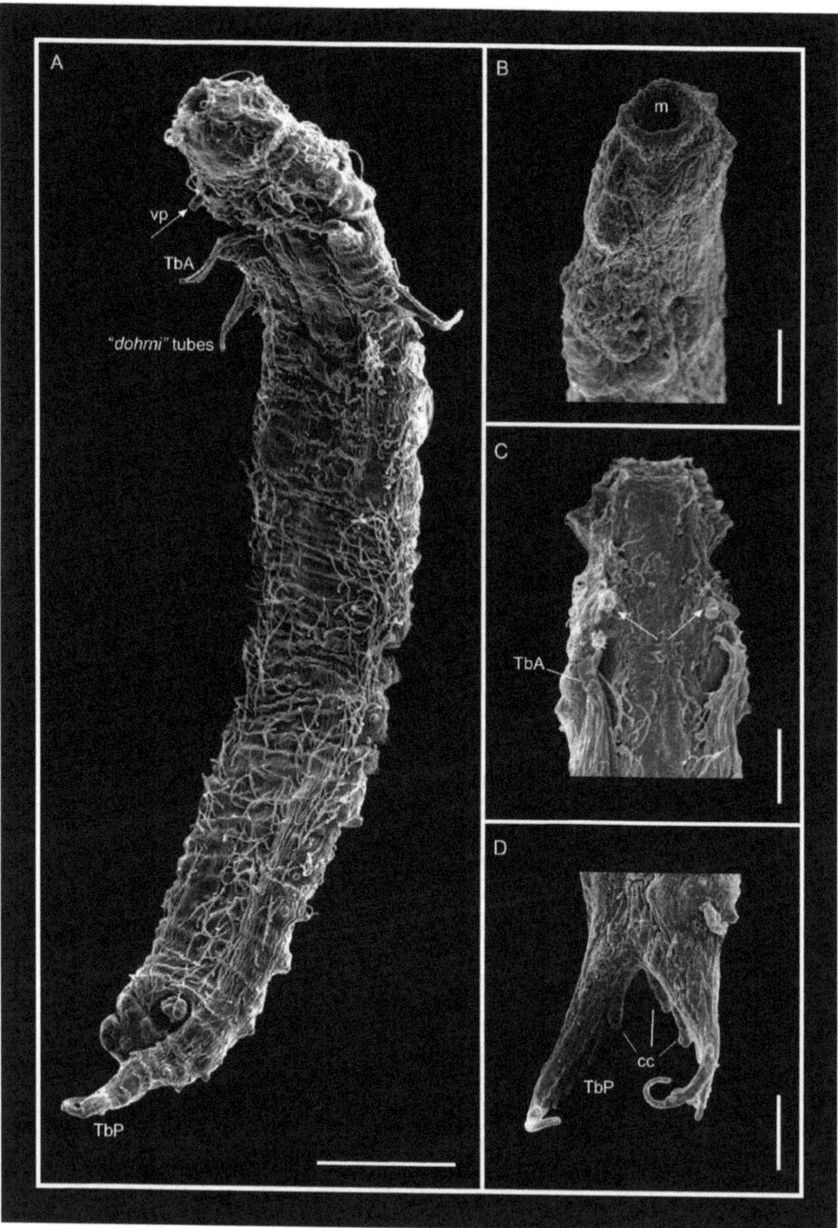

Figure 6. *Paraturbanella tricaudata* sp. nov. (SEM). (**A**) Ventrolateral view of the *Habitus*, showing ventral papillae (vp), anterior adhesive tubes (TbA), "*dohrni*" tubes and ventral ciliation. (**B**) Anterior region showing the terminal mouth (m). (**C**) Ventral view of the anterior region, showing the peribuccal swellings, anterior adhesive tubes (TbA) and ventral papillae (arrows). (**D**) Posterior body region showing the caudal lobe carrying adhesive tubes (TbP) and caudal cones (arrows). Scale bar: **A** = 20; **B**, **C** and **D** = 10 μm.

Examined material: Holotype. Photographs of an adult specimen, collected on 9th November 2017 from Praia do Cachadaço, in the Trindade district in the municipality of Paraty, Rio de Janeiro State, Brazil (23°21′15.8″ S 44°43′41.5″ W). Bare sand of 30 cm depth had the following sediment characteristics: mean = 1.4111 phi (medium sand), SD = 0.8137, skewness = −0.8573, kurtosis = 4.4435, and median = 1.3215.

The specimen was observed alive with a compound microscope, but due to the fragility of the organisms, it was inadvertently destroyed during the study and is no longer available [50]. The holotype is illustrated in Figure 2 (according to the International Code of Zoological Nomenclature, 2017: Article 73, Recommendation 73G, in Declaration 45), and photos are available at the Museum of Zoology, University of Campinas, Brazil, under the accession number ZUEC GCH 61.

Paratypes. Photographs of five adult specimens (adults), collected on 9th November 2017 from Praia do Cachadaço, in Trindade district in the municipality of Paraty, Rio de Janeiro State, Brazil. The specimens were observed alive with a compound microscope, but due to the fragility of the organisms, physical specimens were inadvertently destroyed during the study and are no longer available [50]. Photographs of each specimen are available at the Museum of Zoology, University of Campinas, Brazil, under accession numbers ZUEC GCH 62–66.

Additional material: From the type locality, ten specimens were prepared for SEM and from locations sampled between 2002–2003 [34,35], two specimens are shown in Figures 4 and 5.

Etymology: The specific name refers to the triple caudal cone.

Repository: http://zoobank.org/urn:lsid:zoobank.org:pub:EFA8F92E-7298-4982-B832-54F71E4BB995

Diagnosis: The body is strap-shaped, and its length ranges from 417 to 480 μm. It has a large terminal mouth with a diameter ranging from 13.3 to 17.4 μm. The buccal cavity is heavily cuticularized, and it has a head with noticeable peribuccal swelling and ventral papillae. The pestle organ is absent. The pharyngo-intestinal junction (PhJIn) is between U34 and U38, and the distance of the pharyngeal pores from PhJIn varies from 20.2 to 27.1 μm. Epidermal glands are arranged in a single column on each side. There are five to six anterior adhesive tubes (TbA) on each side, all occurring on fleshy hands. There are two accessory adhesive tubes (*"dohrni"* tubes) of unequal length per side (the longer tube maximum length is 33.8 μm and the shorter is 19.4 μm). There are six posterior adhesive tubes (TbP) per side, the outermost being the longest. Dorsal adhesive tubes (TbD), ventral adhesive tubes (TbV), ventrolateral adhesive tubes (TbVL), and lateral adhesive tubes (TbL) are absent. In the posterior region, body tapering occurs gradually to the caudal base, and the caudum has three caudal cones, one medial and two laterals. Paired testes extend into sperm ducts, which turn anteriorly at U65 and fuse in a mid-ventral pore at U53. The frontal organ is at U71. About 20–30 epidermal glands are distributed along both lateral body margins.

Description: The description is based on both the holotype and ten paratypes (see Table 2). The body is strap-shaped and 417 μm in total length. The head bears noticeable lateral peribuccal swelling (U03) and a pair of papillae ventrally but no pestle organ; the cephalic region is delimited by a neck constriction 34.1 μm wide (U05). Body tapering occurs gradually to the caudal base, and the caudum bilobed has three caudal cones (U96). Widths at the outer oral opening, neck constriction, mid-pharynx, PhJIn, mid-intestine, furcal base, and their locations along the body length are: 16.4, 37.1, 25.2, 22.2, 19.7, and 33.0 μm at U0, U05, U20, U37, U64, and U93, respectively. Epidermal glands are arranged along the body in one column per side of 30 glands, which vary in size (1.6–4.7 μm in diameter).

Adhesive tubes: There are five or six anterior adhesive tubes (TbA) per side, all of which occur on fleshy hands that insert at U15. The innermost, mimicking a thumb, is the shortest, while the second from the inner side is the longest. Accessory adhesive tubes (*"dohrni"* tubes) are posterolaterally directed, and there are two per side of unequal length (the longer tube is 14 μm and the shorter is 9 μm) at U17. Dorsal adhesive tubes (TbD), ventral adhesive tubes (TbV), ventrolateral adhesive tubes (TbVL), and lateral adhesive tubes (TbL) are absent. There are six posterior adhesive tubes (TbP) per side, the outermost being the longest. The distance between the external TbP is greater than the width of the body's caudal base.

Ciliation: The cephalic region has ciliary patches and circumcephalic rows in the mouth. Ventral locomotor cilia are arranged in two longitudinal bands that trace the lateral body sides and join posteriorly near the anal opening. Additional sensory bristles are organized in lateral, dorsolateral and dorsal columns.

Digestive tract: The mouth is terminal and surrounded by the mouth ring, composed of a strengthened cuticle; buccal cavity (6.4 μm wide and 19.1 μm long) mug shaped with walls heavily cuticularized. The pharynx is 142 μm in length and 22.6 μm in wide (U30), with pharyngeal pores near the base at about U33 and PhJIn at U37; the intestine is straight and the anus ventral is at U85.

Reproductive tract: It is hermaphroditic, with paired testes from U52 to U67, which extend into two sperm ducts at U60, turning anteriorly and fusing in a mid-ventral pore at U90. The frontal organ at U70 is vesicular and filled with bundled spermatozoa; the caudal organ is absent; the paired ovaries are at U68, and mature egg dorsal (to the intestine) occurs at U60.

3.1.1. Variability and Remarks on General Morphology

Seven additional adult measured specimens showed six TbP per side and three caudal cones, meaning that these traits are rather constant within and among the investigated populations. On the other hand, the number of TbA is slightly variable but not related to the size of the animal; in fact, one of the specimens attaining a maximum length of 480 μm (Figure 5) possessed only five TbA per side while another attaining 332 μm in total length had six TbA.

3.1.2. Taxonomic Remarks

Currently, there are 23 described species belonging to the genus *Paraturbanella* [6,17,33]. The new species bares closest resemblance to five species: *P. africana* Todaro, Dal Zotto, Bownes & Perissinotto, 2017 [17]; *P. teissieri* Swedmark, 1954 [19]; *P. sanjuanensis* Hummon, 2010 [23]; *P. solitaria* Todaro, 1995 [31]; and *P. xaymacana* Dal Zotto, Leasi & Todaro, 2018 [33], showing similar body and head shape, as well as peribuccal swelling [33].

The species *Paraturbanella tricaudata* sp. nov. is considered new because it has three caudal cones, while the others have just one cone; moreover, the arrangement of the reproductive system, and in particular the location of the male pore, in combination with the extension and anatomical positioning of the testes is unique among congeneric species (two sperm ducts at U60 turning anteriorly and fusing in a mid-ventral pore).

3.1.3. Phylogenetic Analyses

The final alignment of the combined dataset yielded 3525 positions (1693 in 18S rDNA and 1832 in 28S rDNA). The parsimony analysis yields only one most-parsimonious tree with 4626 steps (Figure 7).

Parsimony and Maximum Likelihood analyses yielded congruent topologies. In both analyses the phylogenies showed that the family Turbanellidae (both supported by very high bootstrap value - 100) and the genera *Turbanella* and *Paraturbanella* were monophyletic (Parsimony: *Turbanella* with good bootstrap values - 85; Maximum Likelihood: both genera supported by very high bootstrap value, respectively 100 and 99) and sister groups.

The phylogenetic position of *Paraturbanella tricaudata* sp. nov. was not stable in both analyses. The new species appeared as the sister-group of *Paraturbanella* sp. and both species were closely related with *P. pallida* in Maximum Parsimony analysis (Figure 7). However, *Paraturbanella tricaudata* sp. nov. was sister-group of *P. pallida* and both species were closely related with *Paraturbanella* sp. in Maximum Likelihood analysis (Figure 8).

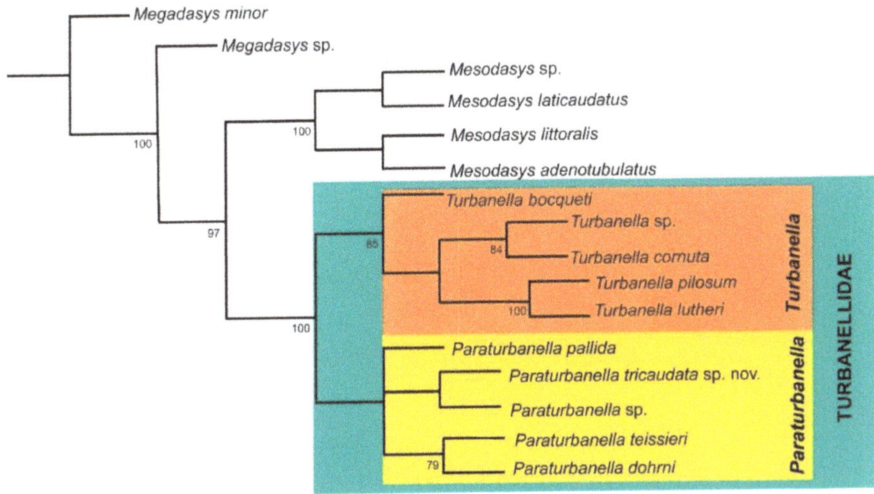

Figure 7. Maximum parsimony reconstruction based on molecular data (18S and 28S rDNA). Molecular dataset aligned in MAFFT. Numbers at nodes indicate bootstrap support (> 75).

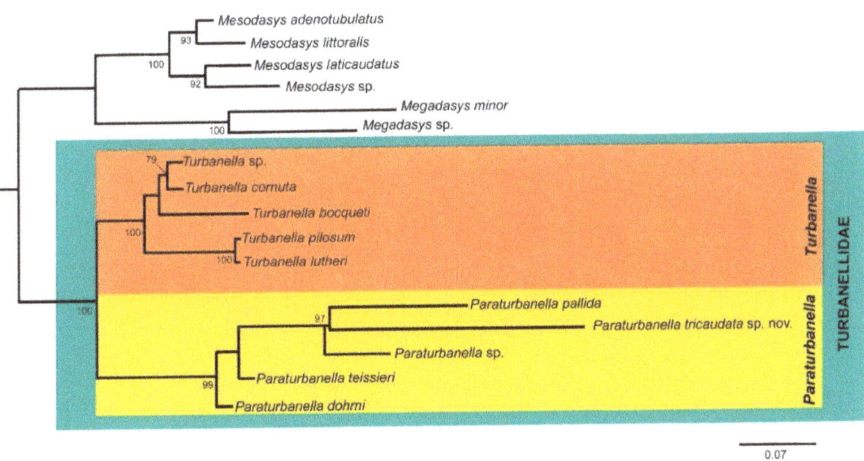

Figure 8. Phylogenetic relationships of Turbanellidae species inferred from Maximum Likelihood analysis of 18S and 28S rRNA. Numbers at nodes represent bootstrap support (1000 bootstrap replicates).

3.1.4. Conclusive Remarks

The majority of marine gastrotrich sampling sites are in the Northern Hemisphere [51], while in tropical countries the investigated areas are very scattered [14,16,17,30,52]. The Brazilian coast is poorly sampled, although it shows a high diversity of marine gastrotrichs, according to current literature [5,51]. In terms of knowledge of Turbanellidae diversity from Brazil, there are at least four candidate species to be formally described (one *Turbanella* species and three *Paraturbanella* species) from the north coast of São Paulo State [34,35].

The Brazilian coast represents a potentially important area for the discovery of new species, and it is worth noting that this region is of particular interest in the study of marine meiofauna and the diversity of Gastrotricha, as recorded in previous studies [2,34,35,51–54]. We emphasize the importance of investigating new geographic areas in order to improve our understanding of morphological diversity and the species richness of gastrotrich species.

Finally, the inclusion of new sequences of data concerning *Paraturbanella tricaudata* sp. nov. and the use of distinct phylogenetic approaches did not change the scenario found by Todaro et al. [44], Kolicka et al. [13], and Garraffoni et al. [4]. However, it is important to highlight that the low number of Turbanellidae species sequenced (from two of the six genera) may result in misleading the phylogenetic relationships within this clade.

Author Contributions: Conceptualization, A.C. and A.R.S.G.; methodology, A.C., M.A.T., A.R.S.G.; validation, A.C., M.A.T., A.R.S.G.; formal analysis, A.C., M.A.T., A.R.S.G.; investigation, A.C., M.A.T., A.R.S.G.; resources, A.R.S.G; data curation, A.C., M.A.T., A.R.S.G.; writing—original draft preparation, A.C., M.A.T., A.R.S.G.; writing—review and editing, A.C., M.A.T., A.R.S.G.; visualization, A.C., M.A.T., A.R.S.G.; supervision, A.R.S.G.; project administration, A.R.S.G.; funding acquisition, A.R.S.G. All authors have read and agreed to the published version of the manuscript.

Funding: This study was financed in part by the Coordenação de Aperfeiçoamento de Pessoal de Nível Superior—Brasil (CAPES) Finance Code 001 and Fundação de Amparo à Pesquisa do Estado de São Paulo (FAPESP-2014/23856-0).

Acknowledgments: The authors would like to thank K.R. Seger for help with molecular analysis. A.C.Z Amaral for providing the nested column of sieves and T.A. Quintão for grain size data analysis. We are also grateful to three anonymous reviewers for their constructive criticism that greatly improving the first version of the manuscript.

Conflicts of Interest: The authors declare no conflict of interest.

References

1. Balsamo, M.; d'Hondt, J.L.; Kisielewski, J.; Todaro, M.A.; Tongiorgi, P.; Guidi, L.; Grilli, P.; de Jong, Y. Fauna Europaea: Gastrotricha. *Biodivers. Data J.* **2015**, *3*, e5800. [CrossRef]

2. Todaro, M.A.; Sibaja-Cordero, J.A.; Segura-Bermúdez, O.A.; Coto-Delgado, G.; Goebel-Otárola, N.; Barquero, J.D.; Cullell-Delgado, M.; Dal Zotto, M. An introduction to the study of Gastrotricha, with a taxonomic key to families and genera of the group. *Diversity* **2019**, *11*, 117. [CrossRef]

3. Remane, A. Neue aberrante Gastrotrichen II: *Turbanella cornuta* nov.spec. und *T. hyalina* M. Schultze, 1853. *Zool. Anz.* **1925**, *64*, 309–314.

4. Garraffoni, A.R.S.; Araújo, T.Q.; Lourenço, A.P.; Guidi, L.; Balsamo, M. Integrative taxonomy of a new *Redudasys* species (Gastrotricha: Macrodasyida) sheds light on the invasion of freshwater habitats by macrodasyids. *Sci. Rep.* **2019**, *9*, 2067. [CrossRef] [PubMed]

5. Campos, A.; Garraffoni, A.R.S. A synopsis of knowledge, zoogeography and an online interactive map of Brazilian marine gastrotrichs. *PeerJ* **2019**, *7*, e7898. [CrossRef]

6. Todaro, M.A. (Ed.) Marine. In *Gastrotricha World Portal*; 2019; Available online: http://www.gastrotricha. unimore.it/marine.htm (accessed on 30 November 2019).

7. Remane, A. Neue Gastrotricha Macrodasyoidea. *Zool. Jahrb. Abt. Anat. Ontog. Tiere* **1927**, *54*, 203–242.

8. Evans, W.A.; Hummon, W.D. A new genus and species of Gastrotricha from the Atlantic coast of Florida, U.S.A. *Trans. Am. Microsc. Soc.* **1991**, *110*, 321–327. [CrossRef]

9. d'Hondt, J.L. Contribution a la connaissance des Gastrotriches intercotidaux du Golfe de Gascogne. *Cah. Biol. Mar.* **1968**, *9*, 387–404.

10. Clausen, C. *Desmodasys phocoides* gen. et. sp. n., family Turbanellidae (Gastrotricha Macrodasyoidea). *Sarsia* **1965**, *21*, 17–21. [CrossRef]

11. Schultze, M. Über *Chaetonotus* und *Ichthydium* (Ehrb.) und eine neue verwandte Gattung *Turbanella*. Müller's Arch. Anat. Physiol. **1853**, *6*, 241–254.

12. Kieneke, A.; Riemann, O.; Ahlrichs, W.H. Novel implications for the basal internal relationships of Gastrotricha revealed by an analysis of morphological characters. *Zool. Scr.* **2008**, *37*, 429–460. [CrossRef]

13. Kolicka, M.; Kotwicki, L.; Dabert, M. Diversity of Gastrotricha on Spitsbergen (Svalbard Archipelago, Arctic) with a description of seven new species. *Ann. Zool.* **2018**, *68*, 609–740. [CrossRef]

14. Todaro, M.A.; Dal Zotto, M.; Perissinotto, R.; Bownes, S.J. First records of Gastrotricha from South Africa, with description of a new species of *Halichaetonotus* (Chaetonotida, Chaetonotidae). *Zookeys* **2011**, *142*, 1–13. [CrossRef] [PubMed]

15. Todaro, M.A.; Leasi, F.; Hochberg, R. A new species, genus and family of marine Gastrotricha from Jamaica, with a phylogenetic analysis of Macrodasyida based on molecular data. *Syst. Biodivers.* **2014**, *12*, 473–488. [CrossRef]
16. Todaro, M.A.; Perissinotto, R.; Bownes, S.J. Two new marine Gastrotricha from the Indian Ocean coast of South Africa. *Zootaxa* **2015**, *3905*, 193–208. [CrossRef]
17. Todaro, M.A.; Dal Zotto, M.; Bownes, S.J.; Perissinotto, R. Two new interesting species of Macrodasyida (Gastrotricha) from KwaZulu-Natal (South Africa). *Proc. Biol. Soc. Wash.* **2017**, *130*, 140–155. [CrossRef]
18. Evans, W.A. Five new species of marine Gastrotricha from the Atlantic coast of Florida. *Bull. Mar. Sci.* **1992**, *51*, 315–328.
19. Swedmark, B. *Turbanella armoricana* n. sp., nouveau gastrotriche macrodasyoide de la côte nord de Bretagne. *Bull. Soc. Zool. Fr.* **1954**, *79*, 469–473.
20. Rao, G.C.; Ganapati, P.N. Some new interstitial gastrotrichs from the beach sands of Waltair coast. *Proc. Sci. Acad. India* **1968**, *67*, 35–53.
21. Rao, G.C. Meiofauna, In: Fauna of Lakshadweep, State Fauna Series. *Zool. Surv. India* **1991**, *2*, 73–75.
22. Maguire, C. Two new species of *Paraturbanella cuanensis* and *P. eireanna*. *Cah. Biol. Mar.* **1976**, *17*, 405–410.
23. Hummon, W.D. Marine Gastrotricha of San Juan Island, Washington, USA, with notes on some species from Oregon and California. *Meiofauna Mar.* **2010**, *18*, 11–40.
24. Wieser, W. Gastrotricha Macrodasyoidea from the intertidal of Puget Sound. *Trans. Am. Microsc. Soc.* **1957**, *76*, 372–381. [CrossRef]
25. Hummon, W.D. Marine Gastrotricha of the Near East: 1. Fourteen new species of Macrodasyida and a redescription of Dactylopodola agadasys Hochberg, 2003. *ZooKeys* **2011**, *94*, 1–59. [CrossRef] [PubMed]
26. Hummon, W.D. Gastrotricha of the North Atlantic Ocean: 1. Twenty-four new and two redescribed species of Macrodasyida. *Meiofauna Mar.* **2008**, *16*, 117–174.
27. Rao, G.C. 3 New interstitial Gastrotrichs from Andhra Coast, India. *Cah. Biol. Mar.* **1970**, *11*, 109–120.
28. Schmidt, P. Interstitielle Fauna von Galapagos. IV. Gastrotricha. *Mikrofauna Meeresbodens.* **1974**, *26*, 1–76.
29. Luporini, P.; Magagnini, G.; Tongiorgi, P. Contribution a la connaissance des gastrotriches des côtes de Toscane. *Cah. Biol. Mar.* **1971**, *12*, 433–455.
30. Clausen, C. Three new species of Gastrotricha Macrodasyida from the Bergen area, western Norway. *Sarsia* **1996**, *81*, 119–129. [CrossRef]
31. Todaro, M.A. *Paraturbanella solitaria*, a new psammic species (Gastrotricha: Macrodasyida: Turbanellidae), from the coast of California. *Proc. Biol. Soc. Wash.* **1995**, *108*, 553–559.
32. Hochberg, R. Two new species of Turbanellidae (Gastrotricha: Macrodasyida) from a high-energy beach on North Stradbroke Island, Australia. *N. Zeal. J. Mar. Fresh.* **2002**, *36*, 311–319. [CrossRef]
33. Dal Zotto, M.; Leasi, F.; Todaro, M.A. A new species of Turbanellidae (Gastrotricha, Macrodasyida) from Jamaica, with a key to species of *Paraturbanella*. *ZooKeys* **2018**, *734*, 105–119. [CrossRef] [PubMed]
34. Todaro, M.A.; Rocha, C.E. Diversity and distribution of marine Gastrotricha along the northern beaches of the State of São Paulo (Brazil), with description of a new species of Macrodasys (Macrodasyida, Macrodasyidae). *J. Nat. Hist.* **2004**, *38*, 1605–1634. [CrossRef]
35. Todaro, M.A.; Rocha, C.E. Further data on marine gastrotrichs from the State of São Paulo and the first records from the State of Rio de Janeiro (Brazil). *Meiofauna Mar.* **2005**, *14*, 27–31.
36. ArcGis. *Paraturbanella* in Brazil [basemap], Scale Not Given. 2019. Available online: https://www.arcgis.com/home/webmap/viewer.html?webmap=fafb1d0942b4483a99ddf828fc24039a (accessed on 17 November 2019).
37. Hummon, W.D.; Todaro, M.A.; Tongiorgi, P. Italian Marine Gastrotricha: II. One new genus and ten new species of Macrodasyida. *Boll. Zool.* **1993**, *60*, 109–127. [CrossRef]
38. Suguio, K. *Introdução à Sedimentologia*; Edgard Blucher: São Paulo, Brazil, 1973; p. 307.
39. Balsillie, J.H.; Donoghue, J.F.; Butler, K.M.; Koch, J.L. Plotting equation for Gaussian percentiles and a spreadsheet program for generating probability plots. *J. Sediment. Res.* **2002**, *72*, 929–933. [CrossRef]
40. Todaro, M.A.; Dal Zotto, M.; Kånneby, T.; Hochberg, R. Integrated data analysis allows the establishment of a new, cosmopolitan genus of marine Macrodasyida (Gastrotricha). *Sci. Rep.* **2019**, *9*, 7989. [CrossRef]
41. Katoh, K.; Standley, D.M. MAFFT multiple sequence alignment software version 7: Improvements in performance and usability. *Mol. Biol. Evol.* **2013**, *30*, 772–780. [CrossRef]
42. Goloboff, P.A.; Farris, J.S.; Nixon, K.C. TNT, a free program for phylogenetic analysis. *Cladistics* **2008**, *24*, 774–786. [CrossRef]

43. Stamatakis, A.; Hoover, P.; Rougemont, J. A rapid bootstrap algorithm for the RAxML web servers. *Syst. Biol.* **2008**, *57*, 758–771. [CrossRef]
44. Todaro, M.A.; Kånneby, T.; Dal Zotto, M.; Jondelius, U. Phylogeny of Thaumastodermatidae (Gastrotricha: Macrodasyida) inferred from nuclear and mitochondrial sequence data. *PLoS ONE* **2011**, *6*, e17892. [CrossRef] [PubMed]
45. Todaro, M.A.; Littlewood, D.T.J.; Balsamo, M.; Herniou, E.A.; Cassanelli, S.; Manicardi, G.; Tongiorgi, P. The interrelationships of the Gastrotricha using nuclear small rRNA subunit sequence data, with an interpretation based on morphology. *Zool. Anz.* **2003**, *242*, 145–156. [CrossRef]
46. Petrov, N.B.; Pegova, A.N.; Manylov, O.G.; Vladychenskaya, N.S.; Mugue, N.S.; Aleshin, V.V. Molecular phylogeny of Gastrotricha on the basis of a comparison of the 18S rRNA genes: Rejection of the hypothesis of a relationship between Gastrotricha and Nematoda. *Mol. Biol.* **2007**, *41*, 445–452. [CrossRef]
47. Paps, J.; Riutort, M. Molecular phylogeny of the phylum Gastrotricha: New data brings together molecules and morphology. *Mol. Phylogenet. Evol.* **2012**, *63*, 208–212. [CrossRef]
48. Rao, G.C.; Clausen, C. *Planodasys marginalis* gen. et sp. nov. and Planodasyidae fam. nov. (Gastrotricha Macrodasyoidea). *Sarsia* **1970**, *42*, 73–82. [CrossRef]
49. Remane, A. Morphologie und Verwandtschaftbeziehungen der aberranten Gastrotrichen I. *Z. Morph. Oekol. Tiere* **1926**, *5*, 625–754. [CrossRef]
50. Garraffoni, A.R.; Kieneke, A.; Kolicka, M.; Corgosinho, P.H.; Prado, J.; Nihei, S.S.; Freitas, A.V. ICZN Declaration 45: A remedy for the nomenclatural and typification dilemma regarding soft-bodied meiofaunal organisms? *Mar. Biodivers.* **2019**, *49*, 2199–2207. [CrossRef]
51. Garraffoni, A.R.S.; Balsamo, M. Is the ubiquitous distribution real for marine gastrotrichs? Detection of areas of endemism using Parsimony Analysis of Endemicity (PAE). *Proc. Biol. Soc. Wash.* **2017**, *130*, 197–210. [CrossRef]
52. Araújo, T.Q.; Balsamo, M.; Garraffoni, A.R.S. A new species of *Pseudostomella* (Gastrotricha, Thaumastodermatidae) from Brazil. *Mar. Biodivers.* **2014**, *44*, 243–248. [CrossRef]
53. Garraffoni, A.R.S.; Di Domenico, M.; Amaral, A.C.Z. Patterns of diversity in marine Gastrotricha from Southeastern Brazilian Coast is predicted by sediment textures. *Hydrobiologia* **2016**, *773*, 105–116. [CrossRef]
54. Hochberg, R.; Atherton, S.; Kieneke, A. Marine Gastrotricha of Little Cayman Island with the description of one new species and an initial assessment of meiofaunal diversity. *Mar. Biodivers.* **2014**, *44*, 89–113. [CrossRef]

Article

DNA Barcoding for Delimitation of Putative Mexican Marine Nematodes Species

Arely Martínez-Arce [1,*], Alberto De Jesús-Navarrete [1] and Francesca Leasi [2,*]

[1] Departamento Sistemática y Ecología Acuática, El Colegio de la Frontera Sur, Unidad Chetumal, Av. Centenario Km. 5.5, Quintana Roo, CP 77014, Mexico; anavarre@ecosur.mx

[2] Department of Biology, Geology and Environmental Science, University of Tennessee at Chattanooga, 615 McCallie Ave, Chattanooga, TN 37403, USA

* Correspondence: armartarce@ecosur.mx (A.M.-A.); francesca-leasi@utc.edu (F.L.); Tel.: +983-8350454 (A.M.-A.); +1-423-425-4797 (F.L.)

Received: 6 February 2020; Accepted: 15 March 2020; Published: 19 March 2020

Abstract: Nematode biodiversity is mostly unknown; while about 20,000 nematode species have been described, estimates for species diversity range from 0.1 to 100 million. The study of nematode diversity, like that of meiofaunal organisms in general, has been mostly based on morphology-based taxonomy, a time-consuming and costly task that requires well-trained specialists. This work represents the first study on the taxonomy of Mexican nematodes that integrates morphological and molecular data. We added eleven new records to the Mexican Caribbean nematode species list: Anticomidae sp.1, *Catanema* sp.1, *Enoploides gryphus*, *Eurystomina* sp.1, *Haliplectus bickneri*, *Metachromadora* sp.1, *Odontophora bermudensis*, *Oncholaimus* sp.1, *Onyx litorale*, *Proplatycoma fleurdelis*, and *Pontonema* cf. *simile*. We improved the COI database with 57 new sequences from 20 morphotypes. All COI sequences obtained in this work are new entries for the international genetic databases GenBank and BOLD. Among the studied sites, we report the most extensive species record (12 species) at Cozumel. DNA barcoding and species delineation methods supported the occurrence of 20 evolutionary independent entities and confirmed the high taxonomic resolution of the COI gene. Different approaches provided consistent results: ABGD and mPTP methods disentangled 20 entities, whereas Barcode Index Numbers (BINs) recovered 22 genetic species. Results support DNA barcoding being an efficient, fast, and low-cost method to integrate into morphological observations in order to address taxonomical shortfalls in meiofaunal organisms.

Keywords: ABGD; BINs; DNA barcoding; meiofauna; mPTP

---

1. Introduction

Nematodes are hyper-diverse, abundant, and distributed worldwide [1]. Free-living species play critical ecological roles in benthic energy flow and contribute to the ecosystem by facilitating mineralization and nutrient cycling [2–7]. In the presence of high inputs of organic matter, their abundance increases, helping to regulate this resource. They are also a source of high-quality food for other animals [8–11]. Currently, about 20,000 nematode species—of which 6500 are marine benthic (= meiofaunal)—have been formally described [12,13], with estimates ranging between 0.1 and 100 million species [14]. The number of existing species is still uncertain because such estimates have been made at the local level, whereas little is known on a global scale. Gathering evidence of nematode diversity and distribution, increasing the record of marine nematodes species, especially in overlooked regions, is nowadays crucial [15].

Nematode taxonomy is an overlooked field of study in Mexico, with only about 119 genera and 183 species known for the country [16–22]. The majority of studies of marine nematodes in this region address ecological questions [17–21,23,24], whereas only five studies are focused on taxonomy [16,25–27]. For this reason, faunistic lists for nematodes in Mexico are at the family or genus level in most cases. The slow advance in the taxonomic knowledge of marine nematodes is due to technical difficulties. The identification of marine benthic nematodes is mostly based on morphological traits of male genital structures in mature individuals [28,29]. However, the occurrence of adult specimens is often rare. For this reason, nematode diversity is commonly disentangled to the family or genus level, especially in ecological studies. Morphology-based taxonomy is a time-consuming task that requires well-trained specialists who are becoming rare [30,31]. The use of morphological traits, in most cases, descriptive and potentially affected by convergent evolution and phenotypic plasticity, could also prevent an accurate quantification of the true nematode diversity [32–34]. Hence, there is an increasing need for methods that can rapidly and cost-effectively estimate nematode diversity in marine sediments. Molecular tools for taxonomic identification, delimitation of species, and an approach to the phylogeny hold the potential to overcome difficulties where morphological studies are painstakingly difficult and/or where the number of species far outweighs the availability of taxonomists. The identification of free-living marine nematodes is particularly difficult [35–39], and an integrated approach including genetic, morphological, and ideally, also ecological and behavioral data is needed [40].

Only one study conducted in Mexico (in Baja, CA, USA) [41] considered genetic tools to investigate the diversity of nematodes. Pereira and collaborators [41] revealed both a wide genetic diversity and geographic distribution of populations of *Mesacanthion* species. They used two molecular markers: 28S ribosomal rRNA gene and 18S, with 28S showing a better taxonomic resolution than 18S in delineating also cryptic species (similarly to [42]). The two markers did not show differences in the phylogenetic relationships among the investigated taxa, except for species of the genus *Rhabdodemania* [41].

In the Caribbean, two relevant studies have been conducted on the nematode fauna and none from Mexico [43,44]. In both studies, two molecular markers were considered: cytochrome oxidase subunit I (COI) and 18S ribosomal RNA gene (18S rRNA). Armenteros and collaborators [43] disentangled species of Desmodorid from Punta Francés, Cuba. They generated 34 sequences for COI across five genera and 27 sequences for 18S rRNA gene across six genera. Either marker could fully resolve the phylogenetic relationships of some lineages (i.e., within the subfamilies Desmodorinae and Spiriniinae). However, COI showed a better resolution than 18S among closely related and cryptic species. Macheriotou et al. [44] generated 18S and COI sequences from nematodes sampled in the equatorial North Pacific, Cuba, Italy (Panarea Island), Papua New Guinea, the Netherlands, Tunisia, and Vietnam. They generated 290 COI and 438 18S sequences; using reference databases for marine nematodes, they identified 39 OTUs (Operational Taxonomic Units with High Throughput Sequencing; HTS). Although the ribosomal marker outperformed the mitochondrial marker in terms of species and genus-level detections., they concluded that, for HTS technologies, it is urgent to continue creating high-quality taxon-specific reference sequence databases.

Free-living marine nematode species are poorly represented in public sequence databases. Limited availability of nematode reference sequences, especially from overlooked both localities and habitats such as the deep-sea, seagrass beds, and tropical coral reefs, hinders biogeographic patterns and characterization of ecosystems. Moreover, although DNA taxonomy is most successful when applied to fast-evolving genes such as the mitochondrial gene COI [42,45–48], genetic reference databases for nematodes mostly include nuclear markers such as 18S and 28S [36,43,48,49]. COI is poorly represented [44,47,50] because of the difficulty in amplifying this gene in a wide range of taxa within the phylum using 'universal' primers. Several studies regularly report low success in the amplification of COI and the necessity to design new specific primers to obtain a robust database [44,50,51]. The limited COI sequence datasets for marine nematodes prevent the establishment of an adequate understanding of intraspecific divergence.

The main objectives of this study are to (i) improve our knowledge of the geographic distribution of meiofaunal nematodes in the Mexican Caribbean; (ii) contribute with new COI sequences to the public genetic databases; and, (iii) apply different delimitation models to test the taxonomic resolution of COI in marine nematodes. Integrating different species delineation models should prevent biased conclusions and disclose patterns of diversity and distribution [52]. We aim to disentangle nematode diversity by applying Automatic Barcode Gap Discovery (ABGD) [53], Barcode Index Number system (BINs) [54], and Poisson Tree Processes model (PTP) [55] on COI sequences.

2. Materials and Methods

2.1. Nematode Sampling and Identification

Marine meiofaunal nematodes were collected in March and September 2011 from the intertidal zone in seven sites along the coast of Quintana Roo State, Mexico (Figure 1, Table 1). Four sediment samples were collected from each site using a Falcon corer (10 × 2 cm) [56]. Individuals were extracted by decantation in the field using two sieves (180 and 63 μm mesh) and fixed with DESS solution [57] or Formalin 10% [58]. In the laboratory, individuals were separated one-by-one and picked up under a stereomicroscope (NIKON SMZ-1). Then, nematodes specimens were individually transferred to temporary slides in a drop of MilliQ water covered by a coverslip and observed with an OLYMPUS BX51 microscope at different magnifications (10×, 40×, and 100×). Well-preserved specimens were identified morphologically and photographed with a camera (Canon G11). When was possible, more than one representative for each morphotype was subsequently selected for further molecular analyses. Morphological identification was carried out using available taxonomic keys for marine nematodes and comparison with original descriptions using several morphological parameters: length and maximum body width, size and position of setae, size and position of amphids, cuticle ornamentation, size and shape of spicula, presence of precloacal supplements, type and tail size and de Man's ratios (a, b and c): a = body length/body width, b = body length/esophagus length and c= body length/tail length [59–61].

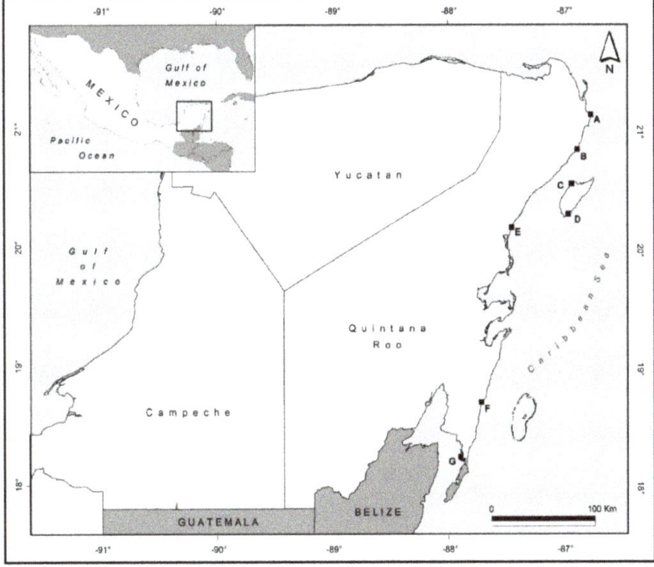

Figure 1. Sampling localities from marine nematodes in the Mexican Caribbean. Site IDs (A, B, C, D, E, F, and G) are explained in Table 1.

Table 1. List of sites where samples were collected. Coordinates are express in decimals *.

Sites	Site ID	Lat	Long	Date of Sampling mm/yy/2011
Cancún	A	21.140	−86.677	03/11
Puerto Morelos	B	20.508	−86.525	03/11
Cozumel (Playa Azul)	C	20.548	−86.929	03/11
Cozumel (Punta sur)	D	20.291	−86.959	03/11
Tulum	E	20.395	−87.315	03/11
Mahahual	F	18.708	−87.712	09/11
Xcalak	G	18.2475	−87.8944	09/11

* Lat, latitude; Long, longitude.

2.2. DNA Extraction, PCR Amplification, and DNA Sequencing

Under the microscope, each selected nematode was removed from the temporary slide using a fine paintbrush, cut in pieces with a scalpel, and preserved in DESS. The DNA was extracted from single individual animals with a HotSHOT technique [62]. DNA was stored at 4 °C for further amplification of the COI gene. PCR reactions were performed in a final volume of 12.5 µL containing 6.25 µL of trehalose 10%, 2 µL of ddH20, 1.25 µL of 10X PCR Buffer, 0.625 µL of MgCl2 50 Mm, 0.125 µL of each primer (10 µM), 0.0625 µL of dNTPs (10 mM), 0.06 µL of Platinum® Taq Polymerase (Invitrogen, Carlsbad, CA, USA), and 2.0 µL of DNA template [63]. We used two primer sets according to [64]: LC01490_t1 and HC02198_t1 (TGTAAAACGACGGCCAGTGGTCAACAAATCATAAAGATATTGG/ CAGGAAACAGCTATGACTAAACTTCAGGGTGACCAAAAAATCA); LC01490 and HC02198 (GGTCAACAAATCATAAAGATATTGG/ TAAACTTCAGGGTGACCAAAAAATCA). Thermocycler conditions, for both primer sets, were: 1 min at 94 °C, 5 cycles of 40 s at 94 °C, 40 s at 45 °C, and 1 min at 72 °C, followed by 35 cycles of 40 s at 94 °C, 40 s at 51 °C and 1 min at 72 °C, and a final extension of 5 min at 72 °C. PCR products were visualized on 2% agarose gels stained with ethidium bromide. Amplicons for LC01490_t1/LC01490_t1 were bidirectionally sequenced with M13F and M13R primers [65]; ABI 3730 capillary sequencer (ABI, Thermo Fisher Scientific, Carlsbad, CA, USA) using the BigDye© Terminator v.3.1 Cycle Sequencing Kit (Applied Biosystems, Foster, CA, USA). Sequences were obtained at the Canadian Centre for DNA Barcoding (CCDB) following standard protocols for high-volume samples [66].

2.3. Data Analysis

Sequences were assembled and edited with Codon Code Aligner v 3.0.3 software. The Clustal W program was used for the sequence alignment with default parameters. Phred score [67] was used to assess the quality of the sequences applying the following categories: no sequences = failed, mean Phred < 30 = low quality, 30 < mean Phred < 40 = medium quality, and mean Phred > 40 = high quality. Sequences with low quality and cross-contamination were removed for the analysis. COI sequences are available as part of the project FMN (Free-Living Marine Nematodes from Quintana Roo, Mexico) on Barcode of Life Data Systems (BOLD, www.boldsystems.org) [68].

Genetic divergence was calculated in MEGA v6 [69] using the Kimura two-parameter (K2P) distance model [70]. The presence of stop codons and indels was verified to discard any contaminants such as NUMTs (nuclear mitochondrial DNA segments) [71,72]. Basic Local Alignment Search Tool (BLAST) [73] and Identification Request, on GenBank and BOLD, respectively, were used to identify matches to the DNA sequences generated in this study.

2.4. Phylogenetic Tree Reconstruction and DNA Taxonomy

A Maximum-Likelihood (ML) tree was reconstructed in MEGA v6 software using 1000 bootstrap replications. The tree was reconstructed with sequences generated in this study and others retrieved from GenBank according to length, quality (stop codons presence), position (concerning barcoding region), and taxonomy (Accession numbers are in Supplement Table S1). The best-fitting substitution model was the General Time Reversible model with nonuniform evolutionary rates and invariant sites (GTR+G+I) and as chosen in MEGA. The COI sequence *Bursaphelenchus* sp. (order Tylenchina) was selected as the outgroup [74].

To delimit evolutionary independent entities of marine nematode species and test the resolution of the sequenced COI gene, we applied three methods: (1) Automatic Barcode Gap Discovery (ABGD) [53] with the following parameters: relative gap (X) of 1.1, minimal intraspecific distance (Pmin) of 0.001, maximal intraspecific distance (Pmax) of 0.1, K2P [70] and JC69 (Jukes-Cantor) [75] as distance metrics, (2) Barcode Index Number (BIN) system [54], and (3) mPTP by selecting single-locus species delimitation with *p*-value 0.001 (http://mptp.h-its.org/#/tree) [55].

3. Results

3.1. Morphological Identification

A total of 30 morphotypes from 107 individuals were identified. The list of morphotypes and sampling locations is available in Table 2. The order Enoplida was represented by the highest number of morphotypes (11), followed by Desmodorida and Chromadorida (7), Monhysterida (3), Aerolaimida (1) and Plectida (Table 2). Families Desmodoridae and Chromadoridae were the best represented with five genera each and six and five species, respectively. Eighteen specimens, all morphological very similar, belonged to the *Spirinia* genus. However, based on de Man´s ratios (different proportions of bulb and tail sizes), we considered the presence of two distinct morphotypes: *Spirinia* sp.1 and *S.* sp.2 (Figure 2). Such morphological differences are supported by the results of integrative taxonomy (see below).

In Tulum, one individual resembling the species *Pontonema simile* (Figure 3) was identified. This species was initially described by Southern as *Oncholaimus similis* [76] and redescribed by Filipjev [77] as *P. simile*. The *Pontomema* cf. *simile* reported in the present study had a shorter size (2100 µm) and a shorter spicule (32 µm) compared to the species described by Filipjev [77].

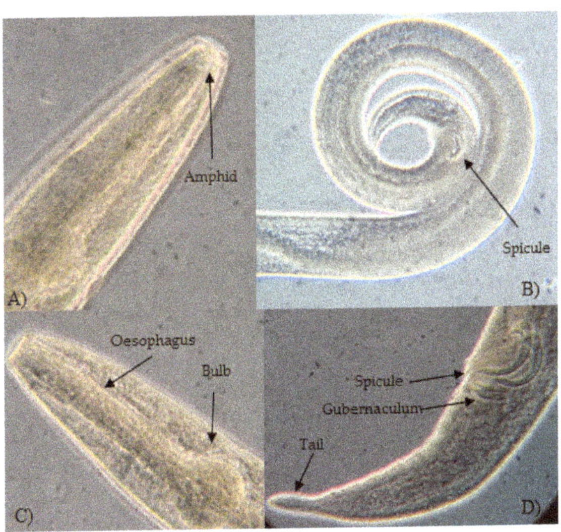

Figure 2. *Spirinia* sp.2. Adult male. (**A**) Head showing circular amphid. (**B**) Posterior part of the body showing tail and spicule. (**C**) Anterior part of the body showing esophagus and esophageal bulb. (**D**) Posterior region showing spicule, gubernaculum, and tail shape.

Table 2. Morphotypes identified at the collection sites. Species are listed by nematode Order. The number of individuals and COI sequences analyzed in this study are indicated. Taxonomy classification is based on Nemys database (www.marinespecies.org) [12]. Sites are indicated as below: Cancún = A; Puerto Morelos= B; Cozumel, Playa Azul = C; Cozumel, Punta Sur = D; Tulum = E; Mahahual = F; and Xcalak = G.

Taxonomy	Identified Specimens (n)	Sites	Sequences (n)
Araeolaimida			
Odontophora bermudensis (Jensen & Gerlach, 1976)	2	D	0
Chromadorida			
Actinonema sp.1	1	B	0
Chromadoridae sp.1	1	G	0
Chromadoridae sp.2	1	E	0
Chromadorida sp.1	2	F	2
Chromadorita sp.1	1	D	0
Rhips sp.1	9	A; B	6
Prochromadorella sp.1	4	E	0
Desmodorida			
Catanema sp.1	1	D	1
Epsilonema sp.1	6	C; E	5
Metachromadora sp.1	2	F	2
Monoposthia mirabilis (Schulz, 1932)	10	A; B; D	8
Onyx litorale (Schulz, 1938)	2	D; F	2
Spirinia sp.1	11	D; F; G	9
Spirinia sp.2	7	G	6
Enoplida			
Anticomidae sp.1	3	B	3
Eurystomina sp.1	1	D	1
Enoploides sp.1	1	D	1
Enoploides sp.2	3	C	2
Enoploides gryphus (Wieser& Hopper, 1967)	2	E	2
Epacanthion sp.1	5	D	0
Halalaimus sp.1	6	B	2
Metaparoncholaimus sp.1	1	E	0
Oncholaimus sp.1	9	E	0
Pontonema cf. *simile* (Southern, 1914) Filipjev, 1921	1	E	1
Proplatycoma fleurdelis (Hope, 1988)	10	C; B; E	1
Monhysterida			
Monhysteridae sp.1	1	F	1
Xyala striata (Cobb, 1920)	2	D	1
Terschellingia longicaudata (de Man, 1907)	1	G	0
Plectida			
Haliplectus bickneri (Chitwood, 1956)	1	B	1

Among the sampled localities, Cozumel harbored the highest species richness (12 species), which all are new for this locality. From the other localities, we found fewer species but a greater abundance of individuals. *Enoploides gryphus* (Wieser & Hopper, 1967) is a new record for the Mexican Caribbean while *Haliplectus bickneri* (Chitwood, 1956), *Proplatycoma fleurdelis* (Hope, 1988), *Pontonema* cf. *simile* (Southern, 1914) Filipjev, 1921, *Rhips* sp.1, *Epsilonema* sp.1, *Metachromadora* sp.1, *Spirinia* sp.2, *Odonthophora bermudensis* (Jensen & Gerlach, 1976), *Onyx litorale* (Schulz, 1938), *Epacanthion* sp.1, *Halalaimus* sp.1, and *Metaparoncholaimus* sp.1, are new for Mexico.

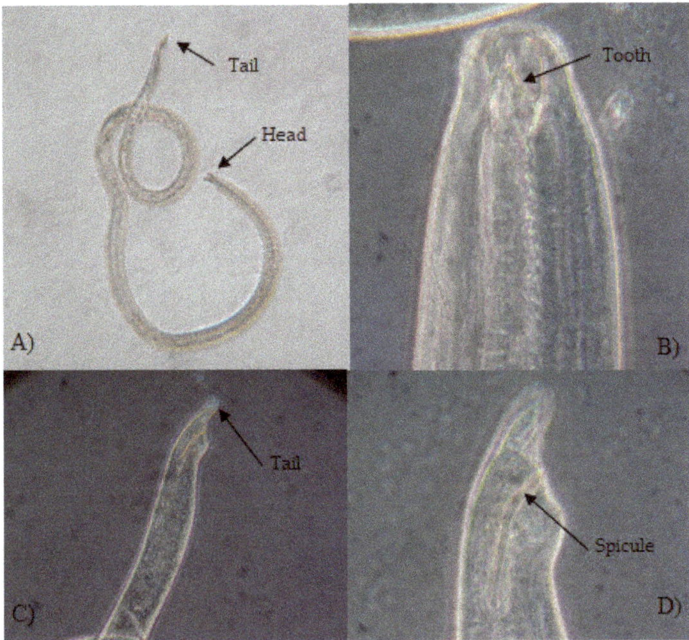

Figure 3. *Pontonema* cf. *simile* (Southern, 1914) Filipjev, 1921. Collected in Tulum locality. Adult male. (**A**) Habitus. (**B**) Head showing dorsal tooth. (**C**) Tail short and rounded. (**D**) Posterior part of the body showing spicule and tail.

3.2. Amplification and Sequencing

Overall, DNA amplification was successful for 67 individuals. However, sequences with low quality were discarded, and 57 sequences from 20 morphological species were considered (Table 2). Individuals of the species *Actinonema* sp.1, *Epacanthion* sp.1, *Metaparoncholaimus* sp.1, *Odontophora bermudensis*, *Oncholaimus* sp.1, *Prochromadorella* sp.1, and some specimens of chromadorids could not be sequenced (Table 2). All COI sequences produced in this study are new for the molecular databases GenBank and BOLD. According to Phred score [67], all sequences were of high quality (mean Phred > 40), longer than 500 bp (502–667 bp), and without internal stop codons. BLAST match values were between 71% and 85% with Nematoda. The mean genetic divergence was 0.43% and 26.45% for intraspecific and interspecific variation, respectively. In most cases, we observed a clear barcoding gap, although low interspecific variation was detected in a few cases (Figure 4). This study provides the first record of mtCOI sequence for *Monoposthia mirabilis*, *Onyx litorale*, *Enoploides gryphus*, *Pontonema* cf. *simile*, *Proplatycoma fleurdelis*, *Xyala striata*, and *Haliplectus bickneri*, and other higher-ranked taxa (Table 2).

3.3. Phylogenetic Analysis and DNA Taxonomy

Maximum-likelihood analysis using the 57 COI sequences obtained in this work and an additional 70 [47,74,78,79] sequences from GenBank showed congruence between DNA barcode and morphological identification. Sequences were grouped in clusters represented by the morphological identity. The tree was supported in recent clades with bootstrap values >90% (Figure 5). The maximum genetic distance values were observed in *Epsilonema* sp.1 clade (4.02%) and *Enoploides* sp.2 (2.24%). The sequences of *Spirinia* sp.1 and *S.* sp.2, with interspecific divergence values of 7.7%, formed two distinct clades.

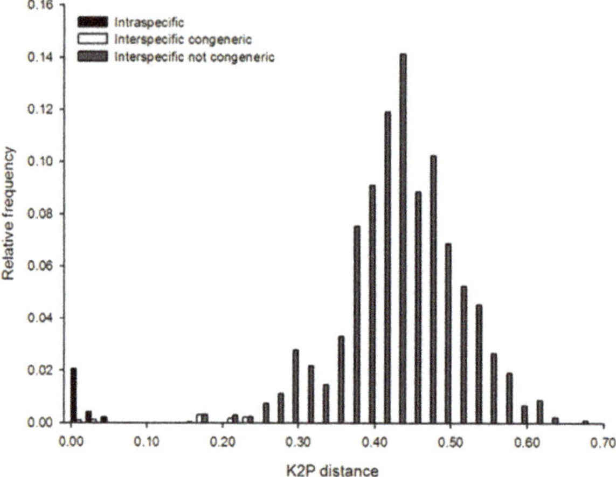

Figure 4. Relative frequencies of Kimura two-parameter (K2P) distances in the COI sequences of marine nematodes analyzed in this study. Frequencies within species are marked in black, among congeneric species in white, and between species from different genera in grey.

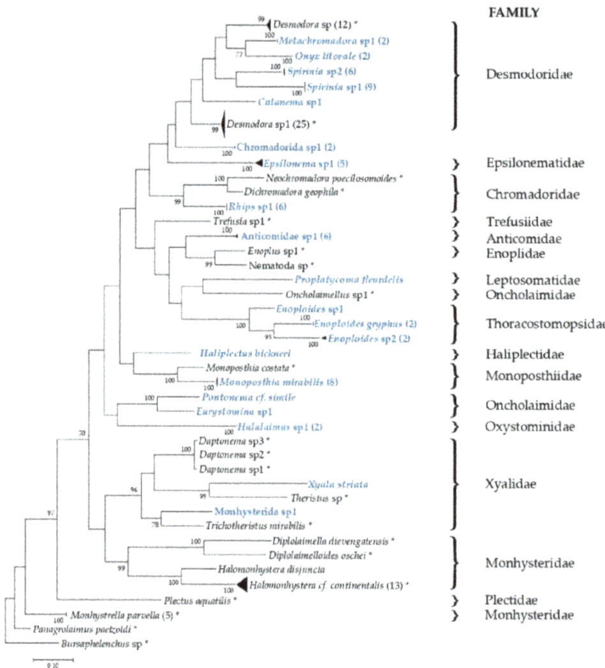

Figure 5. Maximum-Likelihood tree reconstructed from COI DNA sequences based on GTR+G+I model. Sequences of marine nematodes from the present study are annotated according to their morphological identification. The 20 haplotypes generated in this study are highlighted in blue color. Numbers in parentheses refer to the number of specimens sequenced. Reference of 70 sequences obtained from GenBank are indicated with an asterisk. Accession numbers are in Supplementary Table S1. Higher taxa (families) are indicated behind the line. Classification is based on Nemys database in www.marinespecies.org [12]. Arely Martínez 2020 ©

To ascertain the number of entities, sequences obtained in this work were analyzed using three different methods: ABGD, mPTP, and BINs (Table 3). ABGD analysis with K2P showed 20 initial and 25 recursive partitions with prior maximal distances (PMD) of 0.0010; 20 initial and 21 recursive partitions with PMDs ranging from 0.0017 to 0.0028; and 20 initial and 20 recursive partitions with PMDs ranging from 0.0046 to 0.0129. With the JC69 model, the same results were observed (see Table 3). The mPTP method based on our phylogenetic tree recovered 20 evolutionary independent entities (p = 0.001). BINs split *Enoploides* sp.2 and *Epsilonema* sp.1: BOLD: AAU8181 (n = 1) and BOLD: AAU8183 (n = 1) for the first species, and BOLD: AAU8184 (n = 4); BOLD: AAV1242 (n = 1) for the second species. A total of 22 BINs were recovered. Interestingly, results obtained with integrative taxonomy supported that *Spirinia* sp.1 and *S.* sp.2 are two distinct species.

Table 3. Evolutionary independent entities recovered from the COI sequences with the three different delimitation algorithms. Relative gap (X) of 1.1, minimal intraspecific distance (Pmin) of 0.001, maximal intraspecific distance (Pmax) of 0.1, K2P [70], and JC69 (Jukes-Cantor) [75] were selected as distance metrics. I = Initial partition, R= Recursive partition. EiEs = Evolutionary Independent Entities.

	BIN System	mPTP Model	ALGORITHMIC APPROACHES — ABGD																						
p-value		0.001																								
Subst. model			K2P												JC69											
Prior intraspecific divergence (P)			0.0010		0.0017		0.0028		0.0046		0.0077		0.129		0.0010		0.0017		0.0028		0.0046		0.0077		0.129	
Partition			I	R	I	R	I	R	I	R	I	R	I	R	I	R	I	R	I	R	I	R	I	R	I	R
EiEs	22		20	20	25	20	21	20	21	20	20	20	20	20	20	20	25	20	21	20	21	20	20	20	20	20

4. Discussion

4.1. Diversity and Distribution of Meiofaunal Marine Nematodes in Mexico

This study contributes to the current knowledge of the diversity and distribution of meiofaunal marine nematodes in Mexico. The faunistic information in Mexico, and Central America, in general, is traditionally scarce, reflecting the low number of taxonomic experts in the area [80]. Until now, a total of 183 species of marine nematodes are known for Mexico, of which seven have type locality in Mexico [16,22,26,81,82]. In the Mexican Caribbean specifically, there are few ecological and taxonomic studies focusing on marine nematodes [17,18,20] with Isla Mujeres and Banco Chinchorro being the most investigated localities. The Chromadorida order was the most represented in both sites [18,20]. Our results confirm the presence of *Terschellingia longicaudata* (de Man, 1907), *Monoposthia mirabilis* (Schulz, 1932), and *Xyala striata* (Cobb, 1920), all species that have been previously reported for Isla Mujeres [20].

This study adds twelve morphological records to the Mexican list of marine nematodes. Specifically, *Haliplectus bickneri*, *Proplatycoma fleurdelis*, *Pontonema* cf. *simile*, *Rhips* sp.1, *Epsilonema* sp.1, *Metachromadora* sp.1, *Spirinia* sp.2, *Odonthophora bermudensis*, *Onyx litorale*, *Epacanthion* sp.1, *Halalaimus* sp.1, and *Metaparoncholaimus* sp.1 are all new records for the country. Due to specimen immaturity and small sample size, it is necessary to collect additional specimens to obtain proper identification for some of the sampled taxa (e.g., Anticomid, chromadorid, and monhysterid).

The report of Desmodoridae taxon is particularly relevant. This family is globally represented with a high number of species (320 species) [80]. Nevertheless, genetic sequence records for this family are scarce compared with the number of species described [80]. Here, we add 20 COI new sequences from Desmodoridae taxa. However, species identification based only on morphology was particularly tricky due to the lack of well-defined diagnostic traits, weak taxonomic keys reference, and absence of updated databases, including species lists [43,47,80].

We report for the first time meiofaunal nematode species from the Cozumel Island, namely *Chromadorita* sp.1, *Catanema* sp.1, *Epsilonema* sp.1, *Epacanthion* sp.1, *Eurystomina* sp.1, two species of *Enoploides*, *Monoposthia mirabilis*, *Odontophora bermudensis*, *Onyx litorale*, *Proplatycoma fleurdelis*, and *Xyala striata*. Cozumel was represented by the highest number of species, supporting that islands, with a high grade of endemism [83], are particularly relevant areas for the study of biodiversity also for meiofaunal species.

Regarding *P. simile*, specimens found in this work are very similar to the ones described by Southern, 1914 [76], due to comparable buccal cavities with a dorsal tooth, tail shapes, and spicules size. However, we hypothesize that the Mexican record may correspond to a new species for science. First, the original record was found in the intertidal zona of West Ireland, far away from our sample locality. Second, spicule size and total length are appreciably different in individuals from the two populations. Additional investigations comparing specimens from both localities are needed to support our hypothesis.

Individuals of *Monoposthia mirabilis*, *Epsilonema* sp.1, and *Onyx litorale*, were respectively represented by the same COI haplotypes even across different sampling sites, including Cozumel Island and continental sites such as Cancún, Puerto Morelos, Tulum, Mahahual, and Xcalak. This observation suggests that such species have a widespread distribution and a low genetic divergence within and among populations. Therefore, individuals of *M. mirabilis* reported for Isla Mujeres [20] and *Onyx litorale* for Cuba [80], could belong to the same, widely distributed, species.

4.2. COI Amplification

We recommend using HotSHOT technique to extract DNA from marine nematodes [62] because, at least in our study, it provided fairly good results. However, the amplification success rate was difficult (85.07%), confirming previous studies [44,47,79,84]. The problem lies in the high mitochondrial genome diversity among nematodes [85–87] and, consequently, a low success rate using universal primers as well as low availability of specific primers for the Nematoda group [47,88]. In five individuals, the length of COI sequences varied (approximately 150 bp) but these sequences were not found related to any nematode taxa at the rank of genus or family. Specifically, we consistently observed a low success of COI amplification for *Epacanthion* sp.1, *Halalaimus* sp.1, *Prochromadorella* sp.1, *Proplatycoma fleurdelis*, *Odontophora bermudensis*, and *Oncholaimus* sp.1, which highlights the need for new primers specifically designed for these taxa. COI sequences obtained by combining existing protocols developed for zooplankton eggs [62] and by Ivanova et al. for invertebrates [63], allowed us to obtain high-quality DNA barcodes and amplify the complete 'barcode region' (> 500 bp). Obtaining long sequences is particularly important; when COI sequences are amplified in a region without a definite barcoding gap (e.g., JB2/JB3) due to overlapping intra/inter-specific distances, the delineation of species boundaries could be unreliable [44,79,84]. Although not all the sequences could be assigned to a species-level taxon, our results will significantly help the taxonomic identification of nematodes in further studies and when high-throughput sequencing approach on environmental DNA is applied.

4.3. Integrative Taxonomy

COI sequences generated in this work contribute to increasing the record of public genetic repositories and are references for future studies focused on the patterns of diversity and distribution of marine nematodes. In meiofauna and particularly marine nematodes, taxonomic gaps may be compensated only with the integration of DNA and morphological-based taxonomy [31,46]. The combination of both approaches allows us to disentangle diversity at the species level. For example, the two *Spirinias* species showed different morphological traits. However, we could certainly assign them to different species only through integrative taxonomy. In fact, all specimens of *Sprinia* sp.1 were immature, although preliminary observations allowed to ascribe this species to either *S. parasitifera* or *S. septentrionalis*, both previously reported for Mexico [20,22]. *Spirinia* sp.2 was represented by several adult specimens. However, none of the morphometric measurements were sufficient to ascribe them to a known species. *Spirinia* sp.2 could be ascribed to *S. amata* or *S. parasitifera* because of the number of precloacal supplements; however, the larger size of spicules and larger total length size suggest similarities with *S. inaurita*. We do not discard the possibility that *Spirinia* sp.2 could belong to new species to science.

4.4. Application of Species Delineation Models to Disentangle Diversity

Our results confirm the advantage of disentangling species and estimating marine nematode diversity using the COI gene. Fast-evolving genes and relatively short genes such as mtCOI are not expected to resolve the phylogenetic relationships in the deeper nodes [39,89]. The use of a multiple-gene approach is needed to clarify the tree topology, thereby uncovering phylogeographic relationships and historical biogeography within the group [42]. However, as previous studies suggested, mtCOI can resolve relationships among closely related species [79] and, as suggested by Derycke et al. [47], we support that COI is a useful biomarker to disentangle the diversity of nematodes to the species level. The design of new primers for different taxonomic groups of marine nematodes and the increase of COI data sequences will advance our knowledge of the diversity in different ecosystems [84] as well as provide a better estimate of global species diversity [44].

Analytical methods ABGD and mPTP supported the presence of the 20 species identified morphologically. The ABGD method relies on user-selected parameters (distance model and a prior limit on intraspecific divergence). In our case, we took care not to select a high prior intraspecific divergence (P) to avoid combining all sequences in one single group [53]. For this reason, a prior value was set after analyzing the interspecific divergence values within our generated sequences. The value of 0.01 for prior intraspecific divergence showed the strongest congruence between groups recovered and species defined as proposed by Puillandre et al. [53]. Moreover, we obtained an equal number of groups after the third partition (P = 0.0046) with both models (K2P and JC9); this supports that the threshold for the barcoding gap in the sequences considered in this work is well-defined.

The BINs method showed a 90% match for the recovered evolutionary independent entities. Only *Epsilonema* sp.1 and *Enoploides* sp.2 were split into four BIN numbers. It is essential to consider that this method initially employs single linkage clustering, coupled with a 2.2% threshold to establish preliminary OTU boundaries [54]. If the threshold is higher than 2.2%, the method tends to separate a higher number of entities and overestimate diversity. *Epsilonema* sp.1 and *Enoploides* sp.2 both showed the highest intraspecific divergence values > 4% (4.02% and 2.24% respectively), although they are identical morphologically. Based on all the criteria, we recognized *Epsilonema* sp.1 and *Enoploides* sp.2 as a single species, concordantly to the results obtained by applying ABGD and mPTP methods. We should carefully consider the BIN numbers as a tool for disentangling nematode species until a more robust genetic database is developed.

5. Conclusions

Our work contributes towards building a more robust taxonomic and genetic database of meiofaunal nematodes, as well as advancing our knowledge of nematode diversity and distribution. At present, DNA sequencing techniques have opened up new possibilities for taxonomic research in meiofaunal nematodes. However, barcode sequences from marine nematodes are underrepresented in light of the diversity of the phylum [34]. Our study combines classical morphology-based taxonomy with DNA sequences to successfully delimit 20 marine nematode species. Although a new set of primers should be designed for some species, our data support that the COI gene represents an excellent molecular marker to disentangle nematode diversity. Moreover, we validate and support the intraspecific distance value threshold of 5% for nematode species [43,47,84]. Lastly, our study reports new records from an unexplored region and contributes to understanding patterns of diversity and distribution of nematodes in Mexico and worldwide.

Supplementary Materials: The following are available online at http://www.mdpi.com/1424-2818/12/3/107/s1, Table S1: Records downloaded from GenBank and using for reconstructing the Maximum-likelihood tree.

Author Contributions: Conceptualization, A.M.-A. and F.L.; data curation, A.M.-A., A.D.J.-N., and F.L.; formal analysis, A.M.-A. and F.L.; funding acquisition, A.M.-A.; investigation, A.M.-A., A.D.J.-N., and F.L.; methodology,

A.M.-A.; project administration, A.M.-A.; software, F.L.; validation, A.D.J.-N. and F.L.; writing – review & editing, A.M.A., A.D.J.-N., and F.L. All authors have read and agreed to the published version of the manuscript.

Funding: This research was funded by the Consejo Nacional de Ciencia y Tecnología (CONACyT) through the Red MEXBOL Project no. 271108.

Acknowledgments: We thank Jose Angel Cohuo, Mario Yescas, and Blanca Prado for assistance during fieldwork; Jessica Landeros and Josué Puc for their help during editing; Yareli Cota for nematodes mounts; Holger Weissenberger for elaborating the map; Ruth Gingold (sweepandmore.com) for proofreading earlier versions of the manuscript; and, finally, the Canadian Center of DNA Barcoding (CCBD) for support during the sequencing process.

Conflicts of Interest: The authors declare no conflict of interest. The funders had no role in the design of the study; in the collection, analyses, or interpretation of data; in the writing of the manuscript, or in the decision to publish the results.

References

1. Van Den Hoogen, J.; Geisen, S.; Routh, D.; Ferris, H.; Traunspurger, W.; Wardle, D.A.; de Goede, R.G.M.; Adams, B.J.; Ahmad, W.; Andriuzzi, W.S.; et al. Soil nematode abundance and functional group composition at a global scale. *Nature* **2019**, *572*, 194–198. [CrossRef] [PubMed]
2. Heip, C.; Vincx, M.; Vranken, G. Ecolgy of marine nematodes. *Oceanogr. Mar. Biol. Annu. Rev.* **1985**, *23*, 399–489.
3. Vermeeren, H.; Vanreusel, A.; Vanhove, S. Species distribution within the free-living marine nematode genus *Dichromadora* in the Weddell Sea and adjacent areas. *Deep. Res. Part II-Top. Stud. Ocean.* **2004**, *51*, 1643–1664. [CrossRef]
4. Boufahja, F.; Vitiello, P.; Aissa, P. More than 35 years of studies on marine nematodes from Tunisia: A checklist of species and their distribution. *Zootaxa* **2014**, *3786*, 269–300. [CrossRef] [PubMed]
5. Ansari, K.G.M.T.; Pattnaik, A.K.; Rastogi, G.; Bhadury, P. An inventory of free-living marine nematodes from Asia's largest coastal lagoon, Chilika, India. *Wetl. Ecol. Manag.* **2015**, *23*, 881–890. [CrossRef]
6. Semprucci, F. Marine nematodes from the shallow subtidal coast of the Adriatic Sea: Species list and distribution. *Int. J. Biodivers.* **2013**, *2013*, 1–9. [CrossRef]
7. Semprucci, F.; Balsamo, M.; Frontalini, F. La comunidad de nematodos de una laguna costera (Laguna de Varano, Italia meridional): Patrones de la ecología y la biodiversidad. *Sci. Mar.* **2014**, *78*, 579–588. [CrossRef]
8. De Oliveira, D.A.S.; Derycke, S.; Da Rocha, C.M.C.; Barbosa, D.F.; Decraemer, W.; Dos Santos, G.A.P. Spatiotemporal variation and sediment retention effects on nematode communities associated with *Halimeda opuntia* (Linnaeus) Lamouroux (1816) and *Sargassum polyceratium* Montagne (1837) seaweeds in a tropical phytal ecosystem. *Mar. Biol.* **2016**, *163*, 1–13. [CrossRef]
9. Diane, Z.M.; Gouvello, L.; Nel, R.; Harris, L.R.; Bezuidenhout, K. The response of sandy beach meiofauna to nutrients from sea turtle eggs. *J. Exp. Mar. Bio. Ecol.* **2017**, *487*, 94–105. [CrossRef]
10. Ball, S.L.; Hebert, P.D.N.; Burian, S.K.; Webb, J.M. Biological identifications of mayflies (Ephemeroptera) using DNA barcodes. *J. N. Am. Benthol. Soc.* **2006**, *24*, 508–524. [CrossRef]
11. Bonaglia, S.; Nascimento, F.J.A.; Bartoli, M.; Klawonn, I.; Brüchert, V. Meiofauna increases bacterial denitrification in marine sediments. *Nat. Commun.* **2014**, *5*. [CrossRef] [PubMed]
12. World Register of Marine Species. Available online: www.marinespecies.org (accessed on 13 January 2017).
13. Hugot, J.P.; Baujard, P.; Morand, S. Biodiversity in helminths and nematodes as a field of study: An overview. *Nematology* **2001**, *3*, 199–208. [CrossRef]
14. Lambshead, P.J.D. Marine nematode biodiversity. In *Nematology: Advances and perspectives Nematode Morphology, Physiology and Ecology*; Chen, Z.X., Chen, S.Y., Dickson, D.W., Eds.; CABI Publishing: Wallingford, UK, 2004; pp. 436–467.
15. Coomans, A. Present status and future of nematode systematics. *Nematology* **2002**, *4*, 573–582. [CrossRef]
16. Castillo-Fernandez, D.; Lambshead, P.J.D. Revision of the genus *Elzalia* Gerlach, 1957 (Nematoda: Xyalidae) including three new species from an oil producing zone in the Gulf of Mexico, with a discussion of the sibling species problem. *Bull. Br. Museum Natural Hist. Zool. Ser.* **1990**, *56*, 63–71.
17. De Jesús-Navarrete, A.; Herrera-Gómez, J. Vertical distribution and feeding types of nematodes from Chetumal Bay, Quintana Roo, Mexico. *Estuaries* **2002**, *25*, 1131–1137. [CrossRef]
18. De Jesús-Navarrete, A. Diversity of nematoda in a Caribbean atoll: Banco Chinchorro, Mexico. *Bull. Mar. Sci.* **2003**, *73*, 47–56.

19. De Jesús-Navarrete, A. Littoral free living nematode fauna of Socorro Island, Colima, Mexico. *Hidrobiologica* **2007**, *17*, 61–66.

20. De Jesús-Navarrete, A. Nematodos de los arrecifes de Isla Mujeres y Banco Chinchorro, Quintana Roo, México. *Rev. Biol. Mar. Oceanogr.* **2007**, *42*, 193–200. [CrossRef]

21. De Jesús-Navarrete, A. Distribución, abundancia y diversidad de los nematodos (Phylum Nematoda) bénticos de la Sonda de Campeche, México. Enero 1987. *Rev. Biol. Trop.* **1993**, *41*, 57–63.

22. Comision Nacional para el Conocimiento y Uso de la Biodiversidad. Available online: https://www.gob.mx/conabio (accessed on 20 January 2020).

23. De Jesús-Navarrete, A.; Herrera-Gómez, J. Nematofauna asociada a la zona urbana de la bahía de Chetumal, Quintana Roo, México. *Rev. Biol. Trop.* **1999**, *47*, 867–875.

24. Gingold, R.; Mundo-Ocampo, M.; Holovachov, O.; Rocha-Olivares, A. The role of habitat heterogeneity in structuring the community of intertidal free-living marine nematodes. *Mar. Biol.* **2010**, *157*, 1741–1753. [CrossRef] [PubMed]

25. Holovachov, O.; Tandingan De Ley, I.; Mundo-Ocampo, M.; Baldwin, J.G.; Rocha-Olivares, A.; De Ley, P. Nematodes from the Gulf of California. Part 1. The genera *Ceramonema* Cobb, 1920, *Pselionema* Cobb in Cobb, 1933 and *Pterygonema* Gerlach, 1954 (Nematoda: Ceramonematidae). *Nematology* **2008**, *10*, 347–373.

26. Holovachov, O.; Mundo-Ocampo, M.; De Ley, I.T.; De Ley, P. Nematodes from the gulf of California. Part 2. *Ceramonema nasobema* sp. n. (nematoda: Ceramonematidae). *Nematology* **2008**, *10*, 835–844.

27. Mundo-Ocampo, M.; Lambshead, P.J.D.; Debenham, N.; King, I.W.; De Ley, P.; Baldwin, J.G.; De Ley, I.T.; Rocha-Olivares, A.; Waumann, D.; Thomas, W.K.; et al. Biodiversity of littoral nematodes from two sites in the Gulf of California. *Hydrobiologia* **2007**, *586*, 179–189. [CrossRef]

28. De Ley, P. Lost in worm space: Phylogeny and morphology as road maps to nematode diversity. *Nematology* **2002**, *2*, 9–16. [CrossRef]

29. Powers, T. Nematode molecular diagnostics: From bands to barcodes. *Annu. Rev. Phytopathol.* **2004**, *42*, 367–383. [CrossRef]

30. Bik, H.M. Let's rise up to unite taxonomy and technology. *PLoS Biol.* **2017**, *15*, 4–7. [CrossRef]

31. Leasi, F.; Sevigny, J.L.; Laflamme, E.M.; Artois, T.; Curini-Galletti, M.; de Jesus Navarrete, A.; Di Domenico, M.; Goetz, F.; Hall, J.A.; Hochberg, R.; et al. Biodiversity estimates and ecological interpretations of meiofaunal communities are biased by the taxonomic approach. *Commun. Biol.* **2018**, *1*, 1–12. [CrossRef]

32. Derycke, S.; Remerie, T.; Vierstraete, A.; Backeljau, T.; Vanfleteren, J.; Vincx, M.; Moens, T. Mitochondrial DNA variation and cryptic speciation within the free-living marine nematode Pellioditis marina. *Mar. Ecol. Prog. Ser.* **2005**, *300*, 91–103. [CrossRef]

33. De Ley, P.; Tandingan De Ley, I.; Morris, K.; Abebe, E.; Mundo-Ocampo, M.; Yoder, M.; Heras, J.; Waumann, D.; Rocha-Olivares, A.; Burr, A.H.J.; et al. An integrated approach to fast and informative morphological vouchering of nematodes for applications in molecular barcoding. *Philos. Trans. R. Soc. London. Ser. B* **2005**, *360*, 1945–1958. [CrossRef]

34. Da Silva, N.R.R.; Da Silva, M.C.; Genevois, V.F.; Esteves, A.M.; De Ley, P.; Decraemer, W.; Rieger, T.T.; Dos Santos Correia, M.T. Marine nematode taxonomy in the age of DNA: The present and future of molecular tools to assess their biodiversity. *Nematology* **2010**, *12*, 661–672. [CrossRef]

35. Hebert, P.D.N.; Gregory, T.R. The promise of DNA barcoding for taxonomy. *Syst. Biol.* **2005**, *54*, 852–859. [CrossRef] [PubMed]

36. Bhadury, P.; Austen, M.C.; Bilton, D.T.; Lambshead, P.J.D.; Rogers, A.D.; Smerdon, G.R. Development and evaluation of a DNA-barcoding approach for the rapid identification of nematodes. *Mar. Ecol. Prog. Ser.* **2006**, *320*, 1–9. [CrossRef]

37. Bhadury, P.; Austen, M.C. Barcoding marine nematodes: An improved set of nematode 18S rRNA primers to overcome eukaryotic co-interference. *Hydrobiologia* **2010**, *641*, 245–251. [CrossRef]

38. Fontaneto, D.; Flot, J.F.; Tang, C.Q. Guidelines for DNA taxonomy, with a focus on the meiofauna. *Mar. Biodiv.* **2015**, *45*, 433–451. [CrossRef]

39. Blaxter, M.L. The promise of a DNA taxonomy. *Philos. Trans. R. Soc. B* **2004**, *359*, 669–679. [CrossRef] [PubMed]

40. Dayrat, B. Towards integrative taxonomy? *Biol. J. Linn. Soc.* **2005**, *85*, 407–415. [CrossRef]

41. Pereira, T.J.; Fonseca, G.; Mundo-Ocampo, M.; Guilherme, B.C.; Rocha-Olivares, A. Diversity of free-living marine nematodes (Enoplida) from Baja California assessed by integrative taxonomy. *Mar. Biol.* **2010**, *157*, 1665–1678. [CrossRef] [PubMed]

42. Leasi, F.; Andrade, S.C.D.S.; Norenburg, J. At least some meiofaunal species are not everywhere. Indication of geographic, ecological and geological barriers affecting the dispersion of species of Ototyphlonemertes (Nemertea, Hoplonemertea). *Mol. Ecol.* **2016**, *25*, 1381–1397. [CrossRef] [PubMed]

43. Armenteros, M.; Rojas-Corzo, A.; Ruiz-Abierno, A.; Derycke, S.; Backeljau, T.; Decraemer, W. Systematics and DNA barcoding of free-living marine nematodes with emphasis on tropical desmodorids using nuclear SSU rDNA and mitochondrial COI sequences. *Nematology* **2014**, *16*, 979–989. [CrossRef]

44. Macheriotou, L.; Guilini, K.; Bezerra, T.N.; Tytgat, B.; Nguyen, D.T.; Phuong Nguyen, T.X.; Noppe, F.; Armenteros, M.; Boufahja, F.; Rigaux, A.; et al. Metabarcoding free-living marine nematodes using curated 18S and CO1 reference sequence databases for species-level taxonomic assignments. *Ecol. Evol.* **2019**, *9*, 1211–1226. [CrossRef] [PubMed]

45. Mills, S.; Alcántara-Rodríguez, J.A.; Ciros-Pérez, J.; Gómez, A.; Hagiwara, A.; Galindo, K.H.; Jersabek, C.D.; Malekzadeh-Viayeh, R.; Leasi, F.; Lee, J.S.; et al. Fifteen species in one: Deciphering the Brachionus plicatilis species complex (Rotifera, Monogononta) through DNA taxonomy. *Hydrobiologia* **2017**, *796*, 39–58. [CrossRef]

46. Tang, C.Q.; Leasi, F.; Obertegger, U.; Kieneke, A.; Barraclough, T.G.; Fontaneto, D. The widely used small subunit 18S rDNA molecule greatly underestimates true diversity in biodiversity surveys of the meiofauna. *Proc. Natl. Acad. Sci. USA* **2012**, *109*, 16208–16212. [CrossRef] [PubMed]

47. Derycke, S.; Vanaverbeke, J.; Rigaux, A.; Backeljau, T.; Moens, T. Exploring the use of cytochrome oxidase c subunit 1 (COI) for DNA barcoding of free-living marine nematodes. *PLoS ONE* **2010**, *5*, 1–9. [CrossRef] [PubMed]

48. Bhadury, P.; Austen, M.C.; Bilton, D.T.; Lambshead, P.J.D.; Rogers, A.D.; Smerdon, G.R. Evaluation of combined morphological and molecular techniques for marine nematode (*Terschellingia* spp.) identification. *Mar. Biol.* **2008**, *154*, 509–518. [CrossRef]

49. Floyd, R.; Abebe, E.; Papert, A.; Blaxter, M. Molecular barcodes for soil nematodes identification. *Mol. Ecol.* **2002**, *11*, 839–850. [CrossRef] [PubMed]

50. Derycke, S.; De Ley, P.; Tandingan De Ley, I.; Holovachov, O.; Rigaux, A.; Moens, T. Linking DNA sequences to morphology: Cryptic diversity and population genetic structure in the marine nematode *Thoracostoma trachygaster* (Nematoda, Leptosomatidae). *Zool. Scr.* **2010**, *39*, 276–289. [CrossRef]

51. Hebert, P.D.N.; Ratnasingham, S.; DeWaard, J.R. Barcoding animal life: Cytochrome c oxidase subunit 1 divergences among closely related species. *Proc. R. Soc. London. Ser. B* **2003**, *270*, 96–99. [CrossRef]

52. Kekkonen, M.; Hebert, P.D.N. DNA barcode-based delineation of putative species: Efficient start for taxonomic workflows. *Mol. Ecol. Resour.* **2014**, *14*, 706–715. [CrossRef]

53. Puillandre, N.; Lambert, A.; Brouillet, S.; Achaz, G. ABGD, Automatic Barcode Gap Discovery for primary species delimitation. *Mol. Ecol.* **2012**, *21*, 1864–1877. [CrossRef]

54. Ratnasingham, S.; Hebert, P.D.N. A DNA-Based Registry for All Animal Species: The Barcode Index Number (BIN) System. *PLoS ONE* **2013**, *8*. [CrossRef] [PubMed]

55. Zhang, J.; Kapli, P.; Pavlidis, P.; Stamatakis, A. A general species delimitation method with applications to phylogenetic placements. *Bioinformatics* **2013**, *29*, 2869–2876. [CrossRef] [PubMed]

56. Traunspurger, W. The biology and ecology of lotic nematodes. *Freshw. Biol.* **2000**, *44*, 29–45. [CrossRef]

57. Yoder, M.; Tandingan De Ley, I.; King, I.W.; Mundo-Ocampo, M.; Mann, J.; Blaxter, M.; Poiras, L.; De Ley, P. DESS: A versatile solution for preserving morphology and extractable DNA of nematodes. *Nematology* **2006**, *8*, 1–10. [CrossRef]

58. Somerfield, P.J.; Warwick, R.M.; Moens, T. Meiofauna Techniques. In *Methods for the study of marine benthos*; Wiley Online Books: Chichester, UK, 2005; pp. 229–272. ISBN 9780470995129.

59. Platt, H.M.; Warwick, R.M. *Freeliving Marine Nematodes. Part 1: British Enoplids. Pictorial Key to World Genera and Notes for the Identification of British Species*; Cambridge University Press, for the Linnean Society of London and the Estuarine and Brackish-water Sciences Association: Cambridge, UK, 1983; ISBN 0521254221.

60. Platt, H.M.; Warwick, R.M. *Free Living Marine Nematodes: Part II British Chromadorids. Pictoral Key to World Genera and Notes for the Identification of British Species. Synopses of the British Fauna*; Brill, E.J., Backhuys, W., Eds.; Linnean Society of London and the Estuarine and Brackish-water Sciences Association: Leiden, NY, USA, 1988.

61. Warwick, R.M.; Platt, H.M.; Somerfield, P.J. *Free Living Marine Nematodes: Pictorial Key to World Genera and Notes for the Identification of British Species. Part III, Monhysterids*; Bames, R.S.K., Crothers, J.H., Eds.; Linnean Society of London and the Estuarine and Coastal Sciences Association by Field Studies Council: Dorchester, UK, 1998.

62. Montero-Pau, J.; Gómez, A.; Muñoz, J. Application of an inexpensive and high-throughput genomic DNA extraction method for the molecular ecology of zooplanktonic diapausing eggs. *Limnol. Oceanogr. Methods* **2008**, *6*, 218–222. [CrossRef]

63. Ivanova, N.V.; Grainger, C. Increased DNA barcode recovery using Platinum taq. *CCDB Adv.: Methods Release* **2006**, *2*, 1.

64. Folmer, O.; Black, M.; Hoeh, W.; Lutz, R.; Vrijenhoek, R. DNA primers for amplification of mitochondrial cytochrome c oxidase subunit I from diverse metazoan invertebrates. *Mol. Mar. Biol. Biotechnol.* **1994**, *3*, 294–299.

65. Hajibabaei, M.; DeWaard, J.R.; Ivanova, N.V.; Ratnasingham, S.; Dooh, R.T.; Kirk, S.L.; Mackie, P.M.; Hebert, P.D.N. Critical factors for assembling a high volume of DNA barcodes. *Philos. Trans. R. Soc. B* **2005**, *360*, 1959–1967. [CrossRef]

66. Ivanova, N.V.; Dewaard, J.R.; Hebert, P.D.N. An inexpensive, automation-friendly protocol for recovering high-quality DNA. *Mol. Ecol. Notes* **2006**, *6*, 998–1002. [CrossRef]

67. Ewing, B.; Green, P. Base-calling of automated sequencer traces using phred. II. Error probabilities. *Genome Res.* **1998**, *8*, 186–194. [CrossRef]

68. Ratnasingham, S.; Hebert, P.D.N. BOLD: The Barcode of Life Data System (www.barcodinglife.org). *Mol. Ecol. Notes* **2007**, *7*, 355–364. [CrossRef] [PubMed]

69. Tamura, K.; Stecher, G.; Peterson, D.; Filipski, A.; Kumar, S. MEGA6: Molecular evolutionary genetics analysis version 6.0. *Mol. Biol. Evol.* **2013**, *30*, 2725–2729. [CrossRef] [PubMed]

70. Kimura, M. A simple method for estimating evolutionary rates of base substitutions through comparative studies of nucleotide sequences. *J. Mol. Evol.* **1980**, *16*, 111–120. [CrossRef] [PubMed]

71. Richly, E.; Leister, D. NUMTs in sequenced eukaryotic genomes. *Mol. Biol. Evol.* **2004**, *21*, 1081–1084. [CrossRef] [PubMed]

72. Song, H.; Buhay, J.E.; Whiting, M.F.; Crandall, K.A. Many species in one: DNA barcoding overestimates the number of species when nuclear mitochondrial pseudogenes are coamplified. *Proc. Natl. Acad. Sci. USA* **2008**, *105*, 13486–13491. [CrossRef]

73. Altschul, S.; Madden, T.; Schaffer, A.; Zhang, J.; Zhang, Z.; Miller, W.; Lipman, D. Gapped blast and psi-blast: A new generation of protein database search programs. *FASEB J.* **1998**, *12*, 3389–3402. [CrossRef]

74. Hazir, C.; Giblin-Davis, R.M.; Keskin, N.; Ye, W.; Hazir, S.; Scheuhl, E.; Thomas, W.K. Diversity and distribution of nematodes associated with wild bees in Turkey. *Nematology* **2010**, *12*, 65–80.

75. Jukes, T.H.; Cantor, C. Evolution of Protein Molecules. In *Mammalian Protein Metabolism*; Munro, H.N.E., Ed.; Academic Press: New York, NY, USA, 1969; pp. 21–132.

76. Southern, R. Nemathelmaia, Kinorhyncha, and Chaetognatha. *Proc. R. Irish Acad.* **1914**, *31*, 1–80.

77. Filipjev, I.N. Free-living marine nematodes of the Sevastopol area. *Trudy Osob. Zool. Lab. Sebastop. Biol. Sta.* **1921**, *II*, 351–614.

78. Velasco-Castrillón, A.; Stevens, M.I. Morphological and molecular diversity at a regional scale: A step closer to understanding Antarctic nematode biogeography. *Soil Biol. Biochem.* **2014**, *70*, 272–284. [CrossRef]

79. Hauquier, F.; Leliaert, F.; Rigaux, A.; Derycke, S.; Vanreusel, A. Distinct genetic differentiation and species diversification within two marine nematodes with different habitat preference in Antarctic sediments. *BMC Evol. Biol.* **2017**, *17*, 1–14. [CrossRef] [PubMed]

80. Armenteros, M.; Ruiz-Abierno, A.; Decraemer, W. Revision of Desmodorinae and Spiriniinae (Nematoda: Desmodoridae) with redescription of eight known species. *Eur. J. Taxon.* **2014**, *96*, 1–32. [CrossRef]

81. Castillo-Fernandez, D.; Decraemer, W. *Cheironchus paravorax* n. sp. and *Cheironchus vorax* Cobb, 1917 from the Campeche Sound, an oil producing zone in the Gulf of Mexico (Nemata: Selachinematidae). *Bull. l'Institut R. des Sci. Nat. Belgique* **1993**, *63*, 55–64.

82. Holovachov, O.; De Ley, I.T.; Mundo-Ocamnpo, M.; Gingold, R.; De Ley, P. Nematodes from the Gulf of California. Part 3. Three new species of the genus *Diplopeltoides* Gerlach, 1962 (Nematoda: Diplopeltoididae) with overviews of the genera *Diplopeltis* Gerlach, 1962 and *Diplopeltula* Gerlach, 1950. *Russ. J. Nematol.* **2009**, *17*, 43–57.

83. Witt, C.C.; Maliakal-Witt, S. Why are diversity and endemism Linked on islands? *Ecography* **2007**, *30*, 331–333. [CrossRef]

84. Avó, A.P.; Daniell, T.J.; Neilson, R.; Oliveira, S.; Branco, J.; Adão, H. DNA Barcoding and morphological identification of benthic nematodes assemblages of estuarine intertidal sediments: Advances in molecular tools for biodiversity assessment. *Front. Mar. Sci.* **2017**, *4*. [CrossRef]

85. Okimoto, R.; Macfarlene, J.L.; Clary, D.O.; Wolstenholme, D.R. The mitochondrial genomes of two nematodes, *Caenorhabditis elegans* and *Ascaris suum*. *Genetics* **1992**, *130*, 471–498.

86. Ekrem, T.; Willassen, E.; Stur, E. A comprehensive DNA sequence library is essential for identification with DNA barcodes. *Mol. Phylogenet. Evol.* **2007**, *43*, 530–542. [CrossRef]

87. Elsasser, S.C.; Floyd, R.; Hebert, P.D.N.; Schulte-Hostedde, A.I. Species identification of North American guinea worms (Nematoda: *Dracunculus*) with DNA barcoding. *Mol. Ecol. Resour.* **2009**, *9*, 707–712. [CrossRef]

88. Prosser, S.W.J.; Velarde-Aguilar, M.G.; León-Règagnon, V.; Hebert, P.D.N. Advancing nematode barcoding: A primer cocktail for the cytochrome c oxidase subunit I gene from vertebrate parasitic nematodes. *Mol. Ecol. Resour.* **2013**, *13*, 1108–1115. [CrossRef]

89. Leasi, F.; Norenburg, J.L. The necessity of DNA taxonomy to reveal cryptic diversity and spatial distribution of meiofauna, with a focus on nemertea. *PLoS ONE* **2014**, *9*, e104385. [CrossRef] [PubMed]

Article

The Effects of Habitat Heterogeneity at Distinct Spatial Scales on Hard-Bottom-Associated Communities

Fabiane Gallucci [1,*], **Ronaldo A. Christofoletti** [1], **Gustavo Fonseca** [1] and **Gustavo M. Dias** [2]

[1] Institute of Marine Science, Federal University of São Paulo (IMar/UNIFESP),
    Rua Dr. Carvalho de Mendonça 144, Santos SP 11070-100, Brazil; ronaldochristofoletti@gmail.com (R.A.C.);
    gfonseca.unifesp@gmail.com (G.F.)
[2] Center of Natural and Human Sciences, Federal University of ABC (CCNH/UFABC), Rua Santa Adélia, 166,
    Santo Andre SP 09210-170, Brazil; gmunizdias@gmail.com
*   Correspondence: fgallucci@unifesp.br

Received: 30 November 2019; Accepted: 15 January 2020; Published: 20 January 2020

Abstract: For marine benthic communities, environmental heterogeneity at small spatial scales are mostly due to biologically produced habitat heterogeneity and biotic interactions, while at larger spatial scales environmental factors may prevails over biotic features. In this study, we investigated how community structure and β-diversity of hard-bottom-associated meio- and macrofauna varied in relation to small-scale (cm–m) changes in biological substrate (an algae "turf" dominated by the macroalgae *Gelidium* sp., the macroalgae *Caulerpa racemosa* and the sponge *Hymeniacidon heliophile*) in a rocky shore and in relation to larger-scale (10's m) changes in environmental conditions of the same biological substrate (the macroalgae *Bostrychia* sp) in different habitats (rocky shore vs. mangrove roots). Results showed that both substrate identity and the surrounding environment were important in structuring the smaller-sized meiofauna, particularly the nematode assemblages, whereas the larger and more motile macrofauna was influenced only by larger-scale changes in the surrounding ecosystem. This implies that the macrofauna explores the environment in a larger spatial scale compared to the meiofauna, suggesting that effects of spatial heterogeneity on communities are dependent on organism size and mobility. Changes in taxa composition between environments and substrates highlight the importance of habitat diversity at different scales for maintaining the diversity of the associated fauna.

Keywords: meiofauna; macrofauna; associated fauna; biological substrate; species diversity; community ecology; benthic ecology

---

1. Introduction

In natural systems, environmental heterogeneity occurs at varying scales in space and time affecting the diversity of species. However, processes operating at one scale not necessarily scale up or down [1]. For marine benthic communities, environmental heterogeneity at small spatial scales (cm to few meters) are mostly due to biologically produced habitat heterogeneity and biotic interactions, while at larger spatial scales (10's m to kilometers) environmental factors may prevails over biotic features [2]. Regardless of the mechanism behind environmental heterogeneity, it promotes species diversity by increasing the range of resources and reducing niche overlap, which in turn promotes species coexistence [3,4].

In the intertidal zone of rocky shores, sessile organisms such as macroalgae and sponges increase small-scale habitat heterogeneity, harboring a diverse associated fauna [5–7]. The community structure of the associated fauna is affected by several intrinsic properties of the host, such as their physical architecture, the number of microhabitats, sediment deposition, food resources, refuge from predators, and wave damping effects [8–13]. In addition to this small-scale heterogeneity, variations in hydrosedimentary processes at larger scales (meters to kilometers) also affect the structure and diversity of hard-bottom-associated communities, mainly through their effects on larvae supply, dispersal and colonization of organisms [14,15].

Hard-bottom-associated fauna is usually composed by small crustaceans, nematodes, mollusks, annelids and other groups that use the host substratum as shelter and food. These organisms belong to two different ecological compartments, macrofauna and meiofauna, which are distinguished by their different size as well as different life-history traits. Besides being larger and more mobile, most macrobenthic species exhibit a planktonic larval stage so that larval dispersal can be extensive [16]. In contrast, meiofauna have direct benthic development, are smaller and exhibit lower mobility. Additionally, meiofauna is expected to be more specialized in their habitat and food resources [2,17–20], whereas macrofauna are relatively more generalist [21]. As a result, meiofauna, especially nematodes, are generally more affected by within-habitat variability when compared to larger-sized, more mobile macrofauna [16]. The latter, on the other hand, are more prone to be influenced by larger-scale environmental heterogeneity, for instance, in hydrodynamic conditions and position on the shore [22].

In this study, we aim to describe how community structure and diversity of hard-bottom-associated fauna are related to (i) small-scale (cm–m) changes in biological (secondary) substrate in a rocky shore and (ii) larger-scale (10's m) changes in environmental conditions of a same biological substrate in different habitats (rocky shore x mangrove roots). At both scales, we expect habitat heterogeneity to create a mosaic that will contribute to local and regional biodiversity. As such, we have tested the hypotheses that different biological substrates harbor different associated communities (H1) and that the same biological substrate growing in different habitats harbor distinct associated communities (H2). Yet, we expect the meiofauna and macrofauna communities to respond differently to habitat heterogeneity. We expect the meiofauna communities to be more affected by within-habitat environmental variability (i.e., to show differences in composition between the biological substrates) and the macrofauna to be more influenced by larger-scale changes in environmental conditions (i.e., to show differences in species composition between the different habitats).

2. Material and Methods

2.1. Study Area

This study was conducted in the Southeastern Brazil, in the Araçá Bay (São Sebastião, SP; Figure 1). Despite the relatively small size of the bay (~500 m²), this is a heterogeneous bay composed by rocky shores, sandy beaches, small mangrove spots and a larger soft bottom, being a hot spot of marine biodiversity [23]. The variety of different habitats in short distances makes this as a special region to test our hypothesis regarding habitat heterogeneity from smaller to larger scales. In addition to the ecological questions, our study will help to support a discussion regarding anthropogenic impacts at the bay. Araçá Bay is under the threat of the eminent expansion of the port of São Sebastião, which will cover 75% of the bay with a structure suspended by pillars, potentially reducing the occurrence of macroalgae [24], which host diverse associated fauna [25].

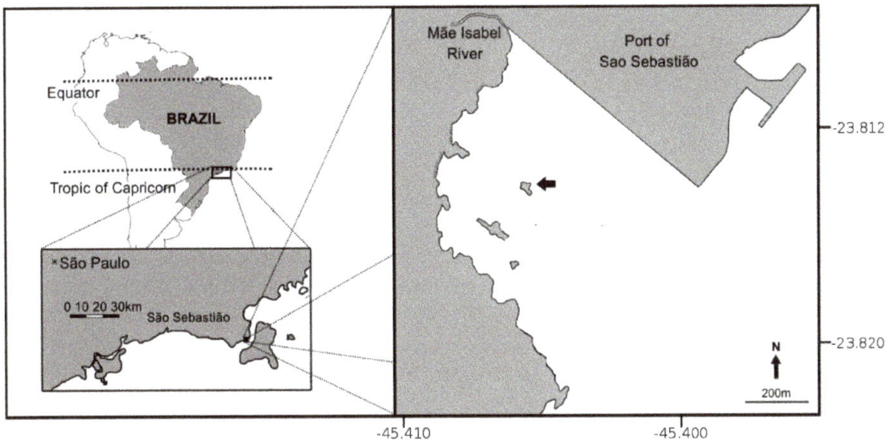

Figure 1. Map indicating the location of the Araçá Bay. The arrow indicates the sampling site.

2.2. Experimental Design

To test the hypothesis that different biological substrates harbor different associated communities increasing local diversity (H1) we have sampled meio- and macrofauna communities from different biological substrates growing in the intertidal zone of a rocky shore from southeastern Brazil. We decided to use the natural heterogeneity present on rocky shores so the natural substrates were (1) a "turf" of different macroalgae species ('Turf') composed mainly by algae of the genus *Gelidium*, (2) the macroalgae *Caulerpa racemosa* ('*Caulerpa*') and (3) the sponge *Hymeniacidon heliophila* ('Sponge'). While the turf substratum was homogeneous, the remaining substrata were dominated by Caulerpa and Sponges, respectively, but turfing algae were still present in abundance underneath the dominant species.

To test the hypothesis that the same biological substrate growing in different habitats harbor distinct associated communities (H2) we sampled the macroalgae *Bostrychia* sp. in two sites that were 20 m apart from each other but were structurally different: a rocky shore and the pneumatophores from the mangrove *Laguncularia racemosa*. In both experiments, four replicate samples were taken for each substrate.

2.3. Sampling and Sample Processing

Four replicate samples were taken at each substrate by scraping a quadrat (10 × 10 cm) using a scraper therefore removing both the hard-bottom surface and the biological substrate. Samples from the different substrates of the rocky shore (H1) were taken within the same tidal level whereas for the comparison of the same substrate (the macroalgae *Bostrychia* sp.) between two different habitats (H2), the tidal level was slightly different. In that case, samples from the pneumatophores from the mangrove *Laguncularia racemose* were taken from an upper tidal level compared to those from the rocky shore. All samples were fixed with 10% formalin and, to better detach organisms from the substrate, were sonicated for 3 min (60 W, 3 times for 1 min with 30 s intervals) [26]. Samples were then sieved through 500 μm and 45 μm mesh sieves to retain macro and meiofauna, respectively. Samples were stored in 4% formalin and stained with Rose Bengal. Meiofauna were further extracted by flotation with Ludox TM 50 (specific density 1.18) [2], divided into four subsamples by using a Folsom plankton splitter. For each sample, one subsample was selected. Macrofauna and meiofauna were counted and identified under a stereomicroscope. Particularly for the meiofauna, 10% of the nematodes per subsample were randomly picked, evaporated slowly in anhydrous glycerol and mounted on permanent slides for identification. Nematodes were identified to the genus level and further separated into morphospecies.

2.4. Statistical Analyses

Because the sampled substrata had distinct growth forms and consequently different biomass, the abundance of both meio- and macrofauna were pondered by the sample volume, which was measured by water dislodgment in a graduated cylinder.

Univariate and multivariate analyses were conducted on log (x + 1) transformed data and presence/absence data, independently for macrofauna, meiofauna community, and for nematode assemblages, as *Nematoda* was the most abundant group in all samples. Abundance data was used to assess differences in community multivariate structure (including taxa composition and taxa abundances) whereas presence/absence data aimed at looking for differences in taxa composition only (i.e., turnover of taxa between the substrates for H1 and the habitats for H2). For each group, density, species richness and Shannon' diversity (H') were compared among substrata (H1) and between habitats (H2) using one-way analyses of variance (ANOVA). Density data was log (x + 1) transformed to equalize the importance of abundant and rare species and used to build a similarity matrix among samples. Differences on species composition were assessed using Sorensen distance for presence/absence, while community structure was assessed using Bray–Curtis distance for log (x + 1) transformed data. For both measures, differences among substrata (H1) and between habitats (H2) were tested independently using Permutational analysis of variance (PERMANOVA) [27] and visually presented by a Non-metric Multidimentional Scaling ordination (NMDS). For all analyses the species that contributed the most for the differences among groups were assessed through Similarity percentage analysis (SIMPER) [28].

3. Results

3.1. Different Biological Substrates Harbor Different Associated Communities

3.1.1. Meiofauna Community and Nematode Assemblage

The meiofauna was represented by 13 higher taxa with mean densities ranging from 238 to 522 individual mL^{-1} (Table S1). Nematodes were the most abundant taxa, corresponding to 58% of all organisms collected, followed by harpacticoid copepods (20%). Ostracods (6%), polychaetes (6%) and amphipods (5%) were the next most abundant groups. Other groups represented less than 2% of the meiofauna.

Meiofauna abundance, the number of meiofaunal groups, diversity and community structure did not differ among substrates (Tables 1 and 2). However, analysis on presence/absence data revealed significant differences in the composition of meiofauna communities and nematode assemblages between the substrata (Table 2). The composition of meiofauna communities associated with the sponge was significantly different from both "Turf" and "*Caulerpa*" communities (post-hoc $p < 0.05$, Figure 2). This difference was mainly due to the exclusive occurrence of oligochaetes in all "sponge" samples (SIMPER analysis: 36.5% and 44.7%, contribution for differences between "Sponge" and "Turf" and "Sponge" and "*Caulerpa*", respectively).

Table 1. Differences in univariate indices from macro- and meiofauna between substrates (H1) using one-way ANOVA.

	df	*F*	*p*
Meiofauna density	2	0.079	0.925
Nematode density	2	0.513	0.610
Number of meiofaunal taxa	2	0.600	0.569
Number of nematode species	2	2.212	0.149
Nematode diversity (H')	2	2.101	0.162
Macrofauna density	2	3.055	0.097
Number of macrofaunal taxa	2	0.694	0.524
Macrofauna diversity (H')	2	2.486	0.138

Table 2. Differences in multivariate structure of macrofauna, meiofauna and nematode assemblages between substrates using PERMANOVA. Bold values indicate $p < 0.05$.

Source	df	Abundance Data (Log (x + 1))		df	Presence/Absence Data	
		Pseudo-*F*	*p* (perm)		Pseudo-F	*p* (perm)
			Macrofauna			
Treatment	2	1.174	0.323	2	0.969	0.441
Residual	9			9		
			Meiofauna			
Treatment	2	1.642	0.115	2	5.021	**0.003**
Residual	9			9		
			Nematodes			
Treatment	2	1.282	0.212	2	2.6989	**0.003**
Residual	9			9		

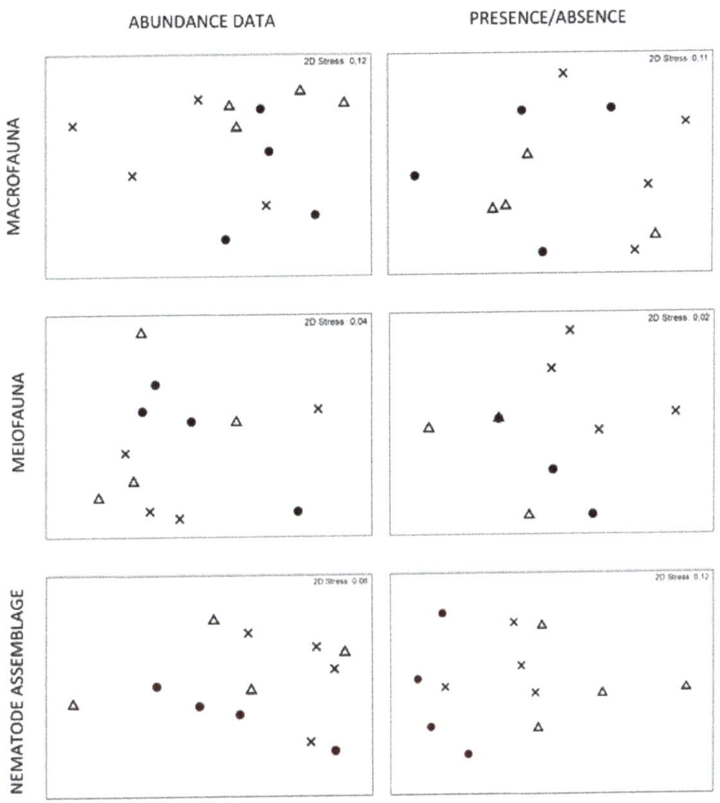

Figure 2. Non-metric multidimentional scaling (NMDS) ordination representing the similarity between samples from the substrates "Turf" (Δ), "Caulerpa" (●) and "Sponge" (x) for macrofauna, meiofauna and nematode assemblages.

Nematodes were represented by 56 species with densities ranging from 115 to 308 ind. mL^{-1} (Table S2). *Spilophorella meyerabichi* was the dominant species in all substrates (45% to 48%). *Enoplus* sp.1 (5%–9%), *Halaphanolaimus* sp.1 (7%–8%) and *Oncholaimellus* sp.1 (5%–8%) were the next most abundant species. All other nematode species occurred in low densities, each accounting for less than 4% of the assemblage. Similarly to the meiofauna, nematode abundances, species richness, diversity and community structure did not differ among substrates (Tables 1 and 2). Analysis on presence/absence data, however, revealed significant differences in the species composition of nematode assemblages between "Caulerpa" and "Turf" assemblages (post-hoc $p < 0.05$, Figure 2). A total of 13 species contributed to 50% of the total dissimilarity between the two substrates (Table S3). Indeed, from a total of 60 species, the different substrates shared from 22 (Caulerpa vs. Turf) to 26 species (Sponge vs. Turf) only. In addition, each substrate showed a relatively high number of exclusive species ("Caulerpa": 11 species, "Sponge": 11 species and "Turf": 7 species).

3.1.2. Macrofauna Community

The macrofauna was represented by 17 higher taxa with densities ranging from 56 to 303 ind. mL^{-1} (Table S4). The most abundant group was Gammaridae (Amphipoda), which accounted for 45% of all organisms collected, followed by Tanaidacea (20%) and Polychaeta (17%). All other taxa contributed with less than 2% each for the total macrofauna sampled. As for the meiofauna, macrofauna densities, number of taxa, diversity and community structure were similar in the three substrata (Table 1). However, differently from meiofauna and nematode assemblages, macrofauna composition as shown by presence/absence data did not differ between the substrata (Table 2, Figure 2).

3.2. The Same Biological Substrate Growing in Different Habitats/Environments Harbor Different Associated Communities

3.2.1. Meiofauna Community and Nematodes Assemblage

Meiofauna was represented by 16 taxa, from which 10 occurred in both environments (Table S5). Mangrove and rocky shore samples each had 3 exclusive taxa. *Nematoda* was the most abundant taxa, representing 73% and 62.5% of total meiofauna collected in the mangrove and rocky-shore samples, respectively. Halocaridina, Rotifera and Copepoda (adult + nauplii) were the next most abundant groups, and accounted for 6%, 9% and 8% in the mangrove and 19%, 5% and 9% in the rocky shore, respectively. Meiofauna densities did not differ between habitats (Table 3), ranging from 44 to 1278 ind. mL^{-1} and from 119 to 374 ind. mL^{-1} in the mangrove and rocky shore, respectively. The number of taxa and community structure did not differ between the two habitats either (Tables 3 and 4, Figure 3).

Nematodes were represented by 28 species with densities ranging from 15 to 982 ind. mL^{-1} (Table S6). Samples from the rocky shore were dominated by *Araeolaimus* sp.1 (35%), followed by *Eleutherolaimus* sp.1 (12%) and *Microlaimus* sp.1 (11%) whereas mangrove samples were dominated by *Thalassomonhystera* sp.1 (46%), followed by *Araeolaimus* sp.1 (19%). Similarly to total meiofauna, nematode densities and species richness did not differ between habitats, but diversity (H') was significantly higher in the rocky shore (Table 3). Contrasting to meiofauna community, nematode assemblages from the two environments were significantly different (Table 4, Figure 3). SIMPER analysis showed that dissimilarities between rocky shore and mangrove samples (aver. dissimilarity = 54.28%) mainly resulted from differences in abundance of a few dominant organisms (Table S7). Rocky-shore samples were characterized by higher abundances of *Eleutherolaimus* sp.1 and *Microlaimus* sp.1, whereas mangrove samples were characterized by higher abundances of *Thalassomonhystera* sp.1 and *Parachanthoncus* sp.3. Altogether these four species accounted for 50% of dissimilarities. In addition to differences in the abundance of species, analysis on presence/absence data revealed significant differences in species composition (PERMANOVA $p = 0.023$). A total of eight species contributed to 50% dissimilarities in species composition between the habitats (Table S8). From the 28 species associated with *Bostrychia* sp., only 9 species were common to both environments (Table S5). A total of 13 species were exclusive to the rocky shore whereas a total of 6 species occurred exclusively at the mangrove environment (Table S5).

Table 3. Differences in univariate indices from macro- and meiofauna between habitat/environment (H2) using one-way ANOVA. Analysis of macro- and meiofauna densities were done on log (x + 1) data.

	df	*F*	*p*
Meiofauna density	1	0.751	0.419
Nematode density	1	0.400	0.550
Number of meiofaunal taxa	1	0.297	0.606
Meiofauna diversity (H′)	1	0.128	0.732
Number of nematode species	1	3.261	0.121
Nematode diversity (H′)	1	11.97	**0.013**
Macrofauna density	1	0.235	0.645
Number of macrofaunal taxa	1	25.00	**0.002**
Macrofauna diversity (H′)	1	53.50	**<0.001**

Table 4. Differences in multivariate structure of macrofauna, meiofauna and nematode assemblages from the different environments using one-way PERMANOVA. Data were log (x + 1) transformed. Bold values indicate $p < 0.05$.

Source	*df*	Pseudo-*F*	*p* (perm)
		Macrofauna	
Treatment	1	5.2506	**0.026**
Residual	6		
		Meiofauna	
Treatment	1	2.3293	0.125
Residual	6		
		Nematodes	
Treatment	1	3.7215	**0.032**
Residual	6		

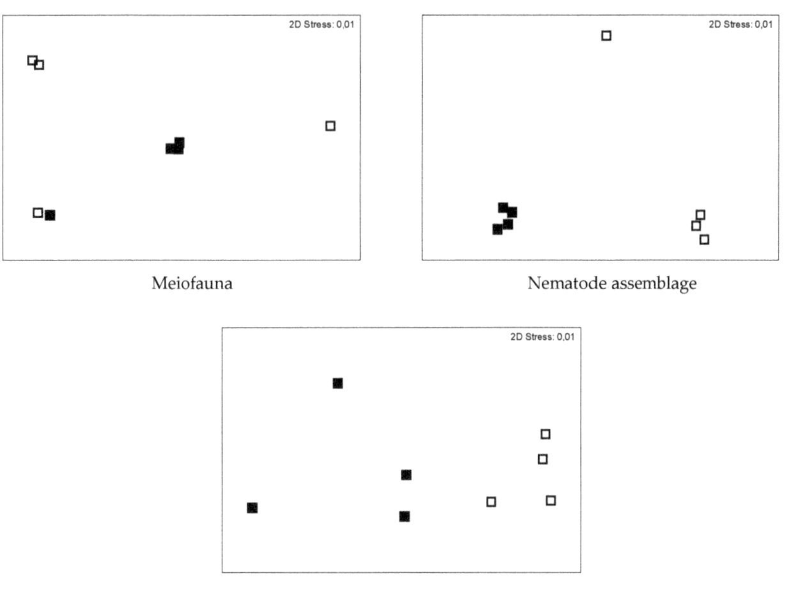

Figure 3. MDS ordination representing the similarity between samples from *Bostrychia* sp. from the rocky shore (■) and mangrove roots (□). Data were log (x + 1) transformed.

3.2.2. Macrofauna Community

Macrofauna associated with *Bostrychia* sp. was represented by 13 taxa, from which 9 were exclusive to the rocky shore and only 4, which were also present in the rocky shore, occurred in mangrove samples (Table S9). In both environments, Chironomidae and Tanaidacea were the most abundant groups. However, whereas in the rocky shore they accounted for 49.5% of the total macrofauna, in the mangrove the two groups represented 98% of all organisms collected.

Total macrofauna densities did not differ between habitats (Table 3) ranging from 5 to 7 and from 2.75 to 9.5 ind. mL^{-1} in the rocky-shore and mangrove samples, respectively. The average number of taxa per sample was significantly lower in the mangrove (2–4 taxonomic groups) than in the rocky-shores communities (5–9 taxonomic groups) (Table 3). Accordingly, diversity was also higher in the rocky shore (Table 3). The difference in species richness and composition resulted in distinct macrofauna communities in the two habitats (Table 4, Figure 3). This difference was mainly due to the absence of Cirripedia, Collembola and Bivalvia and the higher abundances of Chironomidae and Tanaidacea in mangrove samples (SIMPER analyses, Table S10).

4. Discussion

Habitat complexity mediates species coexistence increasing diversity in several communities [29–35]. Here, complexity generated by the biogenic structures and by the surrounding environment creates heterogeneous habitats that support a diverse associated fauna. However, the scale in which environmental changes affected the fauna depended, as expected, on the benthic component analyzed. While the small-sized meiofauna were affected by heterogeneity at both scales, the larger and more mobile macrofauna was unaffected by heterogeneity within habitat but responded to larger-scale differences in the structure of the studied habitats (rocky shore and mangroves).

The influence of biogenic small-scale habitat heterogeneity for meiofauna community structure is well documented for soft sediment environments [36–39]. Evidences for communities from hard-bottom substrates however are limited but corroborate the importance of the complexity for diversity. Meiofauna species from phytal communities are known to be host-specific [40–43] and the increase in algae morphotypes results in a diversified meiofaunal community. Biological substrates can influence the composition and structure of the associated fauna by its architecture [8], the amount of sediment accumulated on it [12,44], the protection it offers against predators [13,45], shelter from extreme physical condition [46,47] and the provision of food resources [10]. In the current study, sponge communities were differentiated by the occurrence of meiofaunal oligochaetes, which are ectocommensals of sponges [48,49]. Oligochaetes inhabiting such filter-feeding organisms benefit from being in an environment with intense water flowing, ensuring high oxygen availability and food (microorganisms and organic detritus), and a lower exposure to the variable conditions of free sediments [48]. In contrast, nematode assemblages associated with sponges did not differ from the other substrates but showed a high turnover of species between "Turf" and "Algae". The species that contributed to the differences occurred in very low densities hampering any conclusion of host association. However, the "Turf" environment creates an intricate mesh of filaments that accumulates much more sediment when compared to the "Algae" environment, probably influencing species abundances and composition. Given the different macroalgal morphologies, it is also possible that differences in irradiance and temperature between the two substrates, especially at daytime and low tide [47], might have played a role. Although the mechanism is not clear, the habitat heterogeneity created by the substrates had an important role in favoring the occurrence of rare species therefore maintaining the regional diversity.

Contrary to the meiofauna and nematode assemblages, differences in substrata identity did not affect macrofauna community therefore agreeing with our expectation, i.e., the meiofauna was more influenced by the smaller-scale heterogeneity investigated in this study than the macrofauna. These results suggest that the larger and relatively more motile macrofauna can possibly move through all studied substrates within habitats and the small-scale variations of environmental conditions offered by the different substrates did not represent contrasting selective pressures for this group. Similar observations were reported for macrofauna inhabiting different type and morphology of sponges [50]. It is possible though that any eventual small-scale variation among biological substrata within the studied zone of the rocky shore is species-specific, so our analyses conducted at lower taxonomic resolution may have prevented us to observe species-specific differences among communities. However, in a region close to our study site, the structure of amphipod assemblages analyzed to the specific level was not affected by algae identity and architecture, but only by its position through the intertidal zone [51]. Even hosts with very distinct architecture, when occurring in the same position of the shore, supported similar associated communities [51]. These findings suggest that factors controlling community assembly of macrofaunal organisms may operate on a wider scale.

Accordingly, the primary substrate and/or the surrounding environment where this occurred influenced both the nematode assemblages and macrofauna. Fauna associated to the same substratum, but under different environmental conditions, are often distinct [52] because hosts do not necessary buffer the distinct selective pressures to which associated fauna is exposed. Mangrove samples were characterized by higher abundances of *Thalassomonhystera* sp.1, which is a genus frequently found in mangrove sediments and mangrove litter (e.g., [53,54], probably as a consequence of both the higher proximity to the sediment and the higher sedimentation in the *Bostrichetum* communities from mangrove roots. In addition, there was a significant overlap in species composition between the two habitats and the number of shared species was relatively low (32%) suggesting that the surrounding environment was highly selective. For the macrofauna, the longer exposure to air to which associated fauna is exposed in mangroves restricted the occurrence of marine groups as Cirripedia, Collembola and Bivalvia reducing diversity, but facilitated the occurrence of typically terrestrial organisms such as Chironomidae larvae. While we did not address the mechanism behind the absence of typically marine species in mangrove habitats, the longer exposure to air can restrict the access of reproductive propagules, but also increase post-recruitment mortality, which probably caused the differences observed here. Differences in taxonomic composition of the associate fauna from the same secondary substrate in different habitats were also observed for fauna associated with mussel beds growing on soft vs. hard substrates [52]. While it has been suggested that changes in wave exposure and the proportion of faeces and pseudo-faeces of mussels deposited within the mussels bed would explain such differences [52]; here we suggest that changes in environmental conditions as distance to the soft-bottom, the amount of sediment deposited over the secondary substrate and aerial exposure would influence the differences in the associated fauna between macroalgae on rocky shores vs. mangrove environments.

5. Conclusions

Changes in species composition between environments and substrates highlight the importance of habitat diversity at different scales in maintaining the diversity of the associated fauna. In addition, whereas both substrate identity and the surrounding environment were important in structuring the smaller-sized meiofauna, the macrofauna was more influenced by larger-scale changes in the surrounding ecosystem. This implies that the larger and more motile macrofauna explores the environment in a larger spatial scale compared to the meiofauna, which are smaller and more sedentary (therefore restricted to a smaller space), suggesting that effects of spatial heterogeneity on communities are dependent on organism size and mobility. Besides, species turnover occurred mainly regarding species with low abundances (i.e., rare species). These results stress the importance of considering the rarer groups in ecological studies and/or impact assessments on local and regional diversity. It is important to note, however, that because our study is limited to a single sampling location, the generalizations of our findings must be further tested in other habitat types and regions. Also, because of the frequency of occurrence and abundance of rare species might change as a function of sampling effort, different rarity patterns could emerge if more samples were taken.

Considering the eminent expansion of the harbor over rocky shores and mangroves from the Araçá Bay and the effects of artificial substrate construction and shading, leading to large modifications on biological substrates on rocky shores [24,55], the anthropogenic impact in the area has the potential to promote large changes to both meiofauna and macrofauna biodiversity in the area. Because both macro- and meiofauna biodiversity are positively related to important ecosystem processes [56,57], such effects can potentially go beyond structural changes in biodiversity and are likely to translate into major impacts to ecosystem functioning.

Supplementary Materials: The following are available online at http://www.mdpi.com/1424-2818/12/1/39/s1, Table S1: Mean density (inds.ml^{-1}) and standard deviation (in parenthesis) of meiofauna higher taxa in the different biological substrates sampled in a rocky-shore from Araçá Bay, Southeast Brazil; Table S2: Mean density (inds. mL^{-1}) and standard deviation (in parenthesis) of nematode morphospecies in the different biological substrates sampled in a rocky shore from Araçá Bay, Southeast Brazil; Table S3: Mean density (inds. mL^{-1}) and standard deviation (in parenthesis) of macrofauna higher taxa in the different biological substrates sampled in a rocky shore from Araçá Bay, Southeast Brazil; Table S4: Mean density (inds. mL^{-1}) and standard deviation (in parenthesis) of meiofauna higher taxa associated with *Bostrychia* sp. sampled from the rocky shore habitats and mangrove roots at Araçá Bay, Southeast Brazil; Table S5: Mean density (inds. mL^{-1}) and standard deviation (in parenthesis) of nematode species associated with *Bostrychia* sp. sampled from the rocky shore habitats and mangrove roots at Araçá Bay, Southeast Brazil; Table S6: Mean density (inds. mL^{-1}) and standard deviation (in parenthesis) of macrofauna taxa associated with *Bostrychia* sp. sampled from the rocky shore habitats and mangrove roots at Araçá Bay, Southeast Brazil; Table S7: SIMPER analysis showing nematode species ranked according to average Sorensen dissimilarity from presence/absence data between the biological substrates 'Caulerpa' and 'Turf'; Table S8: SIMPER analysis showing nematode species ranked according to average Bray–Curtis dissimilarity from log (x + 1) data between the rocky shore habitats and mangrove roots; Table S9: SIMPER analysis showing nematode species ranked according to average Bray–Curtis dissimilarity from log (x + 1) data between the rocky shore habitats and mangrove roots; Table S10: SIMPER analysis showing macrofauna groups ranked according to average Bray–Curtis dissimilarity from log (x + 1) data between the rocky shore habitats and mangrove roots.

Author Contributions: All authors have equally contributed to the conceptualization of the idea and the methods used. Data acquisition (sampling and samples processing) were coordinated by G.M.D., R.A.C., G.F. and F.G. had conducted statistical analysis and data interpretation was done by all authors. F.G. prepared the original draft with considerable contributions from G.M.D. All authors have substantially contributed to the writing of the final text. All authors have read and agreed to the published version of the manuscript.

Funding: This research was supported by research funds granted by São Paulo Research Foundation (FAPESP) to Biota Araçá Project (#2011/50317-5). We thank Felipe Dutra and Viviane Gomes for support with field and laboratory work. We are very grateful to Antônia Cecília Zacagnini Amaral, who acquired funding and wisely administrated the Biota Araçá Project, for this opportunity.

Conflicts of Interest: The authors declare no conflicts of interest. The sponsors had no role in the design, execution, interpretation, or writing of the study.

References

1. Hewitt, J.E.; Thrush, S.F.; Dayton, P.K.; Bonsdorff, E. The effect of spatial and temporal heterogeneity on the design and analysis of empirical studies of scale-dependent systems. *Am. Nat.* **2007**, *169*, 398–408. [CrossRef] [PubMed]

2. Moens, T.; Braeckman, U.; Derycke, S.; Fonseca, G.; Gallucci, F.; Gingold, R.; Guilini, K.; Ingels, J.; Leduc, D.; Vanaverbeke, J.; et al. Ecology of free-living marine nematodes. In *Nematoda*; DE GRUYTER: Berlin, Germany, 2013.

3. Downes, B.J.; Lake, P.S.; Schreiber, E.S.G.; Glaister, A. Habitat Structure and Regulation of Local Species Diversity in a Stony, Upland Stream. *Ecol. Monogr.* **1998**, *68*, 237. [CrossRef]

4. Chesson, P.; Rosenzweig, M. Behavior, Heterogeneity, and the Dynamics of Interacting Species. *Ecology* **1991**, *72*, 1187–1195. [CrossRef]

5. Christie, H.; Norderhaug, K.; Fredriksen, S. Macrophytes as habitat for fauna. *Mar. Ecol. Prog. Ser.* **2009**, *396*, 221–233. [CrossRef]

6. Kelaher, B. Influence of physical characteristics of coralline turf on associated macrofaunal assemblages. *Mar. Ecol. Prog. Ser.* **2002**, *232*, 141–148. [CrossRef]

7. Moreno, T.R.; da Rocha, R.M. Associated Fauna with Eudistoma Carolinense (Tunicata, Ascidiacea) Compared with other Biological Substrates with Different Architectures. *J. Coast. Res.* **2006**, *39*, 1695–1699.

8. Chemello, R.; Milazzo, M. Effect of algal architecture on associated fauna: Some evidence from phytal molluscs. *Mar. Biol.* **2002**, *140*, 981–990.

9. Coull, B.C.; Wells, J.B.J. Refuges from Fish Predation: Experiments with Phytal Meiofauna from the New Zealand Rocky Intertidal. *Ecology* **1983**, *64*, 1599–1609. [CrossRef]

10. Duffy, J.E.; Hay, M.E. Strong Impacts of Grazing Amphipods on the Organization of a Benthic Community. *Ecol. Monogr.* **2000**, *70*, 237. [CrossRef]

11. Gibbons, M.J. Impact of predation by juvenile Clinus superciliosus on phytal meiofauna: Are fish important as predators? *Mar. Ecol. Prog. Ser.* **1988**, *45*, 13–22. [CrossRef]

12. Hicks, G.R.F. Structure of phytal harpacticoid copepod assemblages and the influence of habitat complexity and turbidity. *J. Exp. Mar. Biol. Ecol.* **1980**, *44*, 157–192. [CrossRef]

13. Martin-Smith, K.M. Abundance of mobile epifauna: The role of habitat complexity and predation by fishes. *J. Exp. Mar. Biol. Ecol.* **1993**, *174*, 243–260. [CrossRef]

14. Thiébaut, E.; Cabioch, L.; Dauvin, J.-C.; Retiere, C.; Gentil, F. Spatio-Temporal Persistence of the Abra Alba-Pectinaria Koreni Muddy-Fine Sand Community of the Eastern Bay of Seine. *J. Mar. Biol. Assoc. UK* **1997**, *77*, 1165–1185. [CrossRef]

15. Van Hoey, G.; Vincx, M.; Degraer, S. Temporal variability in the Abra alba community determined by global and local events. *J. Sea Res.* **2007**, *58*, 144–155. [CrossRef]

16. Schratzberger, M.; Maxwell, T.A.D.; Warr, K.; Ellis, J.R.; Rogers, S.I. Spatial variability of infaunal nematode and polychaete assemblages in two muddy subtidal habitats. *Mar. Biol.* **2008**, *153*, 621–642. [CrossRef]

17. Lebreton, B.; Richard, P.; Galois, R.; Radenac, G.; Brahmia, A.; Colli, G.; Grouazel, M.; André, C.; Guillou, G.; Blanchard, G.F. Food sources used by sediment meiofauna in an intertidal *Zostera noltii* seagrass bed: A seasonal stable isotope study. *Mar. Biol.* **2012**, *159*, 1537–1550. [CrossRef]

18. Moens, T.; Verbeeck, L.; de Maeyer, A.; Swings, J.; Vincx, M. Selective attraction of marine bacterivorous nematodes to their bacterial food. *Mar. Ecol. Prog. Ser.* **1999**, *176*, 165–178. [CrossRef]

19. Gallucci, F.; Steyaert, M.; Moens, T. Can field distributions of marine predacious nematodes be explained by sediment constraints on their foraging success? *Mar. Ecol. Prog. Ser.* **2005**, *304*, 167–178. [CrossRef]

20. Vafeiadou, A.-M.; Materatski, P.; Adão, H.; de Troch, M.; Moens, T. Resource utilization and trophic position of nematodes and harpacticoid copepods in and adjacent to *Zostera noltii* beds. *Biogeosciences* **2014**, *11*, 4001–4014. [CrossRef]

21. Species Size Distributions in Marine Benthic Communities on JSTOR. Available online: https://www.jstor.org/stable/4217200?seq=1#metadata_info_tab_contents (accessed on 29 November 2019).

22. Eckman, J.E. Hydrodynamic processes affecting benthic recruitment1. *Limnol. Oceanogr.* **1983**, *28*, 241–257. [CrossRef]

23. Amaral, A.C.Z.; Ciotti, A.M.; Fonseca, G. Biodiversity and functioning of a subtropical coastal ecosystem: Subsidies for integrated management. *Ocean Coast. Manag.* **2018**, *164*, 1–3. [CrossRef]

24. Pardal-Souza, A.L.; Dias, G.M.; Jenkins, S.R.; Ciotti, Á.M.; Christofoletti, R.A. Shading impacts by coastal infrastructure on biological communities from subtropical rocky shores. *J. Appl. Ecol.* **2017**, *54*, 826–835. [CrossRef]

25. Vieira, E.A.; Filgueiras, H.R.; Bueno, M.; Leite, F.P.P.; Dias, G.M. Algas co-ocorrentes e morfologicamente distintas suportam uma diversa fauna associada na zona entremarés na baía do araçá, Brasil. *Biota Neotrop.* **2018**, *18*. [CrossRef]

26. Danovaro, R.; Fraschetti, S. Meiofaunal vertical zonation on hard-bottoms: Comparison with soft-bottom meiofauna. *Mar. Ecol. Prog. Ser.* **2002**, *230*, 159–169. [CrossRef]

27. Anderson, M.J. A new method for non-parametric multivariate analysis of variance. *Austral Ecol.* **2001**, *26*, 32–46.

28. Clarke, K.R. Non-parametric multivariate analyses of changes in community structure. *Austral Ecol.* **1993**, *18*, 117–143. [CrossRef]

29. Taniguchi, H.; Tokeshi, M. Effects of habitat complexity on benthic assemblages in a variable environment. *Freshw. Biol.* **2004**, *49*, 1164–1178. [CrossRef]

30. Gratwicke, B.; Speight, M.R. The relationship between fish species richness, abundance and habitat complexity in a range of shallow tropical marine habitats. *J. Fish Biol.* **2005**, *66*, 650–667. [CrossRef]

31. Sueiro, M.C.; Bortolus, A.; Schwindt, E. Habitat complexity and community composition: Relationships between different ecosystem engineers and the associated macroinvertebrate assemblages. *Helgol. Mar. Res.* **2011**, *65*, 467–477. [CrossRef]

32. Willis, S.C.; Winemiller, K.O.; Lopez-Fernandez, H. Habitat structural complexity and morphological diversity of fish assemblages in a Neotropical floodplain river. *Oecologia* **2005**, *142*, 284–295. [CrossRef]

33. St. Pierre, J.I.; Kovalenko, K.E. Effect of habitat complexity attributes on species richness. *Ecosphere* **2014**, *5*, art22. [CrossRef]

34. Hauser, A.; Attrill, M.J.; Cotton, P.A. Effects of habitat complexity on the diversity and abundance of macrofauna colonising artificial kelp holdfasts. *Mar. Ecol. Prog. Ser.* **2006**, *325*, 93–100. [CrossRef]

35. Öster, M.; Cousins, S.A.O.; Eriksson, O. Size and heterogeneity rather than landscape context determine plant species richness in semi-natural grasslands. *J. Veg. Sci.* **2007**, *18*, 859–868. [CrossRef]

36. Passarelli, C.; Olivier, F.; Paterson, D.; Hubas, C. Impacts of biogenic structures on benthic assemblages: Microbes, meiofauna, macrofauna and related ecosystem functions. *Mar. Ecol. Prog. Ser.* **2012**, *465*, 85–97. [CrossRef]

37. Reise, K. High abundance of small zoobenthos around biogenic structures in tidal sediments of the Wadden Sea. *Helgoländer Meeresunters.* **1981**, *34*, 413–425. [CrossRef]

38. Hasemann, C.; Soltwedel, T. Small-Scale Heterogeneity in Deep-Sea Nematode Communities around Biogenic Structures. *PLoS ONE* **2011**, *6*, e29152. [CrossRef]

39. Bell, S.S.; Watzin, M.C.; Coull, B.C. Biogenic structure and its effect on the spatial heterogeneity of meiofauna in a salt marsh. *J. Exp. Mar. Biol. Ecol.* **1978**, *35*, 99–107. [CrossRef]

40. Veiga, P.; Sousa-Pinto, I.; Rubal, M. Meiofaunal assemblages associated with native and non-indigenous macroalgae. *Cont. Shelf Res.* **2016**, *123*, 1–8. [CrossRef]

41. Trotter, D.B.; Webster, J.M. Feeding preferences and seasonality of free-living marine nematodes inhabiting the kelp Macrocystis integrifolia. *Mar. Ecol. Prog. Ser.* **1984**, *14*, 151–157. [CrossRef]

42. Frame, K.; Hunt, G.; Roy, K. Intertidal meiofaunal biodiversity with respect to different algal habitats: A test using phytal ostracodes from Southern California. *Hydrobiologia* **2007**, *586*, 331–342. [CrossRef]

43. Hicks, G.R.F. Observations on substrate preference of marine phytal harpacticoids (copepoda). *Hydrobiologia* **1977**, *56*, 7–9. [CrossRef]

44. Gibbons, M.J. The impact of sediment accumulations, relative habitat complexity and elevation on rocky shore meiofauna. *J. Exp. Mar. Biol. Ecol.* **1988**, *122*, 225–241. [CrossRef]

45. Russo, A.R. Role of habitat complexity in mediating predation by the gray damselfish Abudefduf sordidus on epiphytal amphipods. *Mar. Ecol. Prog. Ser.* **1987**, *36*, 101–105. [CrossRef]

46. Watt, C.A.; Scrosati, R.A. Bioengineer effects on understory species richness, diversity, and composition change along an environmental stress gradient: Experimental and mensurative evidence. *Estuar. Coast. Shelf Sci.* **2013**, *123*, 10–18. [CrossRef]

47. Umanzor, S.; Ladah, L.; Calderon-Aguilera, L.E.; Zertuche-González, J.A. Intertidal macroalgae influence macroinvertebrate distribution across stress scenarios. *Mar. Ecol. Prog. Ser.* **2017**, *584*, 67–77. [CrossRef]

48. De Ronde, C.E.J.; Stoffers, P.; Garbe-Schönberg, D.; Christenson, B.W.; Jones, B.; Manconi, R.; Browne, P.R.L.; Hissmann, K.; Botz, R.; Davy, B.W.; et al. Discovery of active hydrothermal venting in Lake Taupo, New Zealand. *J. Volcanol. Geotherm. Res.* **2002**, *115*, 257–275. [CrossRef]

49. Rota, E.; Manconi, R. Taxonomy and Ecology of Sponge-Associate *Marionina* spp. (Clitellata: Enchytraeidae) from the Horomatangi Geothermal System of Lake Taupo, New Zealand. *Int. Rev. Hydrobiol.* **2004**, *89*, 58–67. [CrossRef]

50. Ávila, E.; Ortega-Bastida, A.L. Influence of habitat and host morphology on macrofaunal assemblages associated with the sponge *Halichondria melanadocia* in an estuarine system of the southern Gulf of Mexico. *Mar. Ecol.* **2015**, *36*, 1345–1353. [CrossRef]

51. Bueno, M.; Dias, G.M.; Leite, F.P.P. The importance of shore height and host identity for amphipod assemblages. *Mar. Biol. Res.* **2017**, *13*, 870–877. [CrossRef]

52. Thiel, M.; Ullrich, N. Hard rock versus soft bottom: The fauna associated with intertidal mussel beds on hard bottoms along the coast of Chile, and considerations on the functional role of mussel beds. *Helgol. Mar. Res.* **2002**, *56*, 21–30. [CrossRef]

53. Gagarin, V.G.; Thanh, N.V. Four new species of monhysterids (Nematoda: Monhysterida) from mangroves of the Mekong river estuaries of Vietnam. *TAP CHI SINH HOC* **2014**, *30*, 16–25. [CrossRef]

54. Brustolin, M.C.; Nagelkerken, I.; Fonseca, G. Large-scale distribution patterns of mangrove nematodes: A global meta-analysis. *Ecol. Evol.* **2018**, *8*, 4734–4742. [CrossRef] [PubMed]
55. Malerba, M.E.; White, C.R.; Marshall, D.J. The outsized trophic footprint of marine urbanization. *Front. Ecol. Environ.* **2019**, *17*, 400–406. [CrossRef]
56. Schratzberger, M.; Ingels, J. Meiofauna matters: The roles of meiofauna in benthic ecosystems. *J. Exp. Mar. Biol. Ecol.* **2018**, *502*, 12–25. [CrossRef]
57. Belley, R.; Snelgrove, P.V.R. Relative contributions of biodiversity and environment to benthic ecosystem functioning. *Front. Mar. Sci.* **2016**, *3*, 242. [CrossRef]

Article

Sedimentary Organic Matter, Prokaryotes, and Meiofauna across a River-Lagoon-Sea Gradient

Silvia Bianchelli [1,*]**, Daniele Nizzoli** [2]**, Marco Bartoli** [2]**, Pierluigi Viaroli** [2]**, Eugenio Rastelli** [3] **and Antonio Pusceddu** [4]

1   Dipartimento di Scienze della Vita e dell'Ambiente, Università Politecnica delle Marche, Via Brecce Bianche, 60131 Ancona, Italy
2   Dipartimento di Scienze Chimiche, della Vita e della Sostenibilità Ambientale, Università degli studi di Parma, Parco Area delle Scienze 11/A, 43124 Parma, Italy; daniele.nizzoli@unipr.it (D.N.); marco.bartoli@unipr.it (M.B.); pierluigi.viaroli@unipr.it (P.V.)
3   Stazione Zoologica Anton Dorhn, Villa Comunale, 80121 Napoli, Italy; eugenio.rastelli@szn.it
4   Dipartimento di Scienze della Vita e dell'Ambiente, Università degli Studi di Cagliari, Via Fiorelli 1, 09126 Cagliari, Italy; apusceddu@unica.it
*   Correspondence: silvia.bianchelli@univpm.it; Tel.: +39-0712204335

Received: 16 April 2020; Accepted: 6 May 2020; Published: 12 May 2020

Abstract: In benthic ecosystems, organic matter (OM), prokaryotes, and meiofauna represent a functional bottleneck in the energy transfer towards higher trophic levels and all respond to a variety of natural and anthropogenic disturbances. The relationships between OM and the different components of benthic communities are influenced by multiple environmental variables, which can vary across different habitats. However, analyses of these relationships have mostly been conducted by considering the different habitats separately, even though freshwater, transitional, and marine ecosystems, physically linked to each other, are not worlds apart. Here, we investigated the quantity and nutritional quality of sedimentary OM, along with the prokaryotic and meiofauna abundance, biomass, and biodiversity, in two sampling periods, corresponding to high vs. low freshwater inputs to the sea, along a river-to-sea transect. The highest values of sedimentary organic loads and their nutritional quality, prokaryotic and meiofaunal abundance, and biomass were consistently observed in lagoon systems. Differences in the prokaryotic Operational Taxonomic Units (OTUs) and meiofaunal taxonomic composition, rather than changes in the richness of taxa, were observed among the three habitats and, in each habitat, between sampling periods. Such differences were driven by either physical or trophic variables, though with differences between seasons. Overall, our results indicate that the apparent positive relationship between sedimentary OM, prokaryote and meiofaunal abundance, and biomass across the river-lagoon-sea transect under scrutiny is more the result of a pattern of specifically adapted prokaryotic and meiofaunal communities to different habitats, rather than an actually positive 'response' to OM enrichment. We conclude that the synoptic analysis of prokaryotes and meiofauna can provide useful information on the relative effect of organic enrichment and environmental settings across gradients of environmental continuums, including rivers, lagoons, and marine coastal ecosystems.

Keywords: North Adriatic Sea; trophic status; prokaryotes; meiofauna; ecosystem functioning

---

1. Introduction

Pathways and rates of sedimentary organic matter (OM) transfer to higher trophic levels in aquatic ecosystems depend on the OM quantity and nutritional quality [1,2]. In turn, both the quantity and nutritional quality of sedimentary OM depend on its origin (i.e., autotrophic, heterotrophic, and/or detrital), biochemical composition and bioavailability (i.e., refractory vs. labile fraction), and degradation rates [3]. Therefore, the benthic trophic status of an aquatic ecosystem is not only related to the availability of inorganic nutrients, which fuels in situ primary production, but also depends upon the supply rates of OM, including allochthonous and detrital (i.e., not living) sources [3–6]. This, in turn, can influence other ecosystem functions, including nutrient cycling and oxygen availability [3,7]. For instance, the accumulation of huge amounts of detrital OM in marine coastal sediments, triggering increased benthic O_2 consumption and possibly inducing hypoxic and anoxic conditions, can be associated with a preferential accumulation of semi-labile compounds (e.g., the biopolymeric fraction of organic carbon (OC) [8]), particularly enriched in nitrogenous (protein-like) compounds [9]. In such conditions, the decoupling between the production/inputs of OM loads, heterotrophic consumption, and accumulation in sediments can determine strong modifications in the structure and functioning of benthic ecosystems [9–13].

The comprehension of the biogeochemical dynamics in aquatic environments characterized by variable biodiversity levels (e.g., transitional, estuaries, and coastal environments) is strongly limited by the complex and multiple interactions among different biotic components, including microbial and meio- and macrofaunal assemblages [14,15]. In this regard, it is noticeable that changes in the benthic trophic status and OM degradation rates mediated by microbes, through the so-called microbial loop, can be mirrored in changes in the composition and structure of the benthic communities, and vice-versa [16,17].

In benthic ecosystems, OM, prokaryotes, and meiofauna, being trophic resources for higher trophic levels and, at the same time, being responsible, with different roles, for OM cycling, represent a key functional bottleneck in the energy transfer towards a higher trophic level [18]. They also detectably and rapidly respond to a variety of natural and anthropogenic disturbances. Heterotrophic prokaryotes are responsible for detrital OM degradation and transformation and rapidly respond to variations in the quantity and composition of the available OM [19,20]. Meiofauna, due to their strong sensitivity to disturbances, high abundance, lack of pelagic larval dispersion, and short life cycles, rapidly respond to environmental changes in both marine [21,22] and freshwater ecosystems [23].

The relationships between OM and the different components of benthic communities are also influenced by multiple, often interacting, environmental variables, for example, currents and the substrate composition, which, in turn, enhance the levels and variance of natural disturbance, as well as habitat-specific conditions. Analyses of these relationships have mostly been conducted by considering the different benthic components separately. Moreover, freshwater, transitional, and marine coastal ecosystems, though physically linked to each other, have most often been considered as worlds apart, and such a reductive approach especially applies to the analysis of benthic ecosystems.

Here, to provide insights on this topic, we test the null hypothesis that OM quantity, biochemical composition and degradation rates, prokaryotic and meiofaunal biodiversity, and ecosystem functioning do not vary among different ecosystems along a strong salinity gradient in different periods of the year. To test this hypothesis, we investigated the OM quantity, nutritional quality, and degradation rates, along with the prokaryotic and meiofauna biodiversity, in two sampling periods (corresponding to high vs. low freshwater inputs to the sea) along a river-to-sea gradient, comprising Po River (Italy), the North Adriatic Sea, and the associated lagoonal system. The sampling strategy included stations located in the major tributaries of the Po River, in the Po main axis, in a coastal lagoon (Sacca di Goro Lagoon) intercepting Po River outflow, and in the coastal sediments of the North Adriatic Sea facing the lagoon and the Po River delta.

2. Materials and Methods

2.1. Study Area

The Po River is the most important Italian river, with a drainage basin of 71,000 km^2, 44% of which is devoted to agricultural activities, and more than 15 million people. The mean flow discharge is characterized by two major flooding periods, due to snowmelt in the spring and rainfall in the autumn [24]. The Po River outflow of water and sediments is mostly constrained along the western coast of the Adriatic Sea, driven southward by the general circulation of the basin [25,26].

The Sacca di Goro Lagoon is a shallow (average depth 1.5 m) water embayment (27 km^2) of the Po River Delta facing the northern Adriatic Sea from the Italian counterpart (Figure 1). This lagoon is characterized by strong daily variations of salinity and nutrient concentrations due to a microtidal regime (with a maximum amplitude < 1.0 m) and freshwater inputs from the Po River, and saline water input from the adjacent northern Adriatic Sea. The Sacca di Goro Lagoon, being one of the most economically relevant clam farming sites in Europe, whilst at the same time being threatened by dystrophic events [27–29], has largely been investigated in terms of the biogeochemistry [15,30,31], ecophysiology of blooming macroalgae [32], meio- and macrofauna communities [33–35], and ecosystem functions [36,37].

Figure 1. Sampling area: (**A**) the Po River basin, flowing into the North Adriatic sea, Central Mediterranean and (**B**) location of the sampling stations along the Po River basin, the Goro Lagoon, and at sea.

Due to the large inputs from the Po river (which alone accounts for ca. 50% of the terrigenous flux into the whole basin), the sediments of the north western Adriatic are characterized by a strong accumulation of organic loads [9,38–42], which have, for years, triggered hypoxic crises [43–45].

Indeed, the NW Adriatic Sea, which is the most productive basin of the entire Mediterranean Sea, has experienced huge and long-lasting anthropogenic environmental alterations in the last 50 years. Such changes have led to severe consequences for the whole ecosystem functioning, which have ultimately manifested as red tides, mucilage formation, and strong eutrophication along the entire Italian coastline of the basin [46,47].

2.2. Sampling

The sampling stations were located as follows: "Oglio" in the Oglio River (left tributary of the Po River), "Po" (within the Po River, just before the beginning of the Po-di-Goro River, which is part of the Po delta and flows into the Sacca di Goro Lagoon), "Gorino" (at the end of the Po-di-Goro river), "Giralda" (inside the Sacca di Goro Lagoon), "Delta" (located at the mouth of the Sacca di Goro Lagoon), and "Cesenatico" (located south of the lagoon, in the marine coastal environment) (Figure 1). The sampling was carried out in two different seasons: summer (September 2011) and winter (February 2012), identified as representative of periods of low (late summer) vs. high (winter) river outflow, respectively.

The bottom temperature and salinity were measured in situ by means of a multiparametric probe. Sediment samples were collected (three independent replicates per station and sampling period, for both OM and meiofauna) using plexiglass corers operated manually, kept at in situ temperature after being brought to the laboratory (within 4 h), and then immediately frozen once in the laboratory and kept at $-20\,^\circ$C until analysis (within 2 weeks). Only sediment aliquots for the measurement of OM degradation rates were immediately treated, as described below.

2.3. Sedimentary Organic Matter Quantity, Nutritional Quality, and Degradation Rates

Once in the laboratory, the top centimeter from each sediment core was used for analyses of the OM biochemical composition, in terms of the total phytopigment, protein, carbohydrate, and lipid contents. Chlorophyll-a and phaeopigments were analysed fluorometrically [48]. Total phytopigment concentrations were defined as the sum of chlorophyll-a and phaeopigment concentrations and utilized as an estimate of the organic material of algal origin [12]. Sediment phytopigment concentrations were converted into C equivalents using 40 µg C µg phytopigment^{-1} as a conversion factor [3]. Protein, carbohydrate, and lipid analyses were carried out spectrophotometrically [49]. For the analysis of each biochemical class of organic compound, blanks were made with the same sediment samples previously treated in a muffle furnace (450 $^\circ$C, 2 h). Protein, carbohydrate, and lipid concentrations were converted into C equivalents using the conversion factors 0.49, 0.40, and 0.75 mg C mg^{-1}, respectively, and their sum is referred to as the biopolymeric C (BPC) [50].

The fraction of biopolymeric C represented by relatively fresh algal material was assessed as the percentage contribution of phytopigment C to biopolymeric C contents and referred to as the algal fraction of biopolymeric C [3]. The algal and protein fractions of biopolymeric C and the values of the protein to carbohydrate ratio were also used as descriptors of OM nutritional quality (algal and protein fractions) and ageing (protein to carbohydrate ratio) [50,51].

OM degradation rates were estimated from aminopeptidase and beta-glucosidase activities determined by the cleavage of fluorogenic substrates (L-leucine-4-methylcoumarinyl- 7-amide, Leu-; 4-methylumbelliferone-β-D-glucopyranoside, respectively) at saturating concentrations. Briefly, 2.5 mL of sediment subsamples was incubated at in situ temperature in the dark for 2 h with 2.5 mL of filtered, sterile water containing 200 µM L-leucine-4-methylcumarinyl-7-amide and 50 µM 4-methylumbelliferyl β-D-glucopyranoside, respectively, separately for aminopeptidase and β-glucosidase determinations. After incubation, the sediment slurries were centrifuged and the supernatants were analysed fluorometrically [49]. Protease and glucosidase activities (µmol of substrate g^{-1} h^{-1}) were converted into C equivalents using 72 as a conversion factor [13], and their sum is reported as the C degradation rate (µgC g^{-1} h^{-1}).

2.4. Prokaryotic Abundance, Biomass, and Diversity

The prokaryotic abundance and biomass were determined as described by Danovaro (2009) [49]. Briefly, prokaryotic cells were extracted from the sediments according to standard procedures, stained with SYBR Green I, and counted by epifluorescence microscopy. For determination of the prokaryotic biomass, the cell biovolume was converted into the carbon content assuming 310 fg C μm^{-3} as a conversion factor [49]. The prokaryotic abundance and biomass were normalized to the sediment dry weight after desiccation (60 °C, 24 h).

The prokaryotic diversity was assessed according to Danovaro (2009) [49]. DNA was extracted from sediment with the UltraClean soil DNA isolation kit (MoBio Laboratories Inc., Carlsbad, CA USA). In all samples, extracted DNA was determined spectrofluorimetrically using SYBR Green I (Molecular Probes) and quantified vs. standard solutions of genomic DNA from *Escherichia coli*. The extracted DNA was amplified using universal bacterial primers 16S-1392F and 23S-125R, and the latter was fluorescently labeled with the fluorochrome HEX (MWGspa Biotech). PCRs were performed in 50-μL volumes in a thermal cycler (Biometra, Germany) using 30 PCR cycles. PCR products were checked on agarose–Tris-borate-EDTA (TBE) gels (1%). Four different reactions were run for each sample and then combined to form two duplicate PCRs, which were subsequently utilized for Automated Ribosomal Intergenic Spacer Analysis (ARISA). The quality of amplified fragments was checked, and the PCR products were purified and quantified spectrofluorimetrically. For each ARISA, about 5 ng of amplicons was mixed with 14 μL of internal size standard (GS2500-ROX; Applied Biosystems, Foster City, CA, USA) and the automated detection of ARISA fragments was carried out using the ABI Prism 3100 Genetic Analyzer (Applied Biosystems). ARISA fragments in the range of 390 to 1400 bp were determined using GeneScan analytical software version 2.02 (Applied Biosystems). Despite the fact that the DNA fingerprinting approach utilized in the present manuscript does not provide specific taxonomic information for the identified Operational Taxonomic Units (OTUs), it is still a largely utilized approach when assessing the patterns of prokaryotic diversity in environmental samples [52–55].

2.5. Meiofauna

Once in the laboratory, sediment samples for meiofaunal analyses were sliced into five sediment layers (i.e., 0–1, 1–3, 3–5, 5–10, and 10–15 cm), fixed with 4% buffered formalin, and stained with Rose Bengal (0.5 g L^{-1}) until analysis. Sediments were sieved through a 500-μm mesh, and a 20-μm mesh was used to retain the smallest organisms. The fraction remaining on the latter sieve was re-suspended and centrifuged three times with Ludox HS40 [21,49] (diluted with water to a final density of 1.18 g cm^{-3}). All animals remaining in the surnatant were again passed through a 20-μm mesh net, washed with tap water and, after staining with Rose Bengal, sorted under a stereomicroscope [49] (×40 magnification).

The meiofaunal biomass was assessed by bio-volumetric measurements for all specimens encountered. The nematode biomass was calculated from the biovolume, using the formula reported in Andrassy (1956) [56]: V = L × W^2 × 0.063 × 10^{-5} (in which body length, L, and width, W, are expressed in mm). The body volumes of all other taxa were derived from measurements of body length (L, in mm) and width (W, in mm), using the formula V = L × W^2 × C, where C is the approximate conversion factor for each meiofaunal taxon [57]. Each body volume was multiplied by an average density (1.13 g cm^{-3}) to obtain the biomass (mg DW: mg WW = 0.25) and the carbon content was considered to be 40% of the dry weight [58]. The biomass was expressed as μgC 10 cm^{-2}.

2.6. Statistical Analyses

For all of the investigated variables, differences among sampling stations and sampling periods were assessed using distance-based permutational nonparametric analyses of variance (PERMANOVA) in a univariate context [59]. When significant differences were observed, pairwise tests were also carried out to ascertain patterns of differences among stations and/or sampling times. The sampling design included two fixed orthogonal factors: Station (n = 6: Oglio, Po, Gorino, Giralda, Delta, and Cesenatico) and season (n = 2: summer and winter). Although time is typically a continuous source of variation, in this study, we considered the two sampling periods as levels of a fixed factor, assuming that they represented contrasting periods of the Po River discharge regime.

The same experimental design was used to test variations in i) the biochemical composition (in terms of protein, carbohydrate, lipid, and phytopigment contents) and nutritional quality (in terms of the protein to carbohydrate ratio, protein, and chlorophyll-a contributions to biopolymeric C) of sedimentary organic matter, and ii) the composition of OTUs and meiofaunal assemblages (based on abundance data only), again using PERMANOVA, in a multivariate context. The PERMANOVA analyses were based on matrixes of the Euclidean distance after normalization of the data (OM) and Bray Curtis similarity matrixes after square root transformations (prokaryotes and meiofauna) [60].

To visualize differences among stations and seasons in the biochemical composition, the nutritional quality of sedimentary organic matter, and the composition of OTUs and meiofaunal assemblages, bi-plots after a canonical analysis of the principal coordinates (CAP) were also produced [61]. Additionally, a Similarity percentage (SIMPER) analysis was carried out to assess the percentage dissimilarity in the meiofaunal taxonomic composition among systems and seasons. All statistical analyses were performed with the software PRIMER 6+ [62].

To assess whether the sedimentary organic matter content or nutritional quality explained significant differences in the prokaryotic and meiofaunal community composition, non-parametric multivariate multiple regression analyses, based on Euclidean distances, were also carried out using the DISTLM forward routine [60]. The forward selection of predictor variables was carried out with tests by permutation. P values were obtained using 9999 permutations of raw data for the marginal tests (tests of individual variables), whereas, for all of the conditional tests, the routine used 9999 permutations of residuals under a reduced model. Linear regressions were carried out using the Excel software.

3. Results

3.1. Environmental Parameters

The water temperature and salinity measured during the study period are reported in Table 1 A. In summer, the temperature ranged from 17 to 25 °C, with the lowest values at sea (Cesenatico) and highest at the river (Oglio) sampling station, respectively. In winter, the temperature ranged from 8 to 13 °C, with the lowest values at the river and lagoon stations (Oglio, Gorino, and Giralda) and highest at the lagoon mouth station (Delta), respectively. In both summer and winter, the salinity ranged from 0 to 35, at the sampling stations located within the rivers (i.e., Oglio and Po) and at sea (Cesenatico), respectively.

Table 1. Temperature, salinity, organic matter contents (**A**), descriptors of organic matter nutritional quality, and degradation rates (**B**) in the sediments of the investigated stations across the river-to-sea transect in summer and winter.

(A)			Temperature	Salinity	Chlorophyll-a		Phaeopigment		Total Pigment		Protein		Carbohydrate		Lipid		Biopolymeric C	
Sampling season		Sampling station	°C	PSU	$\mu g\,g^{-1}$		$\mu g\,g^{-1}$		$\mu g\,g^{-1}$		$mg\,g^{-1}$		$mg\,g^{-1}$		$mg\,g^{-1}$		$mg\,g^{-1}$	
					avg	sd	avg	sd	avg	sd	avg	sd	avg	sd	avg	sd	avg	sd
Summer	River	Oglio	25	0	0.08	0.00	0.30	0.02	0.35	0.05	0.30	0.02	0.17	0.00	0.03	0.01	0.24	0.02
	River	Po	23	0	1.12	0.09	4.00	0.53	5.12	0.68	0.71	0.08	0.38	0.10	0.14	0.05	0.60	0.12
	Lagoon	Gorino	23	28	3.07	0.59	68.59	19.01	71.66	19.99	14.19	1.60	3.32	0.20	1.51	0.09	9.41	0.93
	Lagoon	Giralda	23	5	6.29	1.03	91.37	9.93	98.66	10.50	8.18	0.92	4.05	0.09	0.64	0.06	6.10	0.53
	Lagoon	Delta	23	28	1.40	0.08	14.03	0.74	15.43	0.88	1.48	0.14	0.44	0.02	0.19	0.02	1.05	0.09
	Sea	Cesenatico	17	35	2.67	1.00	36.45	10.31	39.12	11.31	10.50	0.47	2.55	0.33	0.88	0.04	6.83	0.39
Winter	River	Oglio	8	0	1.38	0.00	19.35	6.24	20.72	3.61	2.45	0.23	3.71	0.99	0.74	0.19	3.24	0.65
	River	Po	10	0	1.31	0.15	8.73	0.17	10.04	0.02	5.80	0.11	7.44	0.27	1.37	0.31	6.85	0.39
	Lagoon	Gorino	8	28	8.74	1.58	125.33	3.10	134.07	4.57	4.60	0.27	7.43	0.60	2.66	0.31	7.22	0.61
	Lagoon	Giralda	8	9	17.28	0.55	93.74	6.60	111.03	7.15	10.34	0.25	6.03	0.68	1.65	0.13	8.72	0.49
	Lagoon	Delta	13	28	9.02	2.36	38.95	10.21	47.97	12.57	4.82	0.82	3.26	0.37	0.91	0.23	4.35	0.72
	Sea	Cesenatico	10	35	1.46	0.15	16.03	1.46	17.00	0.77	5.02	0.62	0.49	0.02	0.32	0.06	2.89	0.35

(B)			Phytopigment: Biopolymeric C		Protein: Biopolymeric C		Protein: Carbohydrate		Aminopeptidase		β-Glucosidase		Alkaline-Phosphatase	
Sampling season		Sampling station	%		%				$\mu g\,C\,g^{-1}\,h^{-1}$		$\mu g\,C\,g^{-1}\,h^{-1}$		$\mu g\,P\,g^{-1}\,h^{-1}$	
			avg	sd	avg	sd	avg	sd	avg	sd	avg	sd	avg	sd
Summer	River	Oglio	1.32	0.03	61.30	0.78	1.25	0.02	14.20	4.38	0.49	0.24	0.67	0.25
	River	Po	7.56	0.91	58.12	4.69	1.19	0.10	20.48	4.81	1.76	0.12	4.82	0.91
	Lagoon	Gorino	1.30	0.12	73.79	1.04	1.51	0.02	13.21	5.06	0.90	0.19	3.23	0.09
	Lagoon	Giralda	4.10	0.33	65.56	1.72	1.34	0.04	29.82	5.11	3.39	1.27	6.53	3.70
	Lagoon	Delta	5.37	0.15	69.35	0.60	1.42	0.01	8.59	2.39	0.68	0.19	0.92	0.31
	Sea	Cesenatico	1.55	0.50	75.41	0.95	1.54	0.02	14.40	3.47	2.01	0.13	5.30	1.29
Winter	River	Oglio	1.75	0.35	37.67	4.19	0.77	0.09	16.60	10.94	3.56	1.14	1.70	0.62
	River	Po	0.76	0.04	41.55	1.57	0.85	0.03	3.51	0.83	3.85	0.77	1.46	0.46
	Lagoon	Gorino	4.82	0.47	31.25	0.78	0.64	0.02	23.31	16.39	1.55	0.52	3.31	0.39
	Lagoon	Giralda	7.94	0.20	58.20	1.87	1.19	0.04	24.91	2.44	3.04	0.16	5.23	1.37
	Lagoon	Delta	8.20	0.83	54.30	0.20	1.11	0.004	20.75	9.63	4.07	1.25	2.01	0.31
	Sea	Cesenatico	2.02	0.04	84.91	0.11	1.73	0.002	8.47	1.49	1.02	0.24	1.98	0.26

3.2. Content, Biochemical Composition, Nutritional Quality, and Degradation Rates of Organic Matter

The chlorophyll-a, phaeopigment, total phytopigment, protein, carbohydrate, lipid, and biopolymeric C contents; algal and protein contributions to biopolymeric C; values of the protein to carbohydrate ratio; and OM degradation rates are reported in Table 1A,B.

The results of one-way PERMANOVA tests reveal a significant effect of the factor Station × Season for organic matter contents (Table S1A). In both seasons, the contents of almost all investigated variables, with only a few exceptions, were significantly the highest at the lagoon stations (Giralda and Gorino). At each station, almost all variables displayed contents in winter that were significantly higher than those in summer, with the exception for those at sea (Cesenatico), where the highest values occurred in summer.

The results of the multivariate PERMANOVA test also show a significant effect of the factor Station × Season on the OM biochemical composition, with significant differences among sampling stations in both seasons and between seasons at each station (Figure 2A).

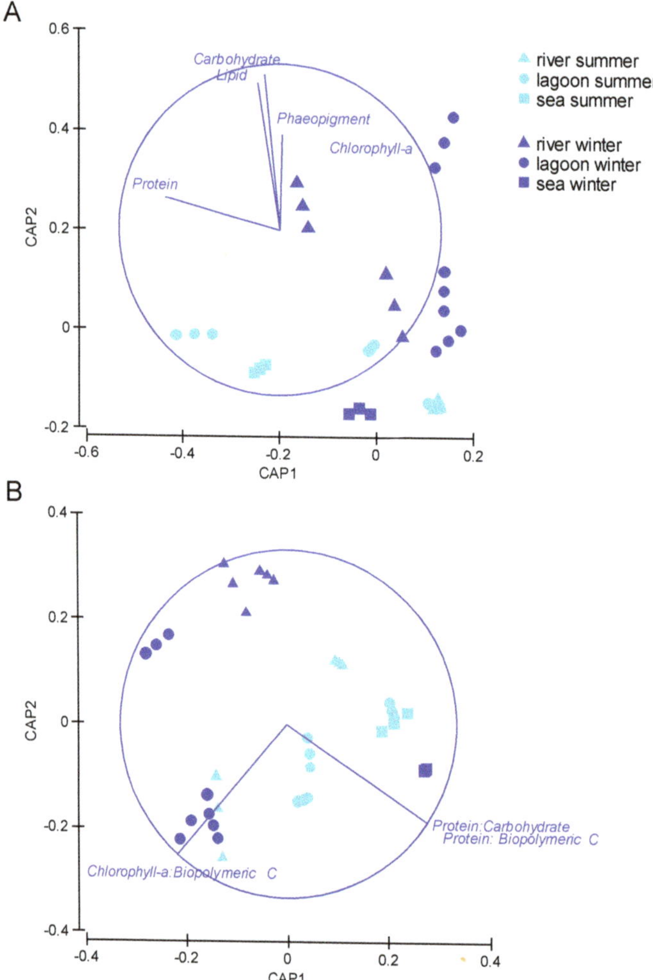

Figure 2. Output of the canonical analysis of principal coordinates (CAP) on the sedimentary organic matter biochemical composition (**A**) and nutritional quality (**B**).

The results of univariate PERMANOVA tests on the nutritional quality and ageing of OM are reported in Table S1B. The factor Station × Season had a significant effect on all descriptors of OM nutritional quality, with values varying significantly among stations in both seasons and between seasons at almost all stations. Specifically, in summer, the algal fraction of biopolymeric C was the highest at the Po River station and lowest at the Oglio River, lagoon (Goro), and sea (Cesenatico) stations; in winter, values at the lagoon stations Delta and Giralda were higher than those in all other stations, whereas the lowest values occurred at the Po River station. The algal fraction of biopolymeric C was only higher in summer than in winter at the Po River station.

In both seasons, the protein fraction of biopolymeric C and the values of the protein to carbohydrate ratio were highest at sea (Cesenatico) and higher in summer than in winter at all sampling stations, except for the marine station (Cesenatico). Overall, the nutritional quality of OM varied significantly among all sampling stations in each season and between seasons at each sampling station (Figure 2B).

The results of one-way PERMANOVA and the consequent pair wise tests carried out on enzymatic activities are reported in Table S1C. In both seasons, the aminopeptidase activity was highest at the lagoon stations Giralda and Gorino, and significantly higher in summer than in winter at the Po River and Cesenatico (sea) stations. The β-glucosidase activity was highest at the lagoon station Giralda in summer and at the lagoon Delta and Po and Oglio River stations in winter. Values of the β-glucosidase activity were higher in winter than in summer at Oglio and Po River stations and at the lagoon Delta station, whereas at sea (Cesenatico), the highest values occurred in summer. In both seasons, the alkaline-phosphatase activity was highest at the lagoon Giralda station. Moreover, it was higher in summer than in winter at Po, Giralda, and Cesenatico stations, with the opposite pattern at Oglio and Delta stations.

3.3. Prokaryotic Abundance, Biomass, and Diversity

The prokaryotic abundance, biomass, richness of OTU, and OTU composition are illustrated in Figure 3A–D. The results of PERMANOVA tests reveal a significant effect of the interaction Station × Season on the prokaryotic abundance, biomass, and OTU composition (Table S2A). Both in summer and winter, the prokaryotic abundance and biomass were the highest at the lagoon (Gorino and Giralda) and sea (Cesenatico) stations. At all stations, with exceptions for the Delta and sea stations, the prokaryotic abundance and biomass were higher in winter than in summer. The richness of prokaryotic OTU was highest at the Po River station in summer and at the Gorino Lagoon station in winter. At all stations, except for the Oglio and Po River and lagoon Delta stations, the richness of prokaryotic OTU was higher in winter than in summer (Figure 3C). The bi-plot produced after the CAP confirms the presence of strong spatial and temporal variations in the OTU composition of prokaryotic assemblages (Figure 3D).

3.4. Meiofaunal Abundance, Richness of Taxa, and Community Structure

The meiofaunal abundance, biomass, richness of taxa, and taxonomic composition are reported in Figure 4A–D. The results of two-way PERMANOVA tests reveal a significant effect of the interaction Station × Season on the meiofaunal abundance, biomass, and taxonomic composition (Table S2B). The results of the pair wise tests (Table S2B) reveal that both the meiofaunal abundance and biomass varied significantly among stations in both seasons and between seasons at almost all sampling stations. Specifically, the highest meiofaunal abundance and biomass occurred at lagoon stations in both seasons (Giralda in summer and Gorino and Giralda in winter). At the Po River and Delta lagoon stations, the meiofaunal abundance was higher in summer than in winter, whereas at the Gorino Lagoon (Gorino) and marine (Cesenatico) stations, values were the highest in winter. At the Delta lagoon station, the meiofaunal biomass was higher in summer than in winter, whereas at sea (Cesenatico), the highest values occurred in winter.

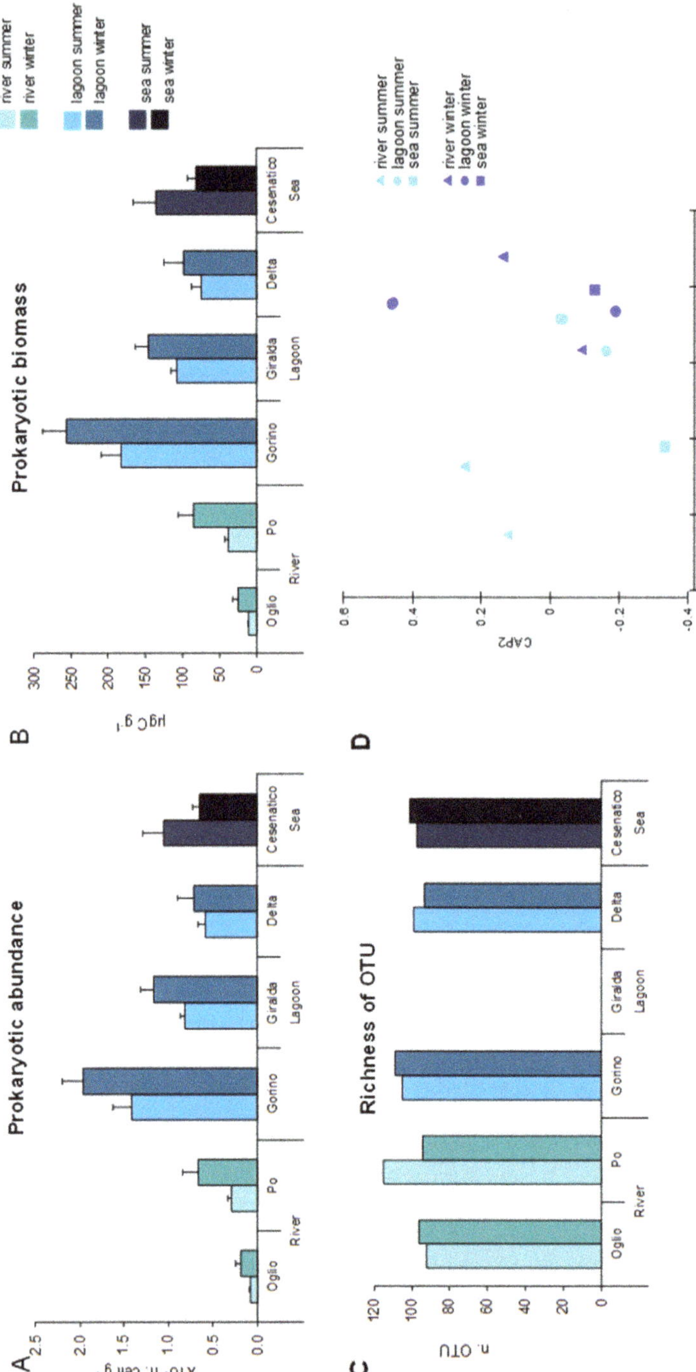

Figure 3. Prokaryotic abundance (**A**), biomass (**B**), richness of Operational Taxonomic Unit (OTU) (**C**), and OTU composition (**D**) in the investigated stations in summer and winter.

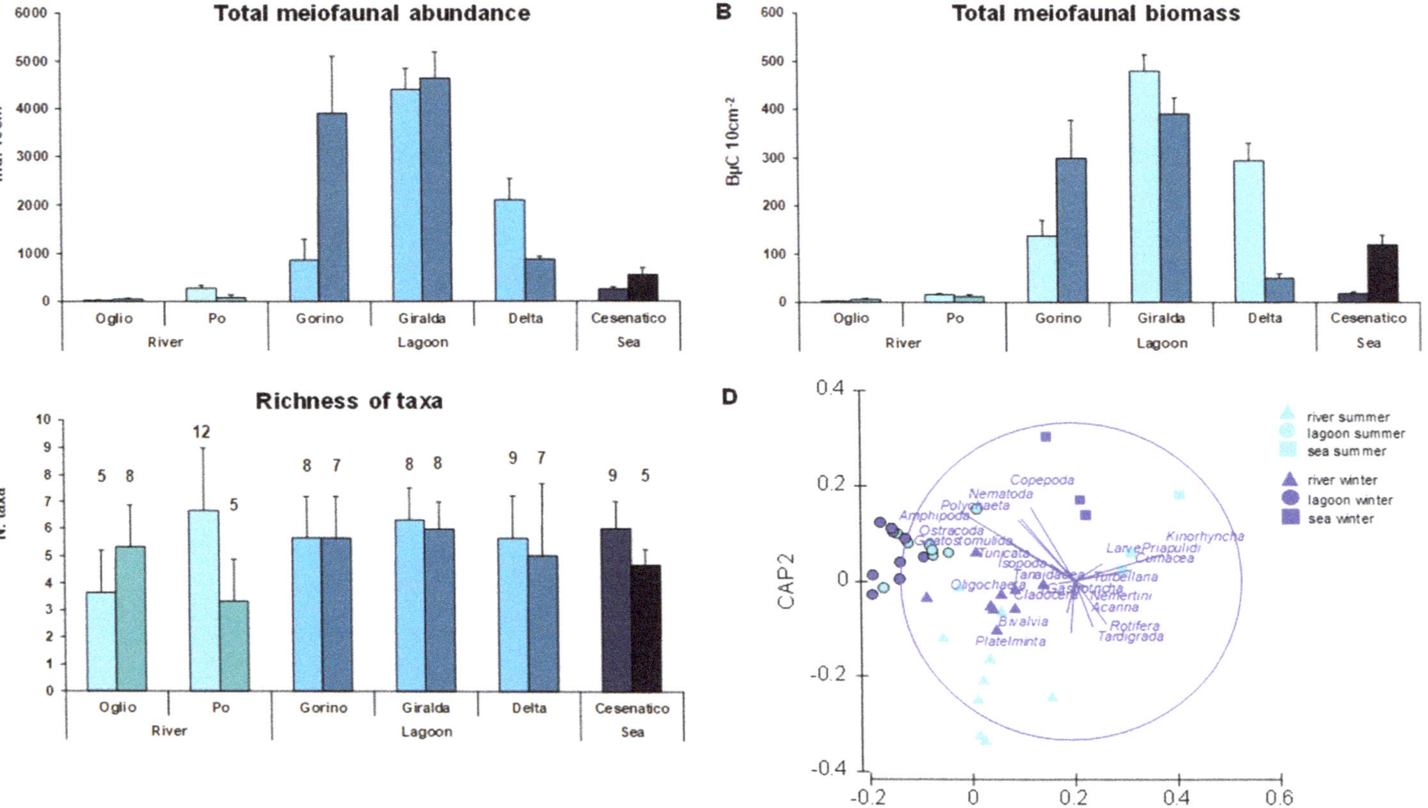

Figure 4. Meiofaunal abundance (**A**), biomass (**B**), richness of higher taxa (**C**), and taxonomic composition (**D**) in the investigated stations in summer and winter.

The richness of meiofaunal taxa did not display significant differences among sampling stations, or between seasons (Table S2B). At all stations and in both seasons, the meiofaunal community was dominated by nematodes (51–98%), followed by copepods (1–22%). "Other" taxa included Bivalvia, Kinorhyncha, Cumacea, Isopoda, Nemertea, Platelminta, Tunicata, and Gnatostomulida, each accounting for <1% of the total meiofaunal abundance.

The results of the multivariate PERMANOVA tests reveal a significant effect of the factors Station and Station × Season on the meiofaunal taxonomic composition (Table S2B). Specifically, the results of the pair wise tests show that the meiofaunal taxonomic composition varied significantly among sampling stations in both seasons and between seasons at four of six sampling stations (Po, Gorino, Delta, and Cesenatico). The bi-plot produced after the CAP analysis confirms the presence of strong spatial and temporal variations in the meiofaunal taxonomic composition (Figure 4D).

SIMPER analysis (Table 2) revealed that the % dissimilarity among different systems was 50–76% in summer and 64–80% in winter, and that the observed % was mainly due to the higher contribution of nematodes in the lagoon community structure and to the absence/reduction of the other taxa, in both seasons.

Table 2. Output of SIMPER analysis (cut off 90%), testing for % dissimilarity among systems in each season and between seasons in each system.

		% Dissimilarity	Responsible Taxa	Presence/Absence % Dissimilarity	Responsible Taxa
In summer	river vs. lagoon	75.3	Nematoda	41.3	Amphipoda, Rotifera, Ostracoda, Oligochaeta, Tardigrada, Acarina, Polychaeta, Nemertini, Bivalvia, Platelminta
	lagoon vs. sea	76.1	Nematoda	36.2	Kinorhyncha, Ostracoda, Rotifera, Priapulida larvae, Polychaeta, Acarina, Cumacea, Oligochaeta, Amphipoda, Nemertea
	river vs. sea	50.0	Nematoda, Copepoda	49.4	Kinorhyncha, Amphipoda, Rotifera, Ostracoda, Polychaeta, Oligochaeta, Tardigrada, Priapulida larvae, Cumacea, Bivalvia
In winter	river vs. lagoon	80.0	Nematoda	41.8	Amphipoda, Rotifera, Ostracoda, Tardigrada, Polychaeta, Oligochaeta, Acarina, Copepoda
	lagoon vs. sea	63.7	Nematoda	22.0	Oligochaeta, Ostracoda, Tardigrada, Acarina, Rotifera, Polychaeta, Amphipoda
	river vs. sea	65.3	Nematoda, Copepoda	40.7	Amphipoda, Rotifera, Oligochaeta, Tardigrada, Polychaeta, Acarina, Copepoda,
In river system	summer vs. winter	63.4	Nematoda, Copepoda, Acarina	37.9	Tardigrada, Polychaeta, Oligochaeta, Rotifera, Ostracoda, Acarina, Copepoda, Amphipoda, Bivalvia, Platelminta
In lagoon system	summer vs. winter	39.3	Nematoda	28.3	Ostracoda, Oligochaeta, Acarina, Amphipoda, Tardigrada, Rotifera, Nemertea, Polychaeta, Bivalvia, Isopoda, Kinorhyncha
In sea system	summer vs. winter	37.0	Nematoda, Copepoda	31.3	Kinorhyncha, Oligochaeta, Polychaeta, Rotifera, Priapulida larvae, Ostracoda

The % dissimilarity between summer and winter was lowest at sea (31%) and highest in river (38%) systems, and such differences were driven by different sets of taxa, depending on the system.

3.5. Relationships between Prokaryotes, Meiofauna, Organic Matter, and Environmental Characteristics

The results of the linear regression analyses indicate a significant and positive relationship between the BPC sedimentary contents and prokaryotic abundance ($p < 0.05$, R^2 0.622; Figure 5A) and biomass ($p < 0.01$, R^2 0.611; Figure 5B), as well as between the chlorophyll-a sedimentary contents and meiofaunal abundance ($p < 0.05$, R^2 0.590; Figure 5C) and biomass ($p < 0.05$, R^2 0.372; Figure 5D).

The results of the multivariate multiple regression analysis (DISTLM forward; Table 3), conducted on the composition of prokaryotic OTUs and meiofaunal assemblages, reveal that, when pooling together all data irrespective of season and station, the sub-set of variables that explained significant variations in the prokaryotic OTU composition explained a total of ca. 85% of variance and included all of the considered variables, with the exception of the protein to carbohydrate ratio.

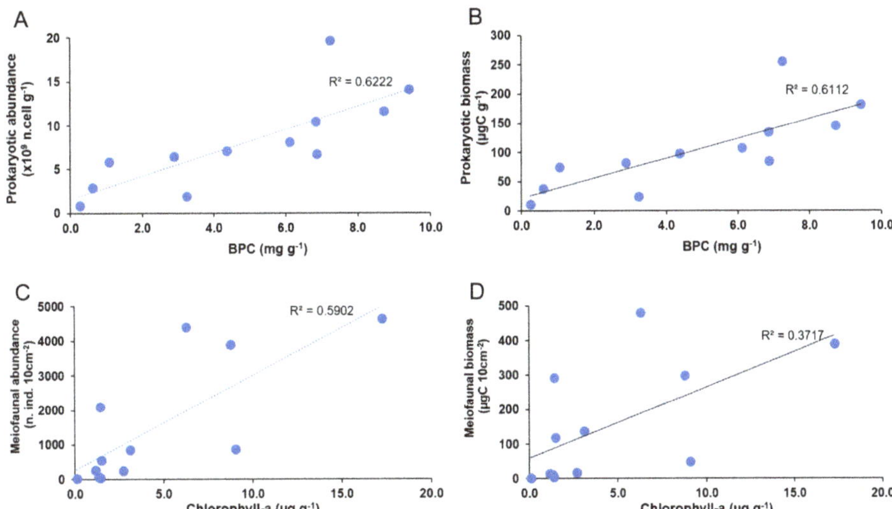

Figure 5. Relationships between biopolymeric C contents and prokaryotic abundance (**A**) and biomass (**B**), as well as between chlorophyll-a and meiofauna abundance (**C**) and biomass (**D**) in the sediments.

Table 3. Results of the multivariate multiple regression analysis carried out to ascertain the effect of the quantity and nutritional quality of sedimentary organic matter on the prokaryotic OTU composition (**A**) and meiofaunal taxonomic composition irrespective of the season (**B**), in summer, (**C**) and in winter (**D**). % Variance = percentage of explained variance (SS = sum of squares; F = F statistic; P = probability level; *** = P < 0.001; ** = P < 0.01; * = P < 0.05; ns = not significant).

	Variable	SS	F	P	% Variance	% Cumulative
(**A**)	Temperature	5505.6	2.8388	***	13.6	13.6
	Phaeopigment	4690.3	3.1907	**	11.6	25.2
	Salinity	4165.5	2.3033	*	10.3	35.5
	Chlorophyll-a	4003.7	2.3956	**	9.9	45.4
	Chlorophyll-a/Biopolymeric C	3940.8	5.3008	**	9.8	55.2
prokaryotic OTU composition	Lipid	3410	3.3767	*	8.4	63.6
	Protein	3293.7	2.4585	*	8.1	71.8
	Carbohydrate	3227.5	2.702	*	8.0	79.8
	Protein/Biopolymeric C	2512.2	4.4342	*	6.2	86.0
	Protein/Carbohydrate	801.55	1.4831	ns	2.0	88.0
(**B**)	Phaeopigment	27,987	14.951	***	30.5	30.5
	Salinity	15,291	10.435	**	16.7	47.2
	Chlorophyll-a/Biopolymeric C	8447.1	6.7732	***	9.2	56.4
	Protein	4192.5	3.639	**	4.6	61.0
meiofaunal taxonomic	Carbohydrate	3872.8	4.4542	**	4.2	65.2
composition	Protein/Carbohydrate	3817.2	3.59	**	4.2	69.4
	Chlorophyll-a	3680.3	3.7822	*	4.0	73.4
	Temperature	2270.2	2.7766	*	2.5	75.9
	Protein/Biopolymeric C	1294.4	1.6557	ns	1.4	77.3
	Lipid	1235.1	1.5408	ns	1.3	78.7
(**C**)	Chlorophyll-a	9912.3	4.7556	**	22.9	22.9
	Temperature	8016.6	4.7468	**	18.5	41.4
	Lipid	7348.9	19.458	**	17.0	58.4
	Salinity	7045.1	5.3933	**	16.3	74.7
meiofaunal taxonomic	Chlorophyll-a/Biopolymeric C	6406.7	7.0101	**	14.8	89.5
composition in summer	Protein	824.86	2.8042	*	1.9	91.4
	Phaeopigment	765.74	2.2364	ns	1.8	93.2
	Protein/Carbohydrate	351.13	1.22	ns	0.8	94.0
	Protein/Biopolymeric C	342.36	1.2184	ns	0.8	94.8
	Carbohydrate	110.93	0.36334	ns	0.3	95.1

Table 3. *Cont.*

	Variable	SS	F	P	% Variance	% Cumulative
(D)	Phaeopigment	20,442	12.098	***	43.1	43.1
	Salinity	13,045	13.987	***	27.5	70.5
	Protein	4212.2	6.0307	***	8.9	79.4
	Chlorophyll-a	2762.1	5.1176	***	5.8	85.2
meiofaunal taxonomic	Carbohydrate	2283.7	5.7903	**	4.8	90.0
composition in winter	Protein/Biopolymeric C	785.55	2.0692	ns	1.7	91.7
	Protein/Carbohydrate	466.83	1.266	ns	1.0	92.7
	Lipid	364.15	0.91691	ns	0.8	93.4
	Temperature	323.36	0.86176	ns	0.7	94.1
	Chlorophyll-a/Biopolymeric C	166.23	0.39556	ns	0.4	94.5

The multivariate multiple regression analysis also reveals that, when pooling together all data irrespective of season and station, a total of 75% of variation in the meiofaunal community composition is significantly explained by the phaeopigment, salinity, chlorophyll-a to biopolymeric C, protein, carbohydrate, protein to carbohydrate ratio, chlorophyll-a, and temperature.

The DISTLM forward analysis carried out separately for the two seasons revealed that two different sub-sets of variables significantly explained the observed variations (ca. 91% and 90% in summer and winter, respectively). In summer, the most important variables explaining the observed variations in the meiofaunal taxonomic composition were chlorophyll-a and temperature, whereas in winter, they were phaeopigment and salinity.

4. Discussion

The rapid increase of human activities has substantially altered the biogeochemical cycles of carbon, nitrogen, and phosphorous, thus becoming, in the last decades, a major issue for most of the freshwater and coastal marine ecosystems worldwide [63,64].

For many years, the North Adriatic Sea has been the most productive region of the whole Mediterranean Sea, with high marine production at all trophic levels, from phytoplankton to fish. Nevertheless, during the last decades, this basin has experienced severe eutrophication, beside hypoxic/anoxic crises and mucilage spreads [47,65]. These events have caused the mass mortality of pelagic and benthic organisms, as well as a deep degradation of the benthic compartment [66]. In the last two decades, the combination of reduced nutrient loads (due to recent regulations limiting detergent use and to a continuously diminished runoff due to climate change [67]) has caused an overall trend of oligotrophication in the basin [68], although some authors recently highlighted that the continental loads of nutrients are still high [63,69].

According to previous studies [3,70], the sedimentary trophic status can be assessed through rankings based on the quantity, biochemical composition, and nutritional quality of organic matter, and their combinations. Such studies have proposed a marine benthic trophic status classification based on the sedimentary contents of protein, carbohydrate, and biopolymeric C and the algal fraction of biopolymeric C [3,70]. In this study, applying the classifications proposed above, all of the investigated sediments can be ranked as meso-eutrophic. Specifically, we observed that, during our study, either the marine or lagoon sediments can be ranked only partially as eutrophic. Our results, when compared with early studies carried out in the same area [3,9,16,33,38,71], pinpoint the decrease of the benthic trophic status, and confirm the "regime shift" of the Adriatic Sea towards progressively more oligotrophic conditions [68,72–74].

Extending the classification proposed by Dell'Anno et al. (2002) [70] and Pusceddu et al. (2009) [3] to the river sediments, the Po and its tributary sediments would be meso-eutrophic. Unfortunately, the lack of similar data from previous study periods in Po River sediments does not allow any inference on temporal changes in the benthic trophic status of the riverine station. Nevertheless, the overall BPC contents in the Po River sediments during our study were in the lowest range observed in other rivers, although in largely different ecological contexts and latitudes [75]. This would suggest that, most likely due to the increased use of inner freshwaters for human usage [76], the sediments of the Po

River could also have recently experienced a decrease in the benthic trophic status. However, most recent studies have demonstrated that excessive simplification of the landscape due to the removal of buffer strips and riparian wetlands has accelerated the nutrient transfer to water bodies [37,63].

The results of our study indicate that, in both seasons, the lagoon sediments (i.e., Gorino, Giralda, and Delta stations) were characterized by the highest sedimentary organic loads and OM nutritional quality. This result is in agreement with the general view of Mediterranean coastal lagoons as transitional water sites of high production [77], where the accumulation of organic matter turns these ecosystems into 'detritus traps' [16,78], and they are net heterotrophic [79,80].

Moreover, according to previous studies [16,20], both the lagoon and marine sites under scrutiny, showing the highest sedimentary BPC concentrations, were characterized by the highest values of prokaryotic abundance, biomass, and diversity. This result confirms the presence of an overwhelming positive effect of OM quantity and bioavailability on benthic prokaryotes [19], but it can also be interpreted on the basis of concurrent physical-chemical gradients along the investigated transect. Indeed, a strong difference in the prokaryote assemblage's composition was observed along the salinity gradient across the river-lagoon-sea transect and was accompanied by an increasing dominance of exclusive taxa in each of the investigated systems. This result corroborates previous investigations showing either an increasing presence of freshwater taxa at stations more influenced by the river discharge or a core microbiome present across all study areas [81]. Unfortunately, as our results do not allow us to provide information on the prokaryotic taxonomic identity (due to the limitations of the ARISA molecular technique, which does not allow DNA sequencing), we cannot make inferences about the specific ecological role of the different prokaryotic groups in each investigated system. Despite this, we observed the highest values of extracellular enzymatic activities in the lagoon sediments, where the highest prokaryotic biomass and diversity were also observed. These results let us hypothesize that lagoons not only behave as 'detritus traps' (sensu [78]), but also as OM degradation hot spots, such as deltas and estuaries, already identified as metabolic reactors for OM and nutrients [63]. The rates of enzymatic OM degradation have been repeatedly used as proxies of benthic ecosystem functioning [12,13]. We report here a clear coupling between OM degradation and prokaryotic and meiofaunal biomass and diversity, so we can infer that the investigated lagoon sediments are hotspots of ecosystem functioning, and that this, at least partially, is promoted by either prokaryotes [19] or meiofauna.

Our results also provided evidence of a concurrent positive relationship between sedimentary OM and meiofaunal abundance and biomass. Meiofauna are considered to be highly sensitive to environmental changes, so they can provide useful information about the benthic component response to ecosystem "regime shifts" in a variety of aquatic ecosystems [16,82–84]. Previous studies carried out in typically oligotrophic conditions (like the coastal W Mediterranean Sea) reported an evident decrease of the abundance, biomass, and richness of higher meiofaunal taxa exposed to high (excess) organic loads [82]. In contrast, our results, obtained in even more eutrophic conditions, show that the higher the organic loads, the higher the values of meiofaunal abundance and biomass. This result suggests that the apparent positive relationship between sedimentary OM and meiofaunal abundance and biomass observed across the river-lagoon-sea transect is more the result of a pattern of prokaryotic and meiofaunal communities specifically adapted to different habitats, rather than an actually 'positive' response to OM enrichment. Accordingly, despite no differences being observed in the richness of prokaryotic OTUs and meiofaunal higher taxa among stations or between seasons, the multivariate analysis revealed that different stations in the two seasons were characterized by very different prokaryotic and meiofaunal assemblages. In this regard, SIMPER analyses also revealed a high % of dissimilarity among systems (particularly between river and lagoon/sea systems) and seasons in each system. These results suggest that meiofaunal assemblages retrieved in the different investigated environments appear well-adapted to the specific trophic and environmental characteristics they face in each environment. Indeed, when the data are presence/absence transformed, different sets of taxa are responsible for the observed dissimilarity (mostly rare taxa, i.e., accounting for <1% of the total assemblage, each, [16]).

In this regard, the results of the multiple multi-regression analysis (DISTLM forward) indicated that both the compositions of prokaryotic OTUs and meiofaunal assemblages were significantly affected by environmental settings (temperature and salinity, cumulatively explaining ca. 24% and 19% of the observed variance in prokaryote and meiofauna assemblages, respectively), the trophic resource quantity (cumulatively 46% and 43%, respectively), and the nutritional quality (cumulatively 16% and 13%, respectively).

5. Conclusions

Our results allow us to reject the null hypothesis that the OM quantity, biochemical composition and degradation rates, prokaryotic and meiofaunal biodiversity, and ecosystem functioning do not vary among different ecosystems along a strong salinity gradient in different periods of the year.

Overall, the results of this study also allow us to conclude that the synoptic analysis of prokaryotes and meiofauna can provide useful information on the relative effects of organic enrichment and environmental settings across gradients of the environmental continuum. Our results also pinpoint that transitional water systems, including rivers, lagoons, and marine coastal ecosystems, represent a sort of end-of-pipe of the watershed continuum connecting the terrestrial and coastal domains, and acting as either a filter or source for nutrients and contaminants [64].

Supplementary Materials: The following are available online at http://www.mdpi.com/1424-2818/12/5/189/s1, Table S1. Results of the one-way PERMANOVA testing for differences in the concentration of biochemical compounds and composition (A), descriptors of nutritional quality (B) of organic matter in the sediment between sampling stations and seasons. Table S2. Results of multivariate PERMANOVA and pair wise test on prokaryotic (A) and meiofaunal (B) abundance, biomass, richness of meiofaunal taxa and taxonomic composition in the sediment among stations and sampling seasons.

Author Contributions: Conceptualization, P.V. and A.P.; methodology, S.B., D.N., M.B., P.V., E.R., and A.P.; writing, S.B. and A.P., revision, S.B., D.N., M.B., P.V., E.R., and A.P. All authors have read and agreed to the published version of the manuscript.

Funding: This study has been carried out in the frame of the project "Origin, composition and fate of organic nitrogen loads in the NW Adriatic: from the Po River to the sea", funded by the Italian Ministry of University and Research under the PRIN 2008 call.

Acknowledgments: The authors are indebted to M. Lo Martire for the precious help during sampling activities.

Conflicts of Interest: The authors declare no conflicts of interest.

References

1. Cebrián, J.; Williams, M.; McClelland, J.; Valiela, I. The dependence of heterotrophic consumption and C accumulation on autotrophic nutrient concentration in ecosystems. *Ecol. Lett.* **1998**, *1*, 165–170. [CrossRef]

2. Huxel, G.R. On the influence of food quality in consumer-resource interactions. *Ecol. Lett.* **1999**, *2*, 256–261. [CrossRef]

3. Pusceddu, A.; Dell'Anno, A.; Fabiano, M.; Danovaro, R. Quantity, biochemical composition and bioavailability of sediment organic matter as complementary signatures of benthic trophic status. *Mar. Ecol. Prog. Ser.* **2009**, *375*, 41–52. [CrossRef]

4. Nixon, S.W. Coastal marine eutrophication: A definition, social causes, and future concerns. *Ophelia* **1995**, *41*, 199–219. [CrossRef]

5. Grall, J.; Chauvaud, L. Marine eutrophication and benthos: The need for new approaches and concepts. *Global Chang. Biol.* **2002**, *8*, 813–830. [CrossRef]

6. Torres, A.I.; Rivera Hernández, J.R.; Giarratano, E.; Faleschini, M.; Green Ruiz, C.R.; Gil, M.N. Potentially toxic elements and biochemical components in surface sediments of NW Mexico: An assessment of contamination and trophic status. *Mar. Pollut. Bull.* **2019**, *149*, 110633. [CrossRef]

7. Cloern, J.E. Our evolving conceptual model of the coastal eutrophication problem. *Mar. Ecol. Prog. Ser.* **2001**, *210*, 223–253. [CrossRef]

8. Fabiano, M.; Danovaro, R.; Fraschetti, S. A three-year time series of elemental and biochemical composition of organic matter in subtidal sandy sediments of the Ligurian Sea (northwestern Mediterranean). *Cont. Shelf Res.* **1995**, *15*, 1453–1469. [CrossRef]

9. Dell'Anno, A.; Pusceddu, A.; Langone, L.; Danovaro, R. Biochemical composition and early diagenesis of organic matter in coastal sediments of the NW Adriatic Sea influenced by riverine inputs. *Chem. Ecol.* **2008**, *24*, 75–85. [CrossRef]

10. Welsh, D.T. It's a dirty job but someone has to do it: The role of marine benthic macrofauna in organic matter turnover and nutrient recycling to the water column. *Chem. Ecol.* **2003**, *19*, 321–342. [CrossRef]

11. Danovaro, R.; Gambi, C.; Manini, E.; Fabiano, M. Meiofauna response to a dynamic river plume front. *Mar. Biol.* **2000**, *137*, 359–370. [CrossRef]

12. Pusceddu, A.; Gambi, C.; Manini, E.; Danovaro, R. Trophic state, ecosystem efficiency and biodiversity of transitional aquatic ecosystems: Analysis of environmental quality based on different benthic indicators. *Chem. Ecol.* **2007**, *23*, 505–515. [CrossRef]

13. Pusceddu, A.; Danovaro, R. Exergy, ecosystem functioning and efficiency in a coastal lagoon: The role of auxiliary energy. *Estuar. Coast. Shelf Sci.* **2009**, *84*, 227–236. [CrossRef]

14. Magri, M.; Benelli, S.; Bondavalli, C.; Bartoli, M.; Christian, R.R.; Bodini, A. Benthic N pathways in illuminated and bioturbated sediments studied with network analysis. *Limnol. Oceanogr.* **2018**, *63*, S68–S84. [CrossRef]

15. Politi, T.; Zilius, M.; Castaldelli, G.; Bartoli, M.; Daunys, D. Estuarine macrofauna affects benthic biogeochemistry in a hypertrophic lagoon. *Water* **2019**, *11*, 1186. [CrossRef]

16. Pusceddu, A.; Bianchelli, S.; Gambi, C.; Danovaro, R. Assessment of benthic trophic status of marine coastal ecosystems: Significance of meiofaunal rare taxa. *Estuar. Coast. Shelf Sci.* **2011**, *93*, 420–430. [CrossRef]

17. Foti, A.; Fenzi, G.A.; Di Pippo, F.; Gravina, M.F.; Magni, P. Testing the saprobity hypothesis in a Mediterranean lagoon: Effects of confinement and organic enrichment on benthic communities. *Mar. Environ. Res.* **2014**, *99*, 85–94. [CrossRef]

18. Zobrist, E.C.; Coull, B.C. Meiobenthic interactions with microbenthic larvae and juveniles: An experimental assessment of the meiofaunal bottleneck. *Mar. Ecol. Prog. Ser.* **1992**, *88*, 1–8. [CrossRef]

19. Danovaro, R.; Pusceddu, A. Biodiversity and ecosystem functioning in coastal lagoons: Does microbial diversity play any role? *Estuar. Coast. Shelf Sci.* **2007**, *75*, 4–12. [CrossRef]

20. Cibic, T.; Franzo, A.; Celussi, M.; Fabbro, C.; Del Negro, P. Benthic ecosystem functioning in hydrocarbon and heavy-metal contaminated sediments of an Adriatic lagoon. *Mar. Ecol. Prog. Ser.* **2012**, *458*, 69–87. [CrossRef]

21. Heip, C.; Vincx, M.; Vranken, G. The ecology of marine nematodes. *Oceanogr. Mar. Biol. Ann. Rev.* **1985**, *23*, 399–489.

22. Semprucci, F.; Losi, V.; Moreno, M. A review of Italian research on free-living marine nematodes and the future perspectives on their use as Ecological Indicators (EcoInds). *Mediterr. Mar. Sci.* **2015**, *16*, 352–365. [CrossRef]

23. Magliozzi, C.; Usseglio-Polatera, P.; Meyer, A.; Grabowski, R.C. Functional traits of hyporheic and benthic invertebrates reveal importance of wood-driven geomorphological processes in rivers. *Funct. Ecol.* **2019**, *33*, 1758–1770. [CrossRef]

24. Tedesco, L.; Socal, G.; Acri, F.; Veneri, D.; Vichi, M. NW Adriatic Sea biogeochemical variability in the last 20 years (1986-2005). *Biogeosciences* **2007**, *4*, 673–687. [CrossRef]

25. Wang, X.H.; Pinardi, N. Modelling the dynamics of sediment transport and resuspension in the northern Adriatic Sea. *J. Geophys. Res. Atmos.* **2002**, *107*, 3225. [CrossRef]

26. Ursella, L.; Poulain, P.M.; Signell, R.P. Surface drifter derived circulation in the northern and middle Adriatic Sea: Response to wind regime and season. *J. Geophys. Res. Oceans* **2006**, *112*, C03S04. [CrossRef]

27. Bartoli, M.; Nizzoli, D.; Viaroli, P.; Turolla, E.; Castaldelli, G.; Fano, E.A.; Rossi, R. Impact of *Tapes philippinarum* farming on nutrient dynamics and benthic respiration in the Sacca di Goro. *Hydrobiologia* **2001**, *455*, 203–212. [CrossRef]

28. Viaroli, P.; Azzoni, R.; Bartoli, M.; Giordani, G.; Tajé, L. Evolution of the trophic conditions and dystrophic outbreaks in the Sacca di Goro lagoon (Northern Adriatic Sea). In *Mediterranean Ecosystems*; Springer: Milano, Italy, 2001; pp. 467–475.

29. Viaroli, P.; Bartoli, M.; Azzoni, R.; Giordani, G.; Mucchino, C.; Naldi, M.; Tajé, L. Nutrient and iron limitation to Ulva blooms in a eutrophic coastal lagoon (Sacca di Goro, Italy). *Hydrobiologia* **2005**, *550*, 57–71. [CrossRef]

30. Nizzoli, D.; Bartoli, M.; Viaroli, P. Nitrogen and phosphorous budgets during a farming cycle of the Manila clam *Ruditapes philippinarum*: An in situ experiment. *Aquaculture* **2006**, *261*, 98–108. [CrossRef]

31. Zilius, M.; Giordani, G.; Petkuviene, J.; Lubiene, I.; Ruginis, T.; Bartoli, M. Phosphorus mobility under short-term anoxic conditions in two shallow eutrophic coastal systems (Curonian and Sacca di Goro lagoons). *Estuar. Coast. Shelf Sci.* **2016**, *164*, 134–146. [CrossRef]

32. Naldi, M.; Viaroli, P. Nitrate uptake and storage in the seaweed *Ulva rigida* C. Agardh in relation to nitrate availability and thallus nitrate content in a eutrophic coastal lagoon (Sacca di Goro, Po River Delta, Italy). *J. Exp. Mar. Biol. Ecol.* **2002**, *269*, 65–83. [CrossRef]

33. Fiordelmondo, C.; Manini, E.; Gambi, C.; Pusceddu, A. Short-term impact of clam harvesting on sediment chemistry, benthic microbes and meiofauna in the Goro lagoon (Italy). *Chem. Ecol.* **2003**, *19*, 173–187. [CrossRef]

34. Mistri, M.; Rossi, R.; Fano, E.A. Structure and secondary production of a soft bottom macrobenthic community in a brackish lagoon (Sacca di Goro, north-eastern Italy). *Estuar. Coast. Shelf Sci.* **2001**, *52*, 605–616. [CrossRef]

35. Ludovisi, A.; Castaldelli, G.; Fano, E.A. Multi-scale spatio-temporal patchiness of macrozoobenthos in the Sacca di Goro lagoon (Po river delta, Italy). *Transit. Waters Bull.* **2013**, *7*, 233–244. [CrossRef]

36. Marinov, D.; Zaldívar, J.M.; Norro, A.; Giordani, G.; Viaroli, P. Integrated modelling in coastal lagoons: Sacca di Goro case study. *Hydrobiologia* **2008**, *611*, 147–165. [CrossRef]

37. Bartoli, M.; Castaldelli, G.; Nizzoli, D.; Viaroli, P. Benthic primary production and bacterial denitrification in a Mediterranean eutrophic coastal lagoon. *J. Exp. Mar. Biol. Ecol.* **2012**, *438*, 41–51. [CrossRef]

38. Dell'Anno, A.; Incera, M.; Mei, M.L.; Pusceddu, A. Mesoscale variability of organic matter composition in NW Adriatic sediments. *Chem. Ecol.* **2003**, *19*, 33–45. [CrossRef]

39. Tesi, T.; Miserocchi, S.; Langone, L.; Boni, L.; Guerrini, F. Sources, fate and distribution of organic matter on the Western Adriatic Continental Shelf, Italy. *Water Air Soil Pollut. Focus* **2006**, *6*, 593–603. [CrossRef]

40. Tesi, T.; Langone, L.; Goni, M.A.; Miserocchi, S.; Bertasi, F. Changes in the composition of organic matter from prodeltaic sediments after a large flood event (Po River, Italy). *Geochim. Cosmochim. Acta* **2008**, *72*, 2100–2114. [CrossRef]

41. Bianchelli, S.; Pusceddu, A.; Buschi, E.; Danovaro, R. Trophic status and meiofauna biodiversity in the Northern Adriatic Sea: Insights for the assessment of good environmental status. *Mar. Environ. Res.* **2016**, *113*, 18–30. [CrossRef]

42. Bianchelli, S.; Pusceddu, A.; Buschi, E.; Danovaro, R. Nematode biodiversity and benthic trophic state are simple tools for the assessment of the environmental quality in coastal marine ecosystems. *Ecol. Indic.* **2018**, *95*, 270–287. [CrossRef]

43. Tahey, T.M.; Duineveld, G.C.A.; de Wilde, P.A.W.J.; Berghuis, E.M.; Kok, A. Sediment O_2 demand, density and biomass of the benthos and phytopigments along the northwestern Adriatic coast: The extent of Po enrichment. *Oceanol. Acta* **1996**, *19*, 117–130.

44. Moodley, L.; Heip, C.H.R.; Middelburg, J.J. Benthic activity in sediments of the northwestern Adriatic Sea: Sediment oxygen consumption, macro- and meiofauna dynamics. *J. Sea Res.* **1998**, *40*, 263–280. [CrossRef]

45. Alvisi, F.; Giani, M.; Ravaioli, M.; Giordano, P. Role of sedimentary environment in the development of hypoxia and anoxia in the NW Adriatic shelf (Italy). *Estuar. Coast. Shelf Sci.* **2013**, *128*, 9–21. [CrossRef]

46. Danovaro, R. Organic inputs and ecosystem efficiency in the deep Mediterranean Sea. *Chem. Ecol.* **2003**, *19*, 391–398. [CrossRef]

47. Danovaro, R.; Fonda Umani, S.; Pusceddu, A. Climate change and the potential spreading of marine mucilage and microbial pathogens in the Mediterranean Sea. *PLoS ONE* **2009**, *4*, e7006. [CrossRef] [PubMed]

48. Lorenzen, C.J.; Jeffrey, S.W. Determination of chlorophyll and phaeopigments spectrophotometric equations. *Limnol. Oceanogr.* **1980**, *12*, 343–346. [CrossRef]

49. Danovaro, R. *Methods for the Study of Deep-Sea Sediments, Their Functioning and Biodiversity*; CRC Press, Taylor & Francis Group: Boca Raton, FL, USA, 2010; pp. 1–428.

50. Pusceddu, A.; Dell'Anno, A.; Fabiano, M. Organic matter composition in coastal sediments at Terra Nova Bay (Ross Sea) during summer 1995. *Polar Biol.* **2000**, *23*, 288–293. [CrossRef]

51. Pusceddu, A.; Bianchelli, S.; Canals, M.; Sanchez-Vidal, A.; De Madron, X.D.; Heussner, S.; Lykousis, V.; de Stigter, H.; Trincardi, F.; Danovaro, R. Organic matter in sediments of canyons and open slopes of the Portuguese, Catalan, Southern Adriatic and Cretan Sea margins. *Deep Sea Res. I Oceanogr. Res. Pap.* **2010**, *57*, 441–457. [CrossRef]

52. Luna, G.M.; Corinaldesi, C.; Rastelli, E.; Danovaro, R. Patterns and drivers of bacterial α-and β-diversity across vertical profiles from surface to subsurface sediments. *Environ. Microbiol. Rep.* **2013**, *201*, 731–739. [CrossRef]

53. Fuhrman, J.A.; Cram, J.A.; Needham, D.M. Marine microbial community dynamics and their ecological interpretation. *Nat. Rev. Microbiol.* **2015**, *13*, 133–146. [CrossRef] [PubMed]

54. Rastelli, E.; Corinaldesi, C.; Dell'Anno, A.; Tangherlini, M.; Lo Martire, M.; Nishizawa, M.; Nomaki, H.; Nunoura, T.; Danovaro, R. Drivers of bacterial α-and β-diversity and functioning in subsurface hadal sediments. *Front. Microbiol.* **2019**, *10*, 2609. [CrossRef] [PubMed]

55. Niederberger, T.; Bottos, E.M.; Sohm, J.A.; Gunderson, T.E.; Parker, A.E.; Coyne, K.J.; Capone, D.G.; Carpenter, E.J.; Cary, S.C. Rapid microbial dynamics in response to an induced wetting event in antarctic dry valley soils. *Front. Microbiol.* **2019**, *10*, 621. [CrossRef] [PubMed]

56. Andrassy, I. The determination of volume and weight of nematodes. *Acta Zool.* **1956**, *2*, 1–15.

57. Feller, R.J.; Warwick, R.M. Energetics. In *Introduction to the Study of Meiofauna*; Higgins, R.P., Thiel, H., Eds.; Smithsonian Institute Press: London, UK, 1988; pp. 181–196.

58. Wieser, W. Benthic studies in Buzzards Bay. II The meiofauna. *Limnol. Oceanogr.* **1960**, *5*, 121–137. [CrossRef]

59. Anderson, M.J. Permutation tests for univariate or multivariate analysis of variance and regression. *Can. J. Fish. Aquat. Sci.* **2001**, *58*, 626–639. [CrossRef]

60. McArdle, B.H.; Anderson, M.J. Fitting multivariate models to community data: A comment on distance-based redundancy analysis. *Ecology* **2001**, *82*, 290–297. [CrossRef]

61. Anderson, M.J.; Willis, T.J. Canonical analysis of principal coordinates: A useful method of constrained ordination for ecology. *Ecology* **2003**, *84*, 511–525. [CrossRef]

62. Clarke, K.R.; Gorley, R.N. PRIMER v6: User Manual/Tutorial. PRIMER-E, Plymouth, 2006, p. 190. Available online: https://www.primer-e.com/ (accessed on 8 May 2020).

63. Viaroli, P.; Nizzoli, D.; Pinardi, M.; Soana, E.; Bartoli, M. Eutrophication of the Mediterranean Sea: A watershed—Cascading aquatic filter approach. *Rend. Fis. Acc. Lincei* **2015**, *26*, 13–23. [CrossRef]

64. Zamparas, M.; Zacharias, I. Restoration of eutrophic freshwater by managing internal nutrient loads. A review. *Sci. Total Environ.* **2014**, *496*, 551–562. [CrossRef]

65. Volf, G.; Atanasova, N.; Kompare, B.; Ožanić, N. Modeling nutrient loads to the northern Adriatic. *J. Hydrol.* **2013**, *504*, 182–193. [CrossRef]

66. Degobbis, D.; Precali, R.; Ivančić, I.; Smodlaka, N.; Fuks, D.; Kveder, S. Long-term changes in the northern Adriatic ecosystem related to anthropogenic eutrophication. *Int. J. Environ. Pollut.* **2000**, *13*, 495–533. [CrossRef]

67. Alcamo, J.; Moreno, J.M.; Nováky, B.; Bindi, M.; Corobov, R.; Devoy, R.J.N.; Giannakopoulos, C.; Martin, E.; Olesen, J.E.; Shvidenko, A. Impacts, Adaptation and Vulnerability. Contribution of Working Group II to the Fourth Assessment Report of the Intergovernmental Panel on Climate Change. In *Climate Change 2007*; Parry, M.L., Canziani, O.F., Palutikof, J.P., van der Linden, P.J., Hanson, C.E., Eds.; Cambridge University Press: Cambridge, UK, 2007; pp. 541–580.

68. Mozetič, P.; Solidoro, C.; Cossarini, G.; Socal, G.; Precali, R.; Francé, J.; Bianchi, F.; De Vittor, C.; Smodlaka, N.; Fonda Umani, S. Recent trends towards oligotrophication of the Northern Adriatic: Evidence from chlorophyll a time series. *Estuar. Coasts* **2010**, *33*, 362–375. [CrossRef]

69. Cozzi, S.; Giani, M. River water and nutrient discharges in the Northern Adriatic Sea: Current importance and long-term changes. *Cont. Shelf Res.* **2011**, *31*, 1881–1893. [CrossRef]

70. Dell'Anno, A.; Mei, M.L.; Pusceddu, A.; Danovaro, R. Assessing the trophic state and eutrophication of coastal marine systems: A new approach based on the biochemical composition of sediment organic matter. *Mar. Pollut. Bull.* **2002**, *44*, 611–622. [CrossRef]

71. Sabbatini, A.; Bonatto, S.; Bianchelli, S.; Pusceddu, A.; Danovaro, R.; Negri, A. Foraminiferal assemblages and trophic state in coastal sediments of the Adriatic Sea. *J. Mar. Syst.* **2012**, *105*, 163–174. [CrossRef]

72. Conversi, A.; Fonda Umani, S.; Peluso, T.; Molinero, J.C.; Santojanni, A.; Edwards, M. The mediterranean sea regime shift at the end of the 1980s, and intriguing parallelisms with other european basins. *PLoS ONE* **2010**, *5*, e10633. [CrossRef]

73. Gašparović, B. Decreased production of surface-active organic substances as a consequence of the oligotrophication in the northern Adriatic Sea. *Estuar. Coast. Shelf Sci.* **2012**, *115*, 33–39. [CrossRef]

74. Giani, M.; Djakovac, T.; Degobbis, D.; Cozzi, S.; Solidoro, C.; Fonda Umani, S. Recent changes in the marine ecosystems of the northern Adriatic Sea. *Estuar. Coast. Shelf Sci.* **2012**, *115*, 1–13. [CrossRef]

75. Kumar, K.S.S.; Nair, S.M.; Salas, P.M.; Cheriyan, E. Distribution and sources of sedimentary organic matter in Chitrapuzha, a tropical tidal river, southwest coast of India. *Environ. Forensics* **2017**, *18*, 135–146. [CrossRef]

76. Delconte, C.A.; Sacchi, E.; Racchetti, E.; Bartoli, M.; Mas-Pla, J.; Re, V. Nitrogen inputs to a river course in a heavily impacted watershed: A combined hydrochemical and isotopic evaluation (Oglio River Basin, N Italy). *Sci. Total Environ.* **2014**, *466–467*, 924–938. [CrossRef]

77. McLusky, D.S.; Elliott, M. Transitional waters: A new approach, semantics or just muddying the waters? *Estuar. Coast. Shelf Sci.* **2007**, *71*, 359–363. [CrossRef]

78. Pusceddu, A.; Dell'Anno, A.; Danovaro, R.; Manini, E.; Sarà, G.; Fabiano, M. Enzymatically hydrolyzable protein and carbohydrate sedimentary pools as indicators of the trophic state of detritus sink systems: A case study in a Mediterranean coastal lagoon. *Estuaries* **2003**, *26*, 641–650. [CrossRef]

79. Viaroli, P.; Christian, R.R. Description of trophic status, hyperautotrophy and dystrophy of a coastal lagoon through a potential oxygen production and consumption index—TOSI: Trophic Oxygen Status Index. *Ecol. Indic.* **2004**, *3*, 237–250. [CrossRef]

80. Duarte, C.M.; Cebrian, J. The fate of marine autotrophic production. *Limnol. Oceanogr.* **1996**, *41*, 1758–1766. [CrossRef]

81. Fazi, S.; Baldassarre, L.; Cassin, D.; Quero, G.M.; Pizzetti, I.; Cibic, T.; Luna, G.M.; Zonta, R.; Del Negro, P. Prokaryotic community composition and distribution in coastal sediments following a Po river flood event (northern Adriatic Sea, Italy). *Estuar. Coast. Shelf Sci.* **2020**, *233*, 106547. [CrossRef]

82. Gambi, C.; Bianchelli, S.; Pérez, M.; Invers, O.; Ruiz, J.M.; Danovaro, R. Biodiversity response to experimental induced hypoxic-anoxic conditions in seagrass sediments. *Biodivers. Conserv.* **2009**, *18*, 33–54. [CrossRef]

83. Mirto, S.; Bianchelli, S.; Gambi, C.; Krzelj, M.; Pusceddu, A.; Scopa, M.; Holmer, M.; Danovaro, R. Fish-farm impact on metazoan meiofauna in the Mediterranean Sea: Analysis of regional vs. habitat effects. *Mar. Environ. Res.* **2010**, *69*, 38–47. [CrossRef]

84. Bianchelli, S.; Buschi, E.; Danovaro, R.; Pusceddu, A. Biodiversity loss and turnover in alternative states in the Mediterranean Sea: A case study on meiofauna. *Sci. Rep.* **2016**, *6*, 34544. [CrossRef]

Article

Meiofauna Life on Loggerhead Sea Turtles-Diversely Structured Abundance and Biodiversity Hotspots That Challenge the Meiofauna Paradox

Jeroen Ingels [1,*], Yirina Valdes [2,*], Letícia P. Pontes [3], Alexsandra C. Silva [3], Patrícia F. Neres [3], Gustavo V. V. Corrêa [3], Ian Silver-Gorges [4], Mariana M.P.B. Fuentes [4], Anthony Gillis [4], Lindsay Hooper [4], Matthew Ware [4], Carrie O'Reilly [4], Quintin Bergman [5], Julia Danyuk [6], Sofia Sanchez Zarate [7], Laura I. Acevedo Natale [1,8] and Giovanni A. P. dos Santos [3]

1. Coastal and Marine Laboratory, Florida State University, St Teresa, FL 32358, USA; lauraisabel1195@gmail.com
2. Department of Biological Sciences, Federal University of Paraíba, 58397-000 Areia-PB, Brazil
3. Zoology Department, Federal University of Pernambuco, 50670-901 Recife-PE, Brazil; leticiapereirapontes@hotmail.com (L.P.P.); alexsandraa.cavalcante@gmail.com (A.C.S.); patricia_neres@yahoo.com.br (P.F.N.); gugaeco@hotmail.com (G.V.V.C.); giopaiva@hotmail.com (G.A.P.d.S.)
4. Department of Earth Ocean and Atmospheric Science, Florida State University, Tallahassee, FL 32306, USA; ims19c@my.fsu.edu (I.S.-G.); mfuentes@fsu.edu (M.M.P.B.F.); anthony.gillis12@gmail.com (A.G.); lhooper@fsu.edu (L.H.); mw15w@my.fsu.edu (M.W.); clo17@my.fsu.edu (C.O.)
5. Public Works-Coastal Division, Indian River County, Vero Beach, FL 32960, USA; qbergman@ircgov.com
6. Florida Department of Environmental Protection, Tallahassee, FL 32399, USA; Yuliya.danyuk@floridadep.gov
7. Department of Computer Science and Engineering, Columbia University, New York, NY 10027, USA; sofia.sanchez@columbia.edu
8. Department of Geosciences and Natural Resource Management, University of Copenhagen, Valby, 2500 Copenhagen, Denmark
* Correspondence: jingels@fsu.edu (J.I.); yirina80@gmail.com (Y.V.)

Received: 3 April 2020; Accepted: 13 May 2020; Published: 20 May 2020

Abstract: Sea turtles migrate thousands of miles annually between foraging and breeding areas, carrying dozens of epibiont species with them on their journeys. Most sea turtle epibiont studies have focused on large-sized organisms, those visible to the naked eye. Here, we report previously undocumented levels of epibiont abundance and biodiversity for loggerhead sea turtles (*Caretta caretta*), by focusing on the microscopic meiofauna. During the peak of the 2018 loggerhead nesting season at St. George Island, Florida, USA, we sampled all epibionts from 24 carapaces. From the subsamples, we identified 38,874 meiofauna individuals belonging to 20 higher taxa. This means 810,753 individuals were recovered in our survey, with an average of 33,781 individuals per carapace. Of 6992 identified nematodes, 111 different genera were observed. To our knowledge, such levels of sea turtle epibiont abundance and diversity have never been recorded. Loggerhead carapaces are without doubt hotspots of meiofaunal and nematode diversity, especially compared to other non-sedimentary substrates. The posterior carapace sections harbored higher diversity and evenness compared to the anterior and middle sections, suggesting increased colonization and potentially facilitation favoring posterior carapace epibiosis, or increased disturbance on the anterior and middle carapace sections. Our findings also shed new light on the meiofauna paradox: "How do small, benthic meiofauna organisms become cosmopolitan over large geographic ranges?" Considering high loggerhead epibiont colonization, the large distances loggerheads migrate for reproduction and feeding, and the evolutionary age and sheer numbers of sea turtles worldwide, potentially large-scale exchange and dispersal for meiofauna through phoresis is implied. We distinguished different groups of loggerhead carapaces based on divergent epibiont communities, suggesting distinct epibiont colonization processes. These epibiont observations hold potential for investigating loggerhead movements and, hence, their conservation.

Keywords: sea turtles; loggerheads; marine biodiversity; meiofauna; epibionts; Florida; Gulf of Mexico; meiofauna paradox; nematodes; Nematoda; hotspots; phoresis

---

1. Introduction

Epibiosis in sea turtles has gained significant attention in recent years to support cryptic migratory and foraging behaviors [1,2]. The majority of epibiotic studies have focused on describing epifaunal diversity. Rarely are community questions and ecological interactions occurring on the turtle carapace addressed [3,4], or relationships between sea turtle epibionts and the environments frequented by the turtles investigated [1]. In addition, the increased focus on sea turtle epibiosis has shown that we are far removed from complete sea turtle epibiont inventories, partly owing to the limited number of taxonomic experts available, and partly because a comprehensive epibiont community analysis is difficult to achieve.

Caretta caretta, the loggerhead sea turtle, occurs in subtropical and temperate waters across continental shelves and estuarine areas in the Atlantic, Pacific and Indian Oceans [5,6]. Throughout this range, loggerheads spend most of their time in nearshore and inshore waters, sometimes associated with reefs and other natural and artificial hard substrates [5]. Loggerheads are opportunistic carnivores, feeding primarily on benthic invertebrates and freshly deceased fish, but also gelatinous plankton [7]. Following an early pelagic developmental period [8], most sea turtles, including loggerheads, transition to more coastal and neritic habitats for feeding and reproduction. In these environments, loggerheads are exposed to intense colonization by marine larvae searching for a hard substrate to start their benthic life stage [1]. So far, over 200 epibiont taxa have been documented on loggerhead carapaces [1]. Turtle carapaces also provide suitable substrates for the smallest of motile invertebrates, including the abundant and diverse meiofauna [3,4]. While prokaryotes and microscopic metazoans are likely the first to colonize carapaces, larger, mostly sessile invertebrates can provide a habitat and structure that substantially facilitate further colonization by motile organisms, and enhance the biodiversity of the fully functional epibiotic carapace ecosystem [4,9]. Loggerhead carapaces have potential as a suitable habitat for microscopic organisms, and because of this, we wish to assess how abundant and diverse such communities are, by focusing on the meiofauna.

Meiofauna (size class between 32 μm and 1 mm, but mesh sizes may vary; identified here as between 63 μm and 1 mm), are rarely the focus of sea turtle or marine mammal epibiont studies. In fact, only a handful of studies have reported the association of meiofaunal organisms and sea turtles [1,3,4,10], yet they are a vast source of biodiversity and fulfill important ecological roles in all marine ecosystems [11]. Currently, 20 metazoan phyla and three protistan (unicellular) phyla have meiofaunal representatives [12], comprising many tens of thousands to millions of species, described and yet undiscovered [13–16]. Given the biomass and diversity of loggerhead macrofaunal epibiont communities, it is likely that loggerhead carapaces have potential to host similarly diverse and abundant meiofaunal communities. Thus, these communities would have the inherent ability to raft with turtles as they migrate, with phoretic dispersal and geographic expansion as potential consequences. Meiofauna distribution patterns from local to global scales offer insights into meiofauna dispersal, migration and phoresis to an extent, but the origins of these patterns remain debated. Meiofauna are often numerically dominant, in shallow waters all the way to the deepest ocean depths. They are also the first metazoans to colonize newly available sediments and substrates [17,18]. Many meiofauna taxa are widespread or even cosmopolitan [19–22], yet meiofauna taxa typically live their entire lives in between the sediment grains, since they do not possess pelagic larval stages; instead, they show direct development or brooding [23]. The assumed contradiction between pronounced dispersal limitations of meiofauna and yet their global presence and ubiquity is captured in the concept of the 'meiofauna paradox' and its central question: "How do small, benthic meiofauna organisms without active means of dispersal become distributed over wide geographic ranges?" However, the presence of pelagic larvae is not the single factor causing dispersal and wide distribution ranges. Many factors may be evoked to understand the distribution ranges of meiofauna, including the underestimated dispersal potential, taxonomic artefacts, biological and evolutionary phenomena like stasis or speciation, as well as geological, oceanographic and climatic processes [24]. Often overlooked in the literature is the role of rafting or phoresis in explaining the distribution of meiofaunal taxa. Rafting implies the unintended travel of small organisms on various substrates, while phoresis is defined as smaller organisms being carried by larger organisms. By assessing the abundance and diversity of meiofaunal epibiotic communities on loggerhead carapaces, we hope to shed light on the potential dispersal capabilities of meiofauna.

Here, we comprehensively document the abundance and diversity of meiofaunal higher taxa and nematode epibionts on loggerhead sea turtles in the northeastern Gulf of Mexico. The few turtle epibiont studies focused on meiofauna have reported high abundances. Thus, we expected to recover high abundances and relatively high diversity on loggerhead sea turtles, with much of it undocumented so far in the literature. To assess how variable loggerhead epibiont communities are, we tested for the similarity of meiofauna and nematode abundance, diversity and community structure among different carapace sections (anterior, middle and posterior, following [25]) and entire carapaces. Loggerheads are highly migratory, capable of traveling hundreds to thousands of kilometers between breeding and foraging areas, where colonization may occur. We therefore also address the potential of loggerhead colonization influencing meiofauna dispersal in a broader context. In addition, differences between epibiont communities may give an indication of where loggerhead turtles have been during epibiont colonization, since distinct epibiont communities from different carapace sections and entire carapaces are likely shaped by different colonization processes (on different spatial and temporal scales) and potentially related to the origin of sea turtle epibionts prior to colonization [1].

2. Material and Methods

2.1. Study Area and Epibiont Sampling

Loggerheads nesting in the northwest Atlantic Ocean comprise the largest loggerhead population globally [26], and are the population from which most epibiont studies have originated [1]. To date, studies of sea turtle epibiosis have focused on turtles nesting on the Atlantic coast of the United States [25], which hosts the majority of nesting in this region [26]. Epibiosis has yet to be documented from loggerheads nesting in the Gulf of Mexico, which forage in different habitats and locations than loggerheads nesting on the Atlantic coast [27,28]. We therefore focused sampling on turtles nesting at St. George Island, Florida, the largest nesting assemblage in the genetically discrete Northern Gulf of Mexico Recovery Unit [26,29,30].

During the peak loggerhead nesting season (June 16th–July 1st, 2018) at St George Island, we carried out a two-week, night-time survey for nesting loggerhead turtles. Teams of scientists patrolled the beach at night, looking for nesting turtles, covering 18 kilometers of the island in four ~4.5 kilometer sections; each section was patrolled several times per night. Upon encountering a turtle, teams assessed the activity of the turtle and only commenced sampling once sampling activity was allowed according to permit conditions (either the turtle did not nest and began to return to the ocean, or the turtle nested and began covering the egg chamber; MTP-18-239). Prior to epibiont collection, turtles were measured for curved carapace length (CCL), both minimum (from the anterior point to the posterior notch along the midline) and standard [notch to tip] (from the anterior point to the posterior tip of the longest supracaudal along the midline), as well as for curved width length (CWL) [31]. Turtles were also checked for Inconel identification tags; if one was not present, then a tag was applied in each front flipper. Twenty-four loggerhead carapaces were sampled, including one recapture, which was sampled a second time one day later. Turtle carapaces were divided into three sections, roughly equal in surface area: anterior (A), middle (M) and posterior (P), following [25] (Figure 1). This resulted in 67 unique samples (some anterior sections could not be sampled, and some sections could not be sampled separately, because of the limited time available before the turtle reached the water). Barnacles were efficiently removed by placing a putty knife at the base of the barnacle and tapping it with a small hammer. This caused the barnacle to come off instantly in one piece and without damage to the carapace. This method was more efficient than pulling barnacles off with pliers or intense scraping, which may cause specimen and carapace damage if the barnacle was encrusted. Each section was then carefully but exhaustively scraped with larger putty knives to collect all visible and invisible organisms. When we were confident all visible organisms were removed, we proceeded by wiping down each section using a separate, uncontaminated wet sponge to clean the carapace entirely. All epibionts, debris, overlaying sand and sponges for each carapace section were placed carefully in 500 mL wide-mouth Nalgene containers. 'Control' sand samples were taken at high (near nest), middle (between nest and tideline) and low (tideline) beach, corresponding with the location for each of seven turtles, by means of 9 cm high, 3 cm diameter plastic cores, to compare meiofauna and nematode abundance and communities found in the beach sand and on the loggerhead carapaces. Three cores were taken from each of the high, middle and low beach locations; these three cores were pooled for analysis, resulting in 21 sand samples (three beach levels x seven turtle locations). These samples were used to assess potential contamination of the carapace by the meiofauna, because during the digging and the covering of the nest, a substantial amount of sand is swept onto the carapace. Fixative (DESS, a DMSO and EDTA salt solution [32], suitable for morphological (present study), as well as molecular analysis (future studies)) was added to the carapace epibiont and beach samples as soon as possible (usually within an hour of sampling). All the samples were kept on ice, until stored in refrigerators within two days of sampling.

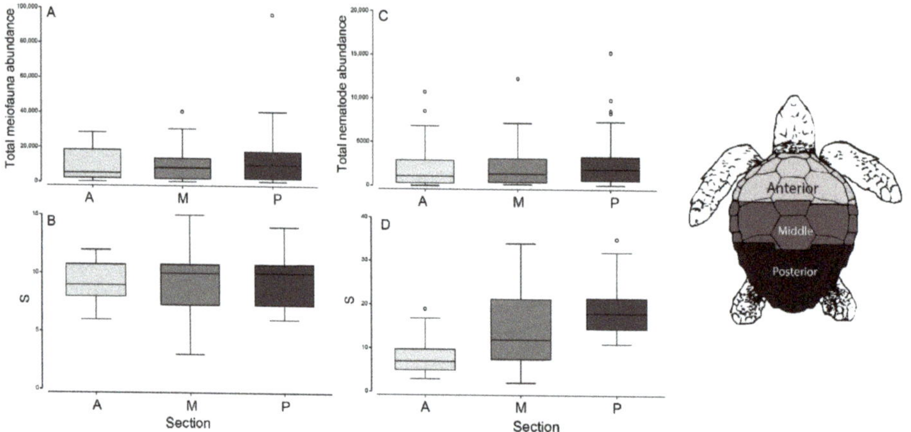

Figure 1. Meiofauna and nematode abundance and taxon richness for each carapace section, represented by Box-whisker plots (median, quartiles and ranges) S = taxon richness. (**A**) Total meiofauna abundance, (**B**) meiofauna taxa richness, (**C**) nematode abundance, (**D**) nematode genera richness.

2.2. Meiofauna and Nematode Processing

Both carapace and sand samples were processed the same way. Samples were washed over a stack of 1 mm and 63 µm sieves to separate the macrofauna (>1 mm) and meiofauna (63 µm–1 mm) size fractions at the Florida State University Coastal and Marine Laboratory. The washed macrofauna were placed in DESS in 500 mL Nalgene bottles, and the rinsed meiofauna were collected in 250 mL Nalgene bottles, also in DESS. To extract the meiofauna organisms from debris and sediments, we applied the decantation method [33]. Each meiofauna sample was placed in 1 L measuring beakers, and topped up with tap water to 550 mL; the volume was vigorously stirred (but avoiding splashes or spills) and left until the sand settled (usually within minutes), after which the supernatant was poured off on a 63 μm sieve, and collected in a 250 mL bottle. This decantation procedure was repeated 10 times for each sample.

We observed that vast numbers of organisms were collected, and that subsampling was necessary. The washed and decanted meiofauna samples were placed in a 1 L beaker (volumes were calibrated) and topped up to 800 mL with tap water. Each sample was homogenized using a magnetic stirrer (600 rpm), and 2% of the total volume was extracted using clean pipettes, ensuring that a representative subsample was taken. Following counting (see below), the subsampling procedure was repeated until at least 300 meiofauna individuals were encountered, unless the entire sample needed to be counted. This yielded a representative subsample of the community present in the sample (11% of total sample size on average). Tap water control samples were taken before, during and after the washing procedure (taps were left running for 10 min on 63 μm sieve several times), to ensure that samples were processed with contamination-free water. No organisms were recovered from these control samples.

Meiofauna were identified and counted in gridded Petri-dishes or purpose-built counting trays to higher taxon level following [34], and specific keys to invertebrates for the Gulf of Mexico, using stereoscopic microscopes (250–500 × magnification). Nematodes were picked out randomly (minimum of 120, or all if less than 120 were present) and placed in embryo dishes for diaphanization in a solution of 50% tap water, 10% glycerol and 40% alcohol, adapted from [35]. Partially covered embryo dishes were left in the oven overnight at 55 °C to allow for the evaporation of water and alcohol, with the nematodes left in pure glycerol. Embryo dishes were then placed in a cabinet desiccator until mounting on glass slides [35]. After mounting, nematodes were identified to genus level (1000× magnification), or family level, where the genus could not be identified (note that family level classifications still comprise at least one unique genus), using [36–38] and specialized nematode taxonomic literature and descriptions.

2.3. Data Analyses

Meiofauna higher taxa and nematode generic analyses were based on total abundance and counts matrices. For two loggerhead turtles, we were not able to sample carapace sections separately; these samples were not considered in analyses where carapace sections were compared, but were included in analyses where only entire carapaces were used. The total abundance data was used for carapace section analyses, while the carapace section data were summed for each turtle individual when conducting entire carapace (i.e., for each individual turtle) analyses. All multivariate analyses were conducted on standardized (by sample totals) and square root transformed data, to reduce the effect of highly variable abundances and highly dominant taxa. For multivariate data (abundance matrices) and univariate data (abundance, density and diversity indices), Bray-Curtis and Euclidean distance were used as the resemblance measures, respectively. Multivariate and univariate PERMANOVAs to test for carapace section differences were conducted using a one-way design (factor: carapace section; anterior, middle, posterior). These were followed with pairwise tests and PERMDISPs, to assess whether dispersion of homogeneity affected the results. Non-metric Multi-Dimensional Scaling ordinations (nMDS) were used to visualize community patterns, overlain with vector plots, bubble plots, and pie bubbles allowing taxa correlations and abundance of different taxa to be represented. These were accompanied by CLUSTER (including SIMPROFs with 5% significance tests) analyses to assess groups of samples that were significantly different from each other. SIMPER was used to assess the dissimilarity of the cluster groups, and to identify the taxa that were mainly responsible for differences between these groups. The DIVERSE routine was used to generate meiofauna higher taxa richness and nematode genera diversity indices (genus richness, ES(51) [39], Hills Indices (N1, N2, Ninf) [40]), covering a range of indices between pure richness (S) and evenness (dominance index = Ninf) [41]. Nematode genus richness was used to calculate cumulative dominance curves to assess diversity and evenness among carapace sections. Finally, coherence plots were generated to assess how specific genera occurrence and abundance changed among different carapace sections [42,43].

Carapace surface area was estimated as the surface area of an ellipse using A = PI × a × b (with a and b being half the minor (CWL) and major (CCL) axes values). For one turtle, width measurements were missing and replaced with calculated width, using the regression equation based on all other complete turtle measurements (y = 0.7214x + 18.198; R^2 = 0.6551). Carapace surface area ranged 5427–8076 cm^2, averaging 6413 cm^2. Surface area for the beach core samples was calculated as 3 × PI × r(1.5 cm)^2 = 21.2 cm^2 (triplicates pooled). The carapace surface areas were used to calculate meiofauna density (ind./10 cm^2) for testing differences among carapace sections. The very low abundances of meiofauna in beach samples relative to the carapace epibiont abundances, the contrast between carapace and core sample surface areas (as calculated above), and the inherent differences in the habitat complexity between hard substrate surfaces and the 3D interstitial sediment matrix made direct comparisons difficult. We therefore used only meiofauna and nematode community structure (nMDS, PERMDISP) and diversity (richness) to compare beach and carapace samples, whereby data were presence-absence transformed. This transformation reduced or eliminated the effect of inherent abundance and habitat structure differences, yet it still allowed us to assess the commonality of taxa between carapace and beach samples. All analyses and plot renderings were conducted using PRIMER v7 [42] and the PERMANOVA+ add-on [44].

3. Results

3.1. Meiofauna Communities

Abundance and density. Meiofauna abundance ranged 353–146,190 per entire carapace (average: 33,781 ± 30,596 SD). The minimum of 353 individuals occurred on the recaptured loggerhead turtle, with the next lowest abundance being 6950 individuals. Meiofauna abundance per carapace section: anterior: 333–28,540, average: 9306 ± 9244 SD; middle: 353–40,570, average: 9660 ± 9450 SD; posterior: 382–96,350, average: 13,561 ± 19,715 SD (Figure 1). There was no significant difference in meiofauna abundance nor meiofauna density (ind./10 cm^2) between the different carapace sections (PERMANOVA, p = 0.567, p = 0.598, respectively).

Community structure. Meiofauna higher taxa community structure did not show any significant differences between the carapace sections (PERMANOVA, p = 0.389), but stacked bar plots in Figure 2 show some variability in taxa composition for each section averaged across all carapaces (e.g., Copepoda, Cirripedia, nauplii and Amphipoda).

Meiofauna higher taxa community cluster analysis indicated two significantly different groups of turtle carapaces (p < 0.05). The nMDS in Figure 3 shows this clustering in relation to taxa that have Pearson correlations with the nMDS axes >0.5 (cf. overlying vectors, size represent strength of correlation), namely nauplii, Cirripedia, Caprellidae and Amphipoda. Caprellidae belong to the Amphipoda but were treated separately here, because they are specifically adapted to attach to substrates by their grasping appendages, called pereopods, unlike the other Amphipoda. Nauplii were mostly, if not all, larval harpacticoid copepod stages. The two carapace groups are distinguished mainly by the differences in nauplii abundance, as indicated by the bubble pie charts in the nMDS of Figure 3. The SIMPER analysis shows that nauplii contribute 28.49% to the differences between the two cluster groups. The smaller cluster in the nMDS has an average similarity value of 75.46%, while the larger cluster has a similarity of 68.34%; their dissimilarity was 42.69%. The prominent role of nauplii in distinguishing the two significant carapace groups is also clear in the stacked bar chart in Figure 4, where a trade off in relative abundance is noticeable between nauplii and other meiofaunal taxa.

Diversity. A total of 20 meiofauna higher taxa were recovered from loggerhead carapaces (Table S1), with 16, 18 and 18 higher taxa appearing on anterior, middle and posterior sections, respectively. Polychaete larvae and kinorhynchs were only found on the middle sections, pycnogonids only on the posterior sections. Average taxon richness per sample was 9.2 ± 2.3 SD, ranging between 3 and 15 taxa. PERMANOVA analyses did not show significant taxon richness differences between sections ($p = 0.995$). For entire carapaces, the average richness was 11.8 ± 2.4, ranging between seven and 16 taxa. Meiofauna taxon richness per section is shown in Figure 1.

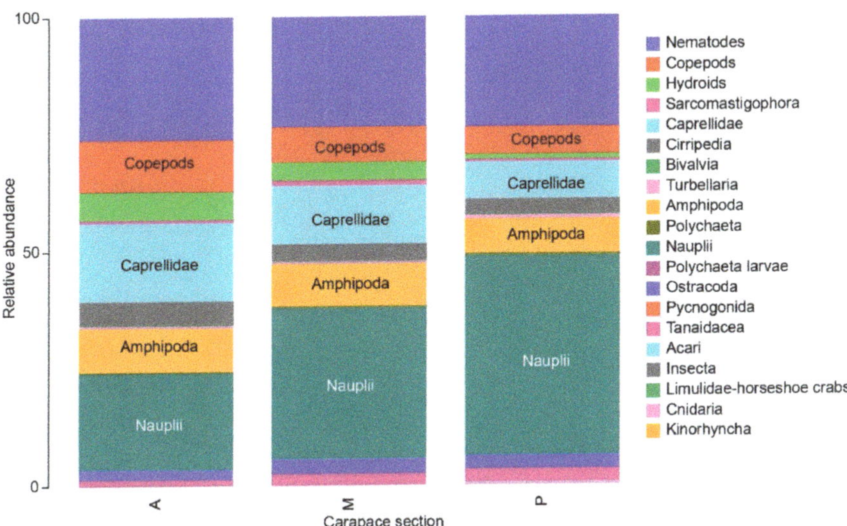

Figure 2. Relative abundance of meiofauna higher taxa and larval forms per carapace section. A: anterior section, M: middle section, P: posterior section.

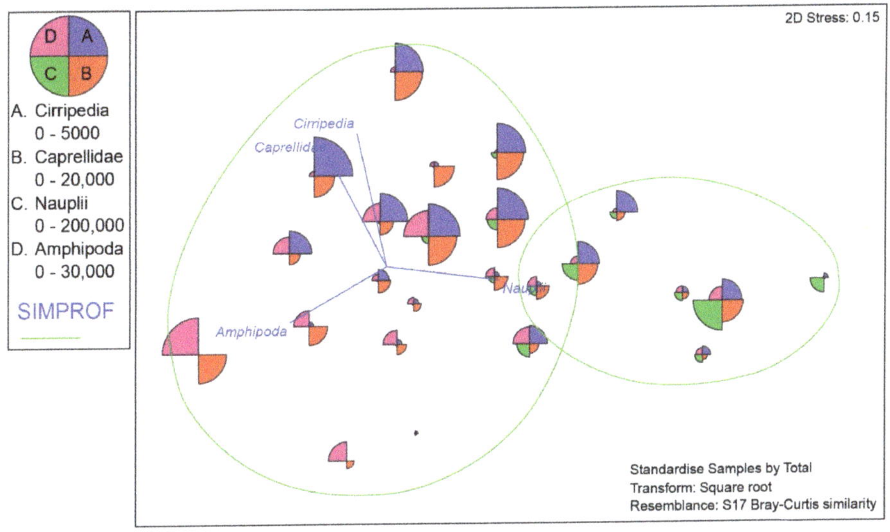

Figure 3. Non-metric Multi-Dimensional Scaling ordinations (nMDS) based on meiofauna higher taxa on loggerhead carapaces. Each carapace is represented by a bubble pie chart based on the four most indicative taxa (Nauplii, Amphipoda, Caprellidae and Cirripedia) showing their total abundance. Superimposed vectors represent Pearson correlations between these four taxa (correlation coefficient >0.5) and the nMDS axes. Green lines separate two main groups based on significant cluster SIMPROF analysis ($p < 0.05$).

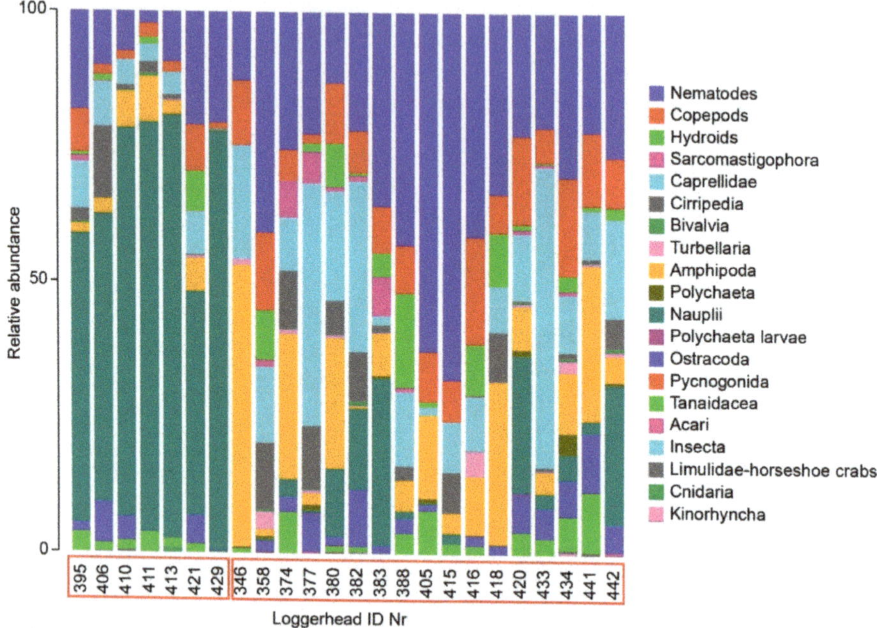

Figure 4. Relative abundance of meiofauna higher taxa for each turtle carapace. Red boxes indicate the significant cluster groups (SIMPROF analysis using 5% significance test).

3.2. Nematode Communities

Abundance and density. Nematode abundance ranged 6–15,300 individuals per carapace section (average 2656 ± 3207), while per entire carapace nematode abundance ranged 240–27,600 (average 8152 ± 7549) (Figure 1). A total of 195,648 nematodes were found on loggerhead carapaces. Nematode abundance and density (ind./10 cm^2) did not differ significantly between the carapace sections (PERMANOVA, $p = 0.561$, 0.522, respectively).

Community structure. Nematode community structure differed significantly between the carapace sections (PERMANOVA, $p = 0.002$), with pairwise tests indicating significant differences between the posterior and anterior sections, as well as between the posterior and middle sections (PERMANOVA, $p = 0.001$, $p = 0.011$, respectively). No significant differences were observed between the anterior and middle sections (PERMANOVA, $p = 0.411$). PERMDISP analyses to assess whether these differences were caused by heterogeneity in dispersions were not significant ($p = 0.986$). These differences can be observed in Figure 5, where a reduced abundance of *Chromadora* and *Theristus*, and increased abundance of *Odontanticoma* occur in the posterior carapace sections. Coherence curves were calculated to assess genera that exhibited significantly similar patterns among carapace sections. This analysis showed that eight different groups of genera (based on the 30 most important genera only) can be distinguished based on their abundance on different sections (Figure S1). Accordingly, five genera showed highest abundance on anterior sections (*Chromadora, Chromadorella, Euchromadora, Theristus* and Xyalidae), and a larger number of genera for the posterior sections, such as *Araeolaimus, Prochromadorella, Synonema, Desmolaimus, Marylynnia, Oncholaimellus* and *Paryeurystomina*, among others. Only three genera showed highest abundance middle sections, *Metachromadora, Metalinhomoeus* and *Microlaimus*.

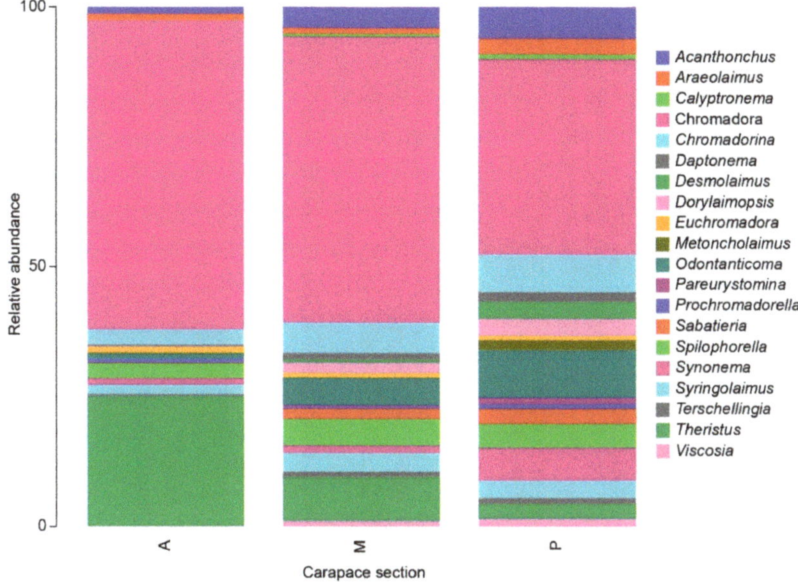

Figure 5. Relative abundance of nematode genera per carapace section. A: anterior section, M: middle section, P: posterior section. Only the 20 most abundant genera are shown.

Nematode genera community cluster analysis indicated four significantly different groups of turtle carapaces ($p < 0.05$). The nMDS in Figure 6 shows this clustering in relation to genera that have Pearson correlations with the nMDS axes > 0.7 (cf. overlying vectors, size represent strength of correlation), namely *Chromadora*, *Daptonema* and *Acanthonchus*. Stacked bar plots in Figure 7 show the structuring of genus composition across the range of entire carapaces, supported by the cluster analyses. The SIMPER similarity and dissimilarity for each of these groups and their pairwise comparisons is given in Table S2, along with the main genera responsible for the (dis)similarities.

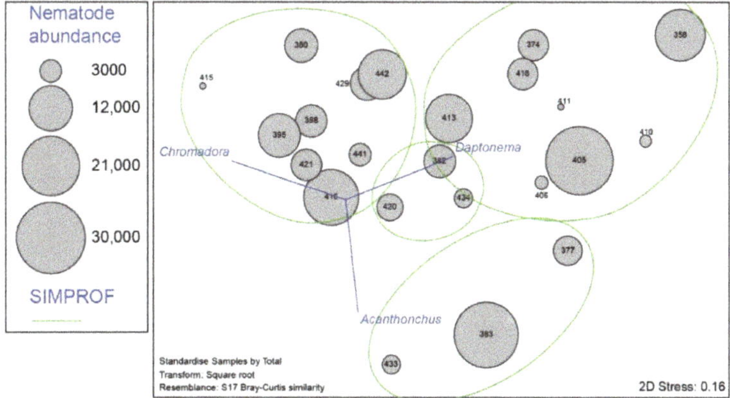

Figure 6. nMDS based on nematode genera on loggerhead carapaces. Each carapace is represented by a bubble representative of total nematode abundance on each carapace. Superimposed vectors represent Pearson correlations between the taxa (correlation coefficient > 0.7) and the nMDS axes, comprising *Chromadora*, *Daptonema* and *Acanthonchus*. Green lines separate four main groups based on significant cluster SIMPROF analysis ($p < 0.05$).

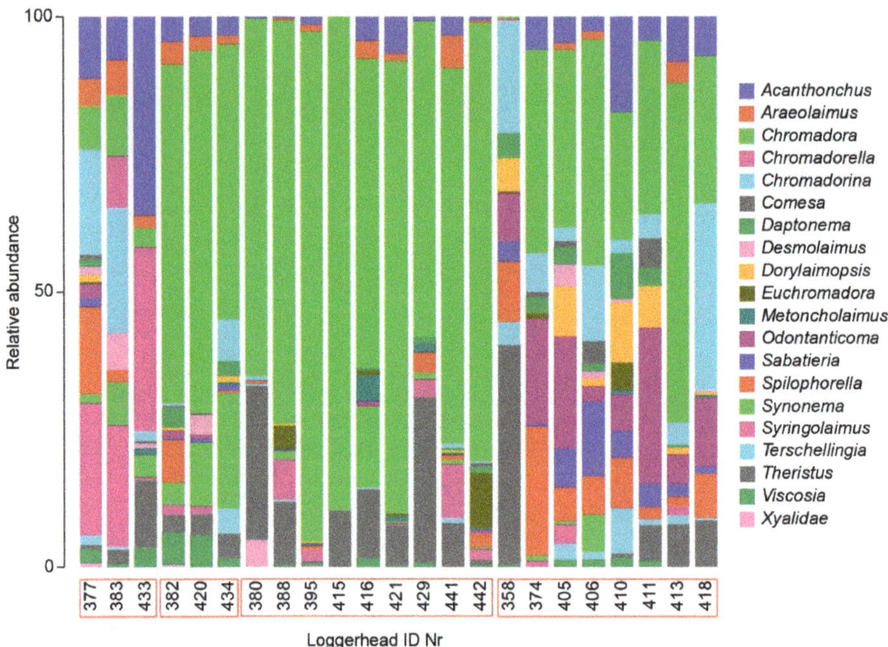

Figure 7. Relative abundance of nematode genera for each turtle carapace. Red boxes indicate the significant cluster groups (SIMPROF analysis using 5% significance test). Only the 20 most abundant genera are shown.

Diversity. A total of 111 nematode genera were observed on loggerhead carapaces. On individual carapace sections, between two and 35 nematode genera were found (average 14.1 ± 8.0), suggesting relatively high turnover between individual samples and carapace sections. The total number of genera on carapace sections were 51, 79 and 92, for anterior, middle and posterior sections, respectively. This suggests greater diversity on posterior sections, as also indicated by: 1) the average genus richness values (Table 1) the cumulative dominance curves, which show evenness and diversity increasing along the anterior, middle, posterior carapace section gradient (Figure S2); and 3) PERMANOVA test ($p = 0.001$,), with pairwise testing showing significant differences ($p < 0.05$) between all the carapace sections (Table 2; PERMDISP non-significant with $p > 0.05$).

Table 1. Summary statistics on the number of nematode genera recovered from samples, sections and entire carapaces collected from loggerhead turtles nesting at St. George Island, Florida during the peak of the 2018 nesting season.

Sample Size	Minimum	Maximum	Average	SD	Total
Anterior	3	19	8.1	4.5	51
Middle	2	34	14.0	8.4	79
Posterior	11	35	19.2	6.5	92
Entire carapace	2	50	26.6	11.0	111

On entire carapaces, between two and 50 genera were observed (average 26.6 ± 11.0), suggesting a reasonably high turnover between turtle individuals. Only two genera were recovered from the recaptured loggerhead, while the minimum number of nematode genera from newly encountered turtles was 16. Further details on number of nematode genera are given in Table 1. Other diversity indices all show the same trend of increasing diversity and evenness from anterior to middle to posterior carapace sections, supported by main PERMANOVA tests with $p < 0.05$ and multiple significant pairwise tests (Table 2, Figure 8).

Table 2. Main and pairwise PERMANOVA test results for nematode genera diversity indices. Comparison between carapace sections, cf. Figure 8. Df: degrees of freedom; SS: sums of squares, MS: means of squares, perm(s): permutation(s), A: anterior, M: middle, P: posterior, mc: Monte Carlo values when the number of permutations were low. Significant values are given in bold.

	df	SS	MS	Pseudo-F	P(Perm)	Perms	Pairwise Groups	t	P(Perm)	Perms
S-sections	2	1285.3	642.66	13.933	**0.001**	883	A, M	2.7304	**0.009/0.01** (mc)	124
Residual	62	2859.7	46.124				A, P	6.3076	**0.001/0.001** (mc)	125
							M, P	2.3529	**0.021/0.027** (mc)	73
ES(51)-sections	2	408.78	204.39	11.713	**0.002**	999	A, M	2.441	**0.017**	998
Residual	62	1081.9	17.45				A, P	5.2855	**0.001**	995
							M, P	2.355	**0.024**	998
N1-sections	2	295.65	147.83	7.4929	**0.001**	999	A, M	2.252	**0.017**	996
Residual	62	1223.2	19.729				A, P	3.7725	**0.001**	998
							M, P	1.9165	0.059	996
N2-sections	2	108.69	54.343	5.6309	**0.009**	999	A, M	1.8712	0.073	998
Residual	62	598.35	9.6508				A, P	3.0995	**0.001**	998
							M, P	1.8111	0.078	997
Ninf-sections	2	15.959	7.9794	4.3242	**0.013**	999	A, M	1.5188	0.153	998
Residual	62	114.41	1.8453				A, P	2.7645	**0.004**	998
							M, P	1.611	0.109	997

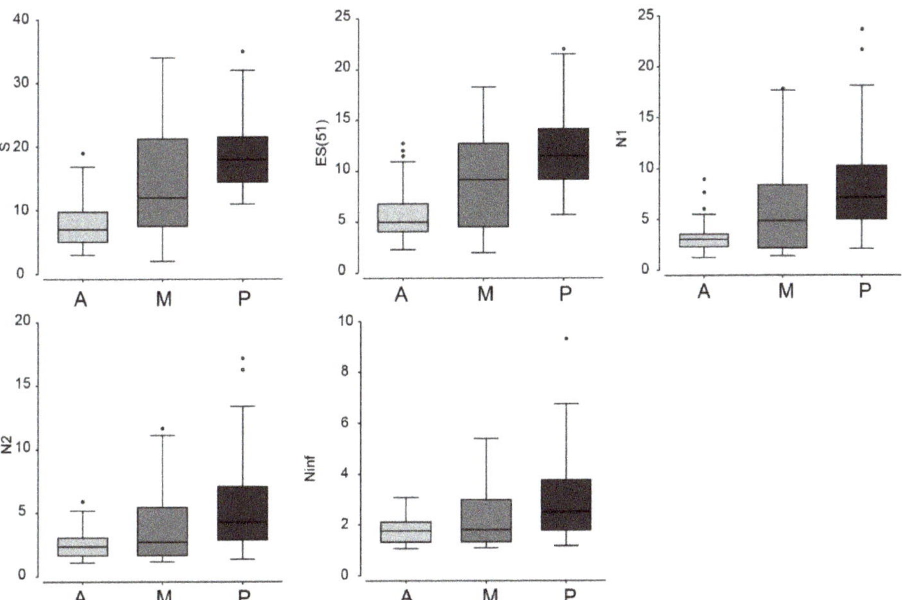

Figure 8. Box and Whisker (median, quartiles and ranges) plots of nematode genera diversity indices for each carapace section. S: nematode genus richness, ES(51): expected number of genera based on 51 individuals, N1: Hill's index 1, N2: Hill's index 2, Ninf: Hill's index inf [40]. A = anterior, M = middle, P = posterior sections.

3.3. Beach vs. Carapace Samples

Sand core samples contained minimal meiofauna abundance compared to carapace samples, with 17–889 individuals only, averaging 136.2 individuals per triplicate sample (low beach samples contained more individuals than high beach samples), which is much lower compared to the epibiont meiofauna and nematode abundance values (Table S1). The nMDS results show a clear distinction in meiofauna and nematode community composition between beach and carapace samples (Figure 9), confirmed by PERMANOVA analysis (meiofauna: $p = 0.001$; nematodes: $p = 0.001$). Within the beach samples, a clear distinction can be made between low, middle and high beach. It is noticeable that the meiofauna higher taxa communities occupy less Bray-Curtis space in the nMDS for the carapace epibionts, compared to the beach samples (PERMDISP, $p = 0.001$), suggesting greater similarities between the carapace meiofauna communities than between the beach meiofauna communities. This is not the case for the nematode genera data (PERMDISP, $p = 0.64$), implying no differences in the nematode community heterogeneity of dispersion between the beach and carapace samples. Carapace samples had a much higher meiofauna diversity and nematode genus richness than the beach samples (20 vs. 11 meiofauna higher taxa, and 111 vs. 25 nematode genera).

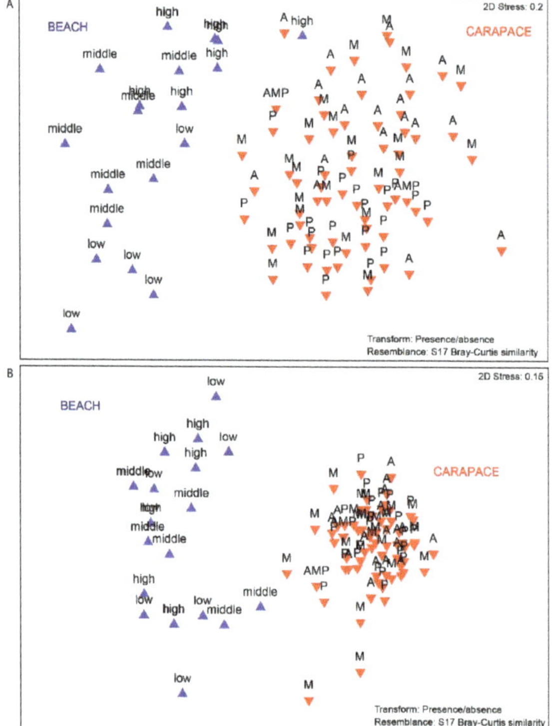

Figure 9. nMDS of (**A**) nematode genera community data and (**B**) meiofauna higher taxa data. Presence–absence transformation was used to eliminate the large abundance differences between the carapace and beach samples, making the comparison based on community composition only. Carapace samples: A = Anterior, M = Middle, P = Posterior; beach samples: high, middle and low beach.

4. Discussion

4.1. Diversity and Structure of Meiofauna on Loggerhead Carapaces

From our results, it is clear that loggerhead carapaces are hotspots of meiofauna and nematode diversity; our observations document more than double the species diversity previously observed on loggerhead carapaces [22,36,37]. Two recent epibiont study on meiofauna and nematodes from Brazilian hawksbill turtles (*Eretmochelys imbricata*) observed 17 meiofauna higher taxa [2] and 80 nematode genera [3], suggesting that loggerheads may harbor between 15% to 30% more diversity than hawksbill turtles. High epibiont diversity for loggerheads has been recorded in other studies (e.g., [22]). For example, dos Santos et al. [3] compared nematode genera richness between hawksbill turtles and non-sedimentary natural and artificial substrates, demonstrating that turtle carapaces are hotspots of meiofauna and nematode biodiversity (Table S3; [45–62]). However, it is worth noting the foraging differences between hawksbill and loggerhead sea turtles. Hawksbills are spongivorous [63] and are generally associated with tropical coral reefs, while loggerheads are benthivores, specialized in consuming benthic invertebrates. There are also distinct differences in the way they consume prey, with evidence that hawksbills can exhibit distinct flipper use to leverage, hold or sweep prey, such as sponges, cnidarians, macroalgae and fishes, while loggerheads can use their flippers to dig into the sediment to gain better access to benthic molluscs [64], in a process coined 'infaunal mining' [65]. In comparison, loggerheads are able to displace much more sediment and, hence, infaunal invertebrates into the water column during feeding activity [64,65], enabling increased colonization of their carapaces by small invertebrates. The difference in general foraging environments (i.e., coral reefs vs. sedimented habitats) and foraging techniques between these two turtle species certainly suggests that epibiont colonization is influenced differently as well.

A number of factors may affect the distribution of epibionts on loggerhead carapaces (and sea turtle carapaces in general), including different water flow patterns, variable desiccation, food accessibility and abrasion or disturbance [22], but also presence of structure-forming, sessile invertebrate species [2,3]. Pfaller et al. [22] reported highest epibiont densities on the posterior carapace section of loggerheads (and vertebral zones, which we did not distinguish in this study), based on distribution patterns of 18 macro-invertebrate taxa on 18 loggerhead carapaces. We did not find such abundance or density differences for meiofauna and nematodes, but our results reveal increased nematode diversity on the posterior section (Table 1, Figures 1–8, Figure S2). Similar to Pfaller et al. [22], our analyses do not allow to discern whether this is consequence of differential colonization or recruitment, or different rates of survival of the epibionts in the different carapace sections. Both are related to complex interactions of processes, including water flow and disturbance or abrasion, turtle behavior and epibiont interactions, as well as the resilience of the epibionts themselves. However, our data suggests that such processes have a limited effect on the abundance and density of meiofauna and nematodes, with no significant difference between the anterior, middle and posterior carapace sections. Dos Santos et al. [3] and Pfaller et al. [22] demonstrated convincingly that the interactions between structure forming macro-invertebrates, such as barnacles, may facilitate an increased abundance/density and diversity of the entire epibiont assemblage, through creating more favorable water flow patterns, increased food settlement and generally providing increased micro-habitat space. Our meiofauna-focused dataset does currently not allow us to answer this question, but we suspect that macrofauna-meiofauna relationships are indeed also facilitative, promoting abundance and diversity as demonstrated for hawksbill epibiont assemblages [3]. Such detailed epibiont ecological interaction studies are few, despite the insights they can provide on colonization processes and dynamics. Future studies should address these ecological relationships between macrofauna and meiofauna epibiont communities on sea turtle carapaces.

With regards to the higher nematode diversity on the posterior loggerhead carapace sections, we follow Pfaller et al.'s [22] reasoning (while necessarily omitting the effect of macrofauna epibiont distributions at this time) that increased flow in the anterior section, and flipper contact with the sides of anterior and middle sections, may affect nematode diversity. In addition, the potential desiccation of the upper middle carapace sections during periods of extended floating at the water surface may have an effect on nematode diversity. Nematode communities respond to many, if not all, types of disturbance, with nematode community shifts likely to occur [8,38], often with reduced diversity and evenness and shifts to more opportunistic species that exhibit greater dominance when conditions deteriorate, or when disturbance precludes the survival or competitiveness of more vulnerable, less resilient species. The much higher abundance of the nematode genus *Theristus*, known to be an opportunistic colonizer [39], on the anterior carapace section compared to the poster section (Figure 5, Figure S1) would be indicative of a more disturbed environment. This is in line with the increasing diversity and evenness from anterior to posterior, with increased dominance in the anterior section (Figure 8). A more detailed, comprehensive analyses of functional diversity and life histories of meiofauna and nematode taxa may provide better insight into meiofauna distribution patterns and their response to variable disturbance on turtle carapaces. Perhaps more notable is the much higher abundance of nauplii on the posterior section compared to the anterior and middle section (with an inclining gradient from anterior to posterior), supporting the notion of increased flow in the anterior section on meiofauna communities. Nauplii larvae are substantially less equipped than copepodites and adult copepods to deal with increased flow [40]. Nauplii have reduced movement ability compared to copepod adults, which have more developed appendages to attach to substrates. However, we do suspect that macro-epifauna patchiness creates suitable habitats able to protect copepods and other taxa from disturbance to an extent.

4.2. Meiofauna Paradox

The wide distribution of many meiofauna species in combination with the assumption of limited dispersal ability because of a lack of pelagic life stages continues to receive much attention [23,24]. In a recent review, Cerca et al. [24] present a number of reasons to why the meiofauna paradox is likely not paradoxical, mainly because of taxonomic issues, and the incongruence between morphological and molecular information to distinguish between species and populations [20,66,67], and the increasing evidence that dispersal of meiofauna individuals is ubiquitous, instead of rare and intermittent [68]. One process that plays potentially a significant role in explaining meiofauna distribution patterns is rafting or phoresis: the ability of meiofauna to 'hitch-hike' with larger organisms, debris or plant material. Other host materials or substrates include macro-algae and microphytobenthos-bacteria mats releasing from the seafloor, or floating ice or sea foam [22–24]. Other meiofauna dispersal can be caused by ships (particularly in sands of ballast water, and fouling), and through marine snow, rich biotic aggregates that can carry diverse meiofauna and are known to travel for long distances. Other examples include different types of floating debris, and even large mammals such as whales, or fouling on human structures, mobile or otherwise, but these are rarely quantified or estimated. For a brief review on the topic we refer to [23].

Molecular evidence precludes the notion that widespread meiofauna species distributions are the result of purely geological processes or vicariance. Instead, genetic exchange can occur resulting in genetic similarities across 1000s of kilometers [20,22,69]. In addition, it has been documented that meiofauna can disperse through active emergence and re-entry into the sediments (during which entrainment by currents or flows can occur), and passively, through erosion resuspension, entrainment and subsequent resettlement (the latter can be active or passive, see for example [17,23,70,71]). Disturbance, especially in high-energy coastal environments, is likely one of the main actors in meiofauna dispersal. Based on our results, we support the notion that the widespread distribution of meiofauna species as a result of phoretic dispersal is likely. The potential for rapid colonization is supported by our observations of epibionts on the recaptured loggerhead; one day, after having returned to the ocean, several hundreds of meiofauna individuals were found on the carapace. However, our comparison of meiofauna higher taxa and nematode genera communities (Figure 9) suggest that epibionts on loggerhead carapaces were not contaminated by sand from the nesting beach itself.

Instead, we must assume these 'hitch hikers' arrive on the carapace during time spent in the water or in close contact with the seafloor, and may have travelled with the turtles over short to long distances, and likely over variable time scales.

Female loggerheads from one nesting assemblage do not migrate to just one foraging area, and individuals may feed at a series of coastal areas en route to resident foraging areas [72]. Distances between loggerhead foraging grounds and their nesting beaches of 100s to 1000s of kilometers have been documented [27,73–76], and studies from the northeastern Gulf of Mexico have shown that post-nesting loggerheads can remain on the Florida continental shelf, or move to long-distance foraging grounds near Mexico and in the Atlantic [27,28,77]. When long-distance loggerhead migrations are accounted for, the potential for meiofaunal dispersal is vast. Considering that we found, on average, >30,000 meiofauna individuals per carapace, and the current global loggerhead sea turtle population is estimated at 314,000 individuals (midpoint using [78,79], with a range between 91,000 and 536,000), billions of meiofauna organisms are transported by loggerheads on a continuous basis with a high potential for dispersal. All sea turtles carry epibionts to a greater or lesser extent [2–4,9,80–82], and sea turtle global abundance is estimated at 6,461,000 (total global midpoint abundance estimate), including green turtles, hawksbills, olive ridleys, leatherbacks, Kemp's ridleys and flatbacks [78,79]. High turtle and meiofauna epibiont numbers, together with the fact that current sea turtle populations, are a fraction of what they were compared to in the 19th century for certain species [83–85], and the >100 million years of evolutionary existence of marine turtles [86–88], suggest a vast potential for coastal meiofauna species dispersal by means of resuspension and phoresis on an ecological, as well as an evolutionary, time scale.

Without an accurate understanding, however, of the meiofauna colonization processes on carapaces, the genetic structuring of epibiont communities and knowledge of the genetic exchange between meiofauna communities of coastal locations frequented by sea turtles, such numbers remain conjecture. Further insight into the colonization patterns of meiofauna epibionts on sea turtle carapaces may be gained through investigating previously captured, tagged and sampled turtles for epibionts when these return for nesting; for loggerheads, every two to four years [89]. Our sampling effort for two weeks, during the peak of the 2018 nesting season, yielded 23 turtles which returned to the ocean, with carapaces that were entirely cleaned of epibionts. Ideally, we would recapture these (by using their tag numbers) in subsequent seasons, to be able to investigate colonization by epibionts over a fixed period of time. This will help us understand better what type of colonization processes have occurred ab initio at foraging grounds.

4.3. Meiofauna and Nematode Epibiont Community Structure across Turtle Individuals

Our results indicated significant meiofauna and nematode community structure differences between loggerhead individuals. Regarding meiofauna higher taxa, two significantly different groups were recognized (Figures 3 and 4), and with respect to nematode genera, four different groups of epibiont communities could be distinguished (Figures 6 and 7). These communities have distinct features, characterized by different diversities and abundances of particular taxa. Whatever the colonization and environmental or ecological processes are that have shaped these communities, there is a very high likelihood that epibiont members within each of these distinct groups have similar colonization patterns that are related to the origin of these colonizers (e.g., [90]). Colonization is dependent upon a spatial and temporal overlap between turtles and colonizers, which should occur at turtle foraging areas [1]. Epibiont species identification and comparison with global or regional coastal species data sets and distributions, as well as trophic (natural stable isotopes) analyses of the epibionts, may not only give information on the origin of the epibionts, but also on the behavior and movements of the turtle hosts (e.g., [91–93]). Few prior studies have attempted to use epibionts to evaluate turtle foraging behaviors, but it proved difficult to distinguish the influence of individual foraging behavior and broad foraging habitat on epibiotic communities [92]. This type of research would be greatly facilitated by molecular analyses on meiofauna epibionts and drawing comparisons between these turtle epibionts and meiofauna from their suspected foraging and breeding areas (as determined via satellite telemetry and stable isotope analysis; see [94]). In the absence of such data, we are reliant on traditional taxonomic identifications, at least for the meiofauna, as we are not aware of molecular meio-epibiont studies for sea turtles. Comprehensive higher-level (e.g., meiofauna higher taxa and nematode genera) taxonomic analyses may reveal clustering patterns and distinguish groups of sea turtles, as is the case in the present study. In a subsequent step, aligning and comparing such epibiont data, as well as their functional diversity and other characteristics with trophic and genetic information from the sea turtles themselves, if available, may reveal corresponding patterns that give insights into the movement of the sea turtles. This would be particularly useful when costly satellite tagging programs are not feasible. Such research could contribute to turtle conservation and management through the identification of the origin of epibiont species or community-types that are likely linked to different sea turtle foraging environments and potentially geographically distinguishable areas.

Supplementary Materials: The following are available online at http://www.mdpi.com/1424-2818/12/5/203/s1, Figure S1: Coherence abundance curves for distinct groups of nematodes, Figure S2: Cumulative dominance curves of nematode genera, Table S1: Total abundance of meiofauna higher taxa on loggerhead carapaces, Table S2: Similarity and dissimilarity values for nematode community cluster groups, Table S3: Nematode genus richness on non-sedimentary substrates from previously published studies.

Author Contributions: Conceptualization, J.I., Y.V., M.M.P.B.F., G.A.P.d.S.; Data curation, J.I., I.S.-G., G.A.P.d.S.; Formal analysis, J.I.; Funding acquisition, J.I., I.S.-G., M.M.P.B.F., G.A.P.d.S.; Investigation, J.I., Y.V., L.P.P., A.C.S., P.F.N., G.V.V.C., I.S.-C., A.G., L.H., M.W., C.O., Q.B., J.D., S.S.Z., L.I.A.N., G.A.P.d.S.; Methodology, J.I., Y.V., L.P.P., A.C.S., I.S.-G., M.M.P.B.F., G.A.P.d.S.; Project administration, J.I., I.S.-G., M.M.P.B.F., G.A.P.d.S.; Resources, J.I., Y.V., L.P.P., A.C.S., I.S.-G., M.M.P.B.F., A.G., M.W., Q.B., G.A.P.d.S.; Software, J.I.; Supervision, J.I., Y.V., I.S.-G., M.M.P.B.F., A.G., L.H., M.W., Q.B., J.D., G.A.P.d.S.; Validation, J.I., G.A.P.d.S.; Visualization, J.I.; Writing—original draft, J.I.; Writing—review & editing, J.I., Y.V., I.S.-G., M.M.P.B.F., G.A.P.d.S. All authors have read and agreed to the published version of the manuscript.

Funding: This research was funded by Florida Sea Turtle License Plate Grant Nr. 18-021R and PADI Grant Nr. 33006.

Acknowledgments: We gratefully acknowledge the funding agencies that supported this work (Florida Sea Turtle License Plate Grant Nr. 18-021R; PADI Grant Nr. 33006). We are indebted to a number of people that have helped with the surveys and project logistics, including Matthew Aguiar, Hector Barrios-Garrido, Janice Becker, Emily Drobes, Andres de la Fe, Katherine Glibowski, Andrew Ibarra, Tayla Lovemore, Natalie Montero, Breanna Muszynski, Dylan Nusche, Tara Wah, Barry Walton, Natalie Wildermann.

Conflicts of Interest: The authors declare no conflict of interest. The funders had no role in the design of the study; in the collection, analyses, or interpretation of data; in the writing of the manuscript, or in the decision to publish the results.

References

1. Frick, M.G.; Pfaller, J.B. Sea Turtle Epibiosis. In *The Biology of Sea Turtles, Volume III*; CRC Press: Boca Raton, FL, USA, 2013; pp. 399–426.

2. Robinson, N.J.; Figgener, C.; Gatto, C.; Lazo-Wasem, E.A.; Paladino, F.V.; Tomillo, P.S.; Zardus, J.D.; Pinou, T. Assessing potential limitations when characterising the epibiota of marine megafauna: Effect of gender, sampling location, and inter-annual variation on the epibiont communities of olive ridley sea turtles. *J. Exp. Mar. Biol. Ecol.* **2017**, *497*, 71–77. [CrossRef]

3. Correa, G.V.V.; Ingels, J.; Valdes, Y.V.; Fonseca-Genevois, V.G.; Farrapeira, C.M.R.; Santos, G.A.P. Diversity and composition of macro- and meiofaunal carapace epibionts of the hawksbill sea turtle (Eretmochelys imbricata Linnaeus, 1822) in Atlantic waters. *Mar. Biodivers.* **2014**, *44*, 391–401. [CrossRef]

4. dos Santos, G.A.P.; Corrêa, G.V.V.; Valdes, Y.; Apolônio Silva de Oliveira, D.; Fonsêca-Genevois, V.G.; Silva, A.C.; Pontes, L.P.; Dolan, E.; Ingels, J. Eretmochelys imbricata shells present a dynamic substrate for a facilitative epibiont relationship between macrofauna richness and nematode diversity, structure and function. *J. Exp. Mar. Biol. Ecol.* **2018**, *502*, 153–163. [CrossRef]

5. Dodd, C.K., Jr. *Synopsis of the Biological Data on the Loggerhead Sea Turtle Caretta caretta (Linnaeus 1758)*; Florida Cooperative Fish and Wildlife Research Unit: Gainesville, FL, USA, 1988.

6. Dodd, C.K.; Byles, R. Post-nesting movements and behavior of loggerhead sea turtles (Caretta caretta) departing from east-central Florida nesting beaches. *Chelonian Conserv. Biol.* **2003**, *4*, 530–536.

7.   Frick, M.G.; Williams, K.L.; Bolten, A.B.; Bjorndal, K.A.; Martins, H.R. Foraging ecology of oceanic-stage loggerhead turtles Caretta caretta. *Endanger. Species Res.* **2009**, *9*, 91–97. [CrossRef]

8.   Carr, A. *So Excellent a Fishe: A Natural History of Sea Turtles*; University Press of Florida: Ganinesville, FL, USA, 1967.

9.   Robinson, N.J.; Majewska, R.; Lazo-Wasem, E.A.; Nel, R.; Paladino, F.V.; Rojas, L.; Zardus, J.D.; Pinou, T. Epibiotic diatoms are universally present on all sea turtle species. *PLoS ONE* **2016**, 11. [CrossRef]

10.   Schärer, M.T. A survey of the epibiota of Eretmochelys imbricata (Testudines: Cheloniidae) of Mona Island, Puerto Rico. *Rev. Biol. Trop.* **2003**, *51*, 87–90.

11.   Schratzberger, M.; Ingels, J. Meiofauna matters: The roles of meiofauna in benthic ecosystems. *J. Exp. Mar. Biol. Ecol.* **2018**, *502*, 12–25. [CrossRef]

12.   Baguley, J.; Coull, B.; Chandler, G. Meiobenthos. In *Encyclopedia of Ocean Sciences*; Cochran, J., Bokuniewicz, H., Yager, P., Eds.; Elsevier: Amsterdam, The Netherlands; Academic Press: Cambridge, MA, USA, 2019; Volume 2.

13.   Appeltans, W.; Ahyong, S.T.; Anderson, G.; Angel, M.V.; Artois, T.; Bailly, N.; Bamber, R.; Barber, A.; Bartsch, I.; Berta, A.; et al. The Magnitude of Global Marine Species Diversity. *Curr. Biol.* **2012**, *22*, 2189–2202. [CrossRef]

14.   Lambshead, P.J.D. Recent developments in marine benthic biodiversity research. *Oceanis* **1993**, *19*, 5–24.

15.   Lambshead, P.J.D. Marine Nematode Biodiversity. In *Nematology: Advances and Perspectives Vol. 1: Nematode Morphology, Physiology and Ecology*; Chen, Z.X., Chen, S.Y., Dickson, D.W., Eds.; CABI Publishing: London, UK, 2004; Volume 1, pp. 436–467.

16.   Lambshead, P.J.D.; Boucher, G. Marine nematode deep-sea biodiversity–hyperdiverse or hype? *J. Biogeogr.* **2003**, *30*, 475–485. [CrossRef]

17.   Ullberg, J.; Olafsson, E. Free-living marine nematodes actively choose habitat when descending from the water column. *Mar. Ecol. Prog. Ser.* **2003**, *260*, 141–149. [CrossRef]

18.   Ullberg, J.; Olafsson, E. Effects of biological disturbance by Monoporeia affinis (Amphipoda) on small-scale migration of marine nematodes in low-energy soft sediments. *Mar. Biol.* **2003**, *143*, 867–874. [CrossRef]

19.   Coomans, A. Nematode systematics: Past, present and future. *Nematology* **2000**, *2*, 3–7. [CrossRef]

20.   Derycke, S.; Remerie, T.; Backeljau, T.; Vierstraete, A.; Vanfleteren, J.; Vincx, M.; Moens, T. Phylogeography of the Rhabditis (Pellioditis) marina species complex: Evidence for long-distance dispersal, and for range expansions and restricted gene flow in the northeast Atlantic. *Mol. Ecol.* **2008**, *17*, 3306–3322. [CrossRef] [PubMed]

21.   Bhadury, P.; Austen, M.; Bilton, D.; Lambshead, P.; Rogers, A.; Smerdon, G. Evaluation of combined morphological and molecular techniques for marine nematode (Terschellingia spp.) identification. *Mar. Biol.* **2008**, *154*, 509–518. [CrossRef]

22.   Worsaae, K.; Kerbl, A.; Vang, Á.; Gonzalez, B.C. Broad North Atlantic distribution of a meiobenthic annelid–against all odds. *Sci. Rep.* **2019**, *9*, 1–13. [CrossRef] [PubMed]

23.   Giere, O. *Meiobenthology: The Microscopic Motile Fauna of Aquatic Sediments*, 2nd ed.; Springer: Berlin, Germany, 2009; p. 527.

24.   Cerca, J.; Purschke, G.; Struck, T.H. Marine connectivity dynamics: Clarifying cosmopolitan distributions of marine interstitial invertebrates and the meiofauna paradox. *Mar. Biol.* **2018**, *165*, 123. [CrossRef]

25.   Pfaller, J.B.; Bjorndal, K.A.; Reich, K.J.; Williams, K.L.; Frick, M.G. Distribution patterns of epibionts on the carapace of loggerhead turtles, Caretta caretta. *Mar. Biodivers. Rec.* **2008**, *1*. [CrossRef]

26.   National Marine Fisheries Service and US Fish and Wildlife Service. *Recovery Plan for the Northwest Atlantic Population of the Loggerhead Sea Turtle (Caretta caretta), Second Revision*; National Marine Fisheries Service: Silver Spring, MD, USA, 2008.

27.   Hart, K.M.; Lamont, M.M.; Fujisaki, I.; Tucker, A.D.; Carthy, R.R. Common coastal foraging areas for loggerheads in the Gulf of Mexico: Opportunities for marine conservation. *Biol. Conserv.* **2012**, *145*, 185–194. [CrossRef]

28.   Hart, K.M.; Lamont, M.M.; Sartain, A.R.; Fujisaki, I. Migration, Foraging, and Residency Patterns for Northern Gulf Loggerheads: Implications of Local Threats and International Movements. *PLoS ONE* **2014**, *9*, e103453. [CrossRef] [PubMed]

29. Shamblin, B.M.; Bolten, A.B.; Abreu-Grobois, F.A.; Bjorndal, K.A.; Cardona, L.; Carreras, C.; Clusa, M.; Monzón-Argüello, C.; Nairn, C.J.; Nielsen, J.T. Geographic patterns of genetic variation in a broadly distributed marine vertebrate: New insights into loggerhead turtle stock structure from expanded mitochondrial DNA sequences. *PLoS ONE* **2014**, 9. [CrossRef] [PubMed]

30. Fuentes, M.M.P.B.; Gredzens, C.; Bateman, B.L.; Boettcher, R.; Ceriani, S.A.; Godfrey, M.H.; Helmers, D.; Ingram, D.K.; Kamrowski, R.L.; Pate, M.; et al. Conservation hotspots for marine turtle nesting in the United States based on coastal development. *Ecol. Appl.* **2016**, *26*, 2708–2719. [CrossRef] [PubMed]

31. Bolten, A.B. Techniques for Measuring Sea Turtles. In *Research and Management Techniques for the Conservation of Sea Turtles*; Eckert, K.L., Bjornal, K.A., Abreu-Grobois, F.A., Donnelly, M., Eds.; IUCN/SSC Marine Turtle Specialist Group Publication No.4: Washington, DC, USA, 1999; pp. 110–114.

32. Yoder, M.; De Ley, I.T.; King, I.W.; Mundo-Ocampo, M.; Mann, J.; Blaxter, M.; Poiras, L.; De Ley, P. DESS: A versatile solution for preserving morphology and extractable DNA of nematodes. *Nematology* **2006**, *8*, 367–376. [CrossRef]

33. Somerfield, P.J.; Warwick, R.M. Meiofauna Techniques. In *Methods for the Study of Marine Benthos*; John Wiley & Sons, Ltd.: Hoboken, NJ, USA, 2013; pp. 253–284. [CrossRef]

34. Higgins, R.P.; Thiel, H. *Introduction to the Study of Meiofauna*; Smithsonian Institution Press: London, UK, 1988.

35. Somerfield, P.; Warwick, R. *Meiofauna in Marine Pollution Monitoring Programmes: A Laboratory Manual*; MAFF Directorate of Fisheries Research Technical Series: Lowestoft, UK, 1996; p. 71.

36. Bain, O.; Baldwin, J.G.; Beveridge, I.; Bezerra, T.C.; Braeckman, U.; Coomans, A.; Decraemer, W.; Derycke, S.; Durette-Desset, M.-C.; Fonseca, G. *Nematoda*; Walter de Gruyter: Berlin, Germany, 2014; Volume 2.

37. Guilini, K.; Bezerra, T.N.; Eisendle-Flöckner, U.; Deprez, T.; Fonseca, G.; Holovachov, O.; Leduc, D.; Miljutin, D.; Moens, T.; Sharma, J.; et al. NeMys: World Database of Free-Living Marine Nematodes. Available online: http://nemys.ugent.be (accessed on 1 May 2018).

38. Platt, H.M.; Warwick, R.M. Free-Living Marine Nematodes. Part II. British Chromadorids: Pictorial Key to World Genera and Notes for the Identification of British Species. *Synop. British Fauna* **1988**, *38*, 502.

39. Hurlbert, S.H. The nonconcept of species diversity: A critique and alternative parameters. *Ecology* **1971**, *52*, 577–586. [CrossRef]

40. Hill, M.O. Diversity and evenness: A unifying notation and its consequences. *Ecology* **1973**, *54*, 427–432. [CrossRef]

41. Heip, C.; Herman, P.; Soetaert, K. Indices of Diversity and Evenness. *Oceanis* **1998**, *24*, 61–87.

42. Clarke, K.R.; Gorley, R.N. *PRIMER v7: User Manual/Tutorial*; PRIMER-E: Plymouth, UK, 2015; p. 296.

43. Somerfield, P.J.; Clarke, K.R. Inverse analysis in non-parametric multivariate analyses: Distinguishing groups of associated species which covary coherently across samples. *J. Exp. Mar. Biol. Ecol.* **2013**, *449*, 261–273. [CrossRef]

44. Anderson, M.J.; Gorley, R.N.; Clarke, K.R. *PERMANOVA+ for PRIMER: Guide to Software and Statistical Methods*; PRIMER-E Ltd.: Plymouth, UK, 2008; p. 214.

45. De Oliveira, D.A.; Derycke, S.; Da Rocha, C.M.; Barbosa, D.F.; Decraemer, W.; Dos Santos, G.A. Spatiotemporal variation and sediment retention effects on nematode communities associated with Halimeda opuntia (Linnaeus) Lamouroux (1816) and Sargassum polyceratium Montagne (1837) seaweeds in a tropical phytal ecosystem. *Mar. Biol.* **2016**, *163*, 102. [CrossRef]

46. Gobin, J.F. Free-living marine nematodes of hard bottom substrates in Trinidad and Tobago, West Indies. *Bull. Mar. Sci.* **2007**, *81*, 73–84.

47. Moore, P. The nematode fauna associated with holdfasts of kelp (Laminaria hyperborea) in north-east Britain. *J. Mar. Biol. Assoc. U. K.* **1971**, *51*, 589–604. [CrossRef]

48. Kito, K. Phytal marine nematode assemblage on Sargassum muticum Agardh with reference to the structure and seasonal. *Mar. Ecol.-Prog. Ser.* **1982**, *37*, 19.

49. Da Rocha, C.M.C.; Venekey, V.; Bezerra, T.N.C.; Souza, J.R.B. Phytal Marine Nematode Assemblages and their Relation with the Macrophytes Structural Complexity in a Brazilian Tropical Rocky Beach. *Hydrobiologia* **2006**, *553*, 219–230. [CrossRef]

50. Derycke, S.; Van Vynckt, R.; Vanoverbeke, J.; Vincx, M.; Moens, T. Colonization patterns of Nematoda on decomposing algae in the estuarine environment: Community assembly and genetic structure of the dominant species Pellioditis marina. *Limnol. Oceanogr.* **2007**, *52*, 992–1001. [CrossRef]

51. Gwyther, J.; Fairweather, P.G. Colonisation by epibionts and meiofauna of real and mimic pneumatophores in a cool temperate mangrove habitat. *Mar. Ecol.-Prog. Ser.* **2002**, *229*, 137–149. [CrossRef]

52. Zhinan, Z. Phytal meiofauna of a rocky shore at the Cape d'Aguilar marine reserve, Hong Kong. *Mar. Flora Fauna Hong Kong South. China IV* **1997**, *4*, 205.

53. Atilla, N.; Wetzel, M.A.; Fleeger, J.W. Abundance and colonization potential of artificial hard substrate-associated meiofauna. *J. Exp. Mar. Biol. Ecol.* **2003**, *287*, 273–287. [CrossRef]

54. Raes, M.; De Troch, M.; Ndaro, S.G.M.; Muthumbi, A.; Guilini, K.; Vanreusel, A. The structuring role of microhabitat type in coral degradation zones: A case study with marine nematodes from Kenya and Zanzibar. *Coral Reefs* **2007**, *26*, 113–126. [CrossRef]

55. Raes, M.; Decraemer, W.; Vanreusel, A. Walking with worms: Coral-associated epifaunal nematodes. *J. Biogeogr.* **2008**, *35*, 2207–2222. [CrossRef]

56. Fonsêca-Genevois, V.D.; Somerfield, P.J.; Neves, M.H.B.; Coutinho, R.; Moens, T. Colonization and early succession on artificial hard substrata by meiofauna. *Mar. Biol.* **2006**, *148*, 1039–1050. [CrossRef]

57. Copley, J.T.P.; Flint, H.C.; Ferrero, T.J.; Van Dover, C.L. Diversity of meiofauna and free-living nematodes in hydrothermal vent mussel beds on the northern and southern East Pacific Rise. *J. Mar. Biol. Assoc. U. K.* **2007**, *87*, 1141–1152. [CrossRef]

58. Jensen, P. Ecology of benthic and epiphytic nematodes in brackish waters. *Hydrobiologia* **1984**, *108*, 201–217. [CrossRef]

59. Zekely, J.; Gollner, S.; Van Dover, C.L.; Govenar, B.; Le Bris, N.; Nemeschkal, H.L.; Bright, M. Nematode communities associated with tubeworm and mussel aggregations on the East Pacific Rise. *Cah. Biol. Mar.* **2006**, *47*, 477–482.

60. Trotter, D.; Webster, J. Distribution and abundance of marine nematodes on the kelp Macrocystic integrifolia. *Mar. Biol.* **1983**, *78*, 39–43. [CrossRef]

61. Coull, B.; Creed, E.; Eskin, R.; Montagna, P.; Palmer, M.; Wells, J. Phytal Meiofauna from the Rocky Intertidal at Murrells Inlet, South Carolina/Colman, J. 1940. On the fauna inhabiting intertidal seaweeds. J. Mar. Biol. Assoc. UK, 24: 129-183. *Trans. Am. Microsc. Soc.* **1983**, *102*, 380–389. [CrossRef]

62. Traunspurger, W.; Rothhaupt, K.-O.; Peters, L.; Wetzel, M. Community development of free-living aquatic nematodes in littoral periphyton communities. *Nematology* **2005**, *7*, 901–916. [CrossRef]

63. Meylan, A. Spongivory in hawksbill turtles: A diet of glass. *Science* **1988**, *239*, 393–395. [CrossRef]

64. Fujii, J.A.; McLeish, D.; Brooks, A.J.; Gaskell, J.; Van Houtan, K.S. Limb-use by foraging marine turtles, an evolutionary perspective. *PeerJ* **2018**, *6*, e4565. [CrossRef]

65. Preen, A.R. Infaunal mining: A novel foraging method of loggerhead turtles. *J. Herpetol.* **1996**, *30*, 94–96. [CrossRef]

66. Derycke, S.; Backeljau, T.; Moens, T. Dispersal and gene flow in free-living marine nematodes. *Front. Zool.* **2013**, *10*, 1. [CrossRef] [PubMed]

67. Derycke, S.; De Ley, P.; Tandingan De Ley, I.; Holovachov, O.; Rigaux, A.; Moens, T. Linking DNA sequences to morphology: Cryptic diversity and population genetic structure in the marine nematode Thoracostoma trachygaster (Nematoda, Leptosomatidae). *Zool. Scr.* **2010**, *39*, 276–289. [CrossRef]

68. Boeckner, M.; Sharma, J.; Proctor, H. Revisiting the meiofauna paradox: Dispersal and colonization of nematodes and other meiofaunal organisms in low-and high-energy environments. *Hydrobiologia* **2009**, *624*, 91–106. [CrossRef]

69. de Oliveira, D.A.S.; Decraemer, W.; Moens, T.; dos Santos, G.A.P.; Derycke, S. Low genetic but high morphological variation over more than 1000 km coastline refutes omnipresence of cryptic diversity in marine nematodes. *BMC Evol. Biol.* **2017**, *17*, 71.

70. Lins, L.; Vanreusel, A.; van Campenhout, J.; Ingels, J. Selective settlement of deep-sea canyon nematodes after resuspension—An experimental approach. *J. Exp. Mar. Biol. Ecol.* **2013**, *441*, 110–116. [CrossRef]

71. Ullberg, J.; Zoologiska, I.; Stockholms, U. *Dispersal in Free-Living, Marine, Benthic Nematodes: Passive or Active Processes?* Zoologiska Institutionen University: Stockholm, Sweden, 2004.

72. Plotkin, P.T.; Spotila, J.R. Post-nesting migrations of loggerhead turtles Caretta caretta from Georgia, USA: Conservation implications for a genetically distinct subpopulation. *Oryx* **2002**, *36*, 396–399. [CrossRef]

73. Marcovaldi, M.Â.; Lopez, G.G.; Soares, L.S.; Lima, E.H.; Thomé, J.C.; Almeida, A.P. Satellite-tracking of female loggerhead turtles highlights fidelity behavior in northeastern Brazil. *Endanger. Species Res.* **2010**, *12*, 263–272. [CrossRef]

74. Hawkes, L.A.; Witt, M.J.; Broderick, A.C.; Coker, J.W.; Coyne, M.S.; Dodd, M.; Frick, M.G.; Godfrey, M.H.; Griffin, D.B.; Murphy, S.R. Home on the range: Spatial ecology of loggerhead turtles in Atlantic waters of the USA. *Divers. Distrib.* **2011**, *17*, 624–640. [CrossRef]

75. Foley, A.M.; Schroeder, B.A.; Hardy, R.; MacPherson, S.L.; Nicholas, M. Long-term behavior at foraging sites of adult female loggerhead sea turtles (Caretta caretta) from three Florida rookeries. *Mar. Biol.* **2014**, *161*, 1251–1262. [CrossRef]

76. Griffin, D.B.; Murphy, S.R.; Frick, M.G.; Broderick, A.C.; Coker, J.W.; Coyne, M.S.; Dodd, M.G.; Godfrey, M.H.; Godley, B.J.; Hawkes, L.A. Foraging habitats and migration corridors utilized by a recovering subpopulation of adult female loggerhead sea turtles: Implications for conservation. *Mar. Biol.* **2013**, *160*, 3071–3086. [CrossRef]

77. Girard, C.; Tucker, A.D.; Calmettes, B. Post-nesting migrations of loggerhead sea turtles in the Gulf of Mexico: Dispersal in highly dynamic conditions. *Mar. Biol.* **2009**, *156*, 1827–1839. [CrossRef]

78. SWOT. The State of the World's Sea Turtles. 2020, p. 54. Available online: www.seaturtlestatus.org (accessed on 1 April 2020).

79. Wallace, B.P.; DiMatteo, A.D.; Bolten, A.B.; Chaloupka, M.Y.; Hutchinson, B.J.; Abreu-Grobois, F.A.; Mortimer, J.A.; Seminoff, J.A.; Amorocho, D.; Bjorndal, K.A.; et al. Global Conservation Priorities for Marine Turtles. *PLoS ONE* **2011**, *6*, e24510. [CrossRef] [PubMed]

80. Robinson, N.J.; Lazo-Wasem, E.A.; Paladino, F.V.; Zardus, J.D.; Pinou, T. Assortative epibiosis of leatherback, olive ridley and green sea turtles in the Eastern Tropical Pacific. *J. Mar. Biol. Assoc. U. K.* **2017**, *97*, 1233–1240. [CrossRef]

81. Pereira, S.; Lima, E.; Ernesto, L.; Mathews, H.; Ventura, A. Epibionts associated with Chelonia mydas from northern Brazil. *Mar. Turt. Newsl.* **2006**, *111*, 17–18.

82. Casale, P.; D'Addario, M.; Freggi, D.; Argano, R. Barnacles (Cirripedia, Thoracica) and Associated Epibionts from Sea Turtles in the Central Mediterranean. *Crustaceana* **2012**, *85*, 533–549. [CrossRef]

83. Limpus, C. Global overview of the status of marine turtles: A 1995 viewpoint. *Biol. Conserv. Sea Turt.* **1995**, *2*, 605–609.

84. Ross, J.P. Historical decline of loggerhead, ridley, and leatherback sea turtles. In *Biology and Conservation of Sea Turtles*; Smithsonian Institution Press: Washington, DC, USA, 1982; pp. 189–195.

85. Hays, G.C. Good news for sea turtles. *Trends Ecol. Evol.* **2004**, *19*, 349–351. [CrossRef]

86. Hirayama, R. Oldest known sea turtle. *Nature* **1998**, *392*, 705–708. [CrossRef]

87. Harvey, V.L.; LeFebvre, M.J.; deFrance, S.D.; Toftgaard, C.; Drosou, K.; Kitchener, A.C.; Buckley, M. Preserved collagen reveals species identity in archaeological marine turtle bones from Caribbean and Florida sites. *R. Soc. Open Sci.* **2019**, *6*, 191137. [CrossRef]

88. Bowen, B.W.; Nelson, W.S.; Avise, J.C. A molecular phylogeny for marine turtles: Trait mapping, rate assessment, and conservation relevance. *Proc. Natl. Acad. Sci. USA* **1993**, *90*, 5574–5577. [CrossRef]

89. Miller, J.D. Reproduction in sea turtles. *Biol. Sea Turt.* **1997**, *1*, 51–82.

90. Pfaller, J.B.; Alfaro-Shigueto, J.; Balazs, G.H.; Ishihara, T.; Kopitsky, K.; Mangel, J.C.; Peckham, S.H.; Bolten, A.B.; Bjorndal, K.A. Hitchhikers reveal cryptic host behavior: New insights from the association between Planes major and sea turtles in the Pacific Ocean. *Mar. Biol.* **2014**, *161*, 2167–2178. [CrossRef]

91. Hosono, T.; Minami, H. Stable isotope analysis of epibiotic caprellids (Amphipoda) on loggerhead turtles provides evidence of turtle's feeding history. In *New Frontiers in Crustacean Biology*; Brill: Leiden, The Netherlands, 2011; pp. 299–309.

92. Reich, K.J.; Bjorndal, K.A.; Frick, M.G.; Witherington, B.E.; Johnson, C.; Bolten, A.B. Polymodal foraging in adult female loggerheads (Caretta caretta). *Mar. Biol.* **2010**, *157*, 113–121. [CrossRef]

93. Detjen, M.; Sterling, E.; Gómez, A. Stable isotopes in barnacles as a tool to understand green sea turtle (Chelonia mydas) regional movement patterns. *Biogeosci. Discuss.* **2015**, *12*, 4655–4669. [CrossRef]

94. Ceriani, S.A.; Weishampel, J.F.; Ehrhart, L.M.; Mansfield, K.L.; Wunder, M.B. Foraging and recruitment hotspot dynamics for the largest Atlantic loggerhead turtle rookery. *Sci. Rep.* **2017**, *7*, 1–13. [CrossRef]

Article

The Influential Role of the Habitat on the Diversity Patterns of Free-Living Aquatic Nematode Assemblages in the Cuban Archipelago

Maickel Armenteros [1,2,*], José Andrés Pérez-García [2], Diana Marzo-Pérez [2] and Patricia Rodríguez-García [2]

[1] Instituto de Ciencias del Mar y Limnología, Universidad Nacional Autónoma de México, Circuito Exterior S/N, Coyoacán 04510, Ciudad de Mexico, Mexico
[2] Centro de Investigaciones Marinas, Universidad de La Habana, Calle 16 # 114, Playa 11300, Habana, Cuba; jose.andres@cim.uh.cu (J.A.P.-G.); diana.marzo@cim.uh.cu (D.M.-P.); patricia.rodriguez@cim.uh.cu (P.R.-G.)
[*] Correspondence: maickel.armenteros@gmail.com; Tel.: +52-55-6069-6240

Received: 25 July 2019; Accepted: 11 September 2019; Published: 16 September 2019

Abstract: Free living nematodes are the most abundant and diverse metazoans in aquatic sediments. We used a framework of habitat types to reveal quantitative patterns in species richness (SR), β-diversity, and biological traits (BT). Meiofauna was quantitatively collected from 60 sites within nine habitat types and 24,736 nematodes were identified to species level. We reported a regional richness of 410 ± 12 species for the Cuban archipelago; however, caves and deep waters need to be sampled more intensively. Relationships between SR and abundance supported the dynamic equilibrium model with habitats ordered across gradients of resource availability and physical disturbance. Seagrass meadows were the most specious and freshwater/anchihaline caves the least diverse habitats. Differences in β-diversity likely were due to habitat heterogeneity and limitations for dispersal. The assemblage composition was unique in some habitats likely reflecting the effects of habitat filtering. However, coastal habitats shared many species reflecting high connectivity and dispersal capability of nematodes due to hydrodynamics. The BTs "life strategy", "trophic group", and "tail shape" reflected ecological adaptations; but "amphidial fovea" and "cuticle", likely reflected phylogenetic signatures from families/genera living in different habitats. Habitat type played an influential role in the diversity patterns of aquatic nematodes from taxonomic and functional points of view.

Keywords: species richness; β-diversity; biological traits; tropical; marine; freshwater; meiofauna; Caribbean

---

1. Introduction

The study of the habitat, the place where organisms live [1], is a tenet of community ecology. The pioneer studies [2,3] highlighted the importance of habitat type as template for ecological strategies. More recently, the habitat type has been reintroduced as a main organizing principle to study communities [4]. In theory, the habitat type may act as a selective force driving the community assemblage, but mediated by other processes such as dispersal, speciation, and drift [5]. A critical step in the endeavor of linking habitat type with community ecology is the definition of the habitat itself. For instance, aquatic habitats are often defined a priori on the basis of overarching abiotic structures/variables such as confined spaces (e.g., caves) or extreme abiotic conditions (e.g., trenches). But, the delineation of habitat types based uniquely on abiotic factors has been termed as environmental filter and it has often failed to explain the community assembly and dynamics since biological interactions may also play a key role [6]. Habitat types are sometimes defined from an operational point of view (e.g., reef versus seagrass bed) that does not always match with the spatial and temporal scales at which organisms perceive the environment (e.g., a reef contains several microhabitats) [7]. Further, the very continuity of the environment in aquatic ecosystems make the test of the habitat type as a main driver of community assemblage challenging. Despite previous shortcomings, the diversity and distribution patterns of marine communities can be successfully approached from a habitat perspective giving generality and predictive power. The habitat type has been used as a successful predictor in studies of marine benthos, e.g., [8–10], and may be linked to one of the most important components of the diversity: β-diversity (i.e., variation in the identities of species among sites [11]).

Free living nematodes are the most abundant and diverse metazoan assemblages in aquatic sediments and the most studied taxon within the meiobenthos [12,13]. However, several knowledge gaps remain in the ecology of nematode assemblages. Three of the most important gaps are: (a) the incomplete knowledge about the species richness of the group, for instance, the known fraction of marine nematode species is about 12% of the estimated total of species [14]; (b) the ecological processes driving the β-diversity patterns of nematode assemblages are known mostly from temperate ecosystems (e.g., [15], and references therein), and even from deep-sea ecosystems, e.g., [16–18], but patterns across tropical habitats and their ecological drivers are substantially understudied; and (c) the link between nematode biological traits and ecosystem functioning based on evolutionary adaptations. In particular, the problem is to identify the traits and environmental factors that are most responsible for the most striking and important patterns in the field [19].

Gradients in physical disturbance and resource availability constitute environmental axes of importance to define habitats relevant for nematode assemblages [15]. Latter authors proposed a conceptual model between nematode abundance and species richness/diversity in relation to habitat types. The general model was originally proposed by Huston [20] and termed as the Dynamic Equilibrium Model (DEM). In some extension, DEM built on previous theories relating diversity to frequency of disturbance (Intermediate Disturbance Hypothesis, [21]) and energy availability (Species–Energy Theory, [22]). However, DEM as analyzed in [16] did not include tropical habitats (e.g., coral degradation zones, seagrass meadows). Actually, most of the recent surveys on diversity patterns of nematode assemblage structure have included polar, temperate or deep-sea habitats. Comparatively less research has been done in tropical ecosystems resulting in a gap of knowledge about diversity patterns (but see studies [23,24] in Kenya, and [25,26] in the Maldives). Recently, some other understudied tropical habitats have recently been surveyed at the species level, such as volcanic sands [27], mangrove sediments [28], and rivers [29].

Diversity may be studied using descriptive, functional and/or phylogenetic approaches. From a descriptive point of view, diversity may be effectively measured by the species richness (SR) and complementarity [30]. For hyperdiverse assemblages (e.g., nematodes), the use of non-parametric estimators (e.g., Chao 1) besides observed SR may bring higher accuracy in the assessment of the richness [31]. The measurement and interpretation of complementarity between assemblages (or samples) has been the subject of much debate. The partition of the β-diversity in α- and γ-components, using additive or multiplicative relationships, is a feasible approach to disentangling the effects of scale or other nested units of analysis. However, the alternative use of the concept of β-diversity, measured as Bray–Curtis dissimilarity index, constitutes a simple and effective approach to assess how similar two sets of species are [11]. In addition to the approach to diversity, which constitutes the basis for further hypotheses about processes, a functional approach may be applied using the biological traits of organisms.

A biological trait is any morphological, physiological or phenological feature measurable at the individual level [32]. The biological trait analysis (BTA) gives generality and predictability in ecological studies [19] because the spatio-temporal distribution of organisms and their functional role in ecosystems depend on their traits rather than on their taxonomical affiliation [33,34]. The BTA was brought into the marine realm by Bremner et al. [35,36] focusing on epibenthic invertebrates. Later, BTA was successfully applied to the ecology of nematode assemblages [37–39]. However, studies on tropical assemblages are scarce (but see [40,41]). Currently, there is a need for empirical studies exploring the BTA across a variety of taxa in order to increase the predictability of assemblage patterns and/or reveal their roles in ecosystem functioning [33].

Studies on free-living nematodes in the Cuban archipelago are rather scarce. Andrassy's pioneering work [42] described several new species of nematodes from a variety of habitats from freshwater caves to intertidal sandy beaches. It constituted the first taxonomic list of free-living nematodes from Cuba. However, the poor description of sampling methods and habitat types prevented the quantitative analysis of his data. López-Canovas and Pastor de Ward [43] published a taxonomic list of nematodes living in seagrass meadows of the northcentral part of the island, but no information about location of the sampling points or quantitative data. Thus, only the most recent studies in several habitats from the Cuban archipelago during the last 10 years are available for quantitative studies.

In this contribution, we addressed the above-mentioned knowledge gaps using a framework of tropical aquatic habitats to reveal patterns in species richness, β-diversity, and biological traits. We propose two objectives: (1) to describe the patterns of species richness and β-diversity of nematode assemblages across nine aquatic habitats; and (2) to explore the distribution of five biological traits in relationship with the habitat type.

2. Materials and Methods

2.1. Study Sites and Habitats

Nematodes were collected from 60 sites around the Cuban archipelago (Figure 1, Supplementary Table S1). The sites were classified according to nine habitat types and they were always submerged (i.e., not influenced by tides). The selection of habitat types was based on the perceived importance in terms of diversity for Caribbean aquatic ecosystems, although some important habitats (e.g., mangroves, beaches) could not be covered. The descriptions of the habitats were based on the original publications and/or personal observations and we have highlighted those abiotic variables that may directly affect nematode assemblages (e.g., grain size, physical disturbance, and organic content). These variables in general match with environmental axes (e.g., productivity, disturbance) defined by other authors, e.g., [7,15].

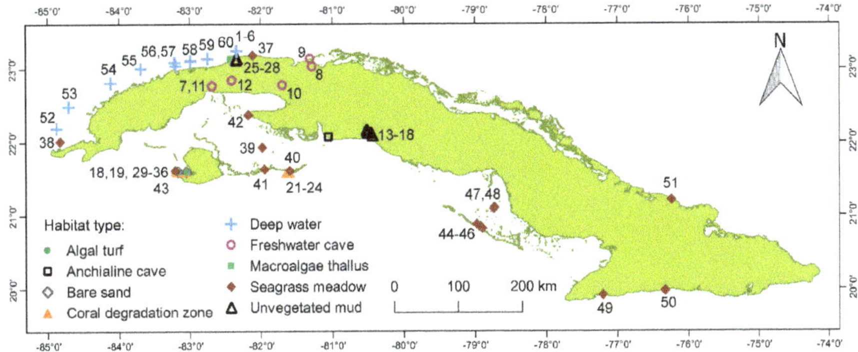

Figure 1. Map indicating the 60 sampling sites of aquatic nematode assemblages around the Cuban archipelago. Symbols indicate the nine habitat types.

Freshwater cave (FC, five sites) referred to ponds within caves, permanently in darkness, with very fine sediment (mostly silt/clay), hypoxic environment (dissolved oxygen 1.0–3.5 mg O_2 L^{-1}), negligible physical disturbance, and an oligotrophic environment due to absence of primary production and limited input of allochthonous organic matter through terrestrial runoff.

Anchihaline cave (AC, one site) was a limestone cave connected with the ocean through subterranean passages. The cave was opened to the sky through a portion of the collapsed roof (i.e., cenote). Bottom water at ~60 m deep had full marine salinity and surface waters had freshwater from rainfall and runoff.

The bottom is permanently in darkness, with very fine sediment (mostly silt/clay), negligible physical disturbance, an oligotrophic environment due to absence of primary production, but allochthonous sources of organic matter existed through terrestrial runoff and potentially subterranean currents from the adjacent sea.

Bare sand (BS, two sites) sites were located between the lagoon and the reef crest, there was a fringe of ~10 m wide unvegetated coralline sand deposited on rocky substrate. The vertical thickness of the sand layer fluctuated between 4 and 12 cm. The water was well oxygenated, and depth varied between 2 and 3 m. The percentage of organic carbon in the sand was ~5% and productivity was assumed as high due to primary production by microphytobenthos (e.g., diatoms). Sites were subjected to high hydrodynamic disturbance by currents and wave action.

Unvegetated mud (UM, 10 sites) referred to subtidal bottoms devoid of vegetation, located in semi-enclosed bays (Cienfuegos and Havana), with depths of 5–12 m, and an estuarine regime due to river discharges, rainfall and terrestrial runoff. Bottoms mostly with fine sediment (silt/clay) and high accumulation of organic matter as detritus because of eutrophication are subjected to seasonal hypoxia/anoxia, and industrial and urban pollution.

Algal turf (AT, two sites) sites were located immediately shoreward of the reef crest (mostly consisting of elkhorn coral (*Acropora palmata*) and fire coral (*Millepora complanata*)), comprised of a hard bottom covered by seaweeds, and depths ranging from 0.5 to 3 m. There was an intense hydrodynamic regime because of wave turbulence created in the vicinity of the reef crest. Dense turfs of filamentous algae were abundant and mixed, with the most common genera being *Galaxaura, Jania, Lobophora, Padina, Sargassum,* and *Stypopodium.*

Macroalgae thallus (MT, six sites) samples actually came from a single site within a polygon of 100 m × 100 m, but the proximate substrates were different macroalgae species living on rocky bottoms at 9–21 m depth. Six species with different degrees of structural complexity were collected: *Amphiroa* sp. 1, *Bryothamnion* sp. 1, *Bryothamnion* sp. 2, *Dictyopteris* sp. 1, *Halimeda* sp. 1, and *Lobophora* sp. 1. There was some sediment deposited within the macroalgae thallus. The site was subjected to strong hydrodynamic regime by currents and wave action.

Seagrass meadow (SM, 17 sites) sites were subtidal bottoms with muddy/sandy sediments covered in variable degree by marine seagrasses, mostly by turtle seagrass *Thalassia testudinum.* Seagrass canopy provides protection from physical disturbance by hydrodynamic forces. The content of organic matter was high (5–10%) because of the intense primary production (seagrasses, epiphytic microalgae, and macroalgae) and the accumulation of detritus in sediments.

Coral degradation zone (CDZ, eight sites) is a biotope within coral reefs where the coral fragments and rubble accumulate and degrade to coralline sand at depths of 2–6 m. Dead coral fragments were covered by biofilm (observable at naked eye), likely reflecting a high level of food availability due to the productivity of the microphytobenthos; but there was minimum accumulation of organic matter due to the intense hydrodynamic regime by currents and wave action.

Deep water (DW, nine sites) sites refer to the bottom along the shelf slope around 1500 m depth, where hydrodynamic regime is supposed intense because of the steep slope and presence of large areas of hard substrate devoid of sediments, with very fine grain size (silt/clay), and an oligotrophic environment due to lack of primary production and reduced input of organic particles from the water column.

2.2. Collection and Processing of Samples

This survey summarizes the results of several papers published or in preparation during a period of about 10 years. Therefore, the techniques for the collection and processing of the nematodes varied and specific details can be found in the published papers and in the Supplementary Table S1. Hereafter, we synthetized the most important features of the sampling. Collection in shallow soft-bottoms (i.e., FC, AC, BS, UM, and SM) was done using hand-held plastic cores (2.6 cm inner diameter); usually 3–10 sampling units were collected in an area smaller than 50 m × 50 m. In AT, 20 thalli ((six macroalgae species × four replicates) minus four missing replicates) were carefully collected into nylon bags; strictly speaking, the nematode assemblages were epiphytes. We kept the different macroalgae species separate as sites although they were sampled in the same location. The CDZ were sampled by hand collection of dead coral rubble within a quadrat of 10 cm × 10 cm. The dead coral fragments were carefully placed within plastic jars underwater. In deep waters, 27 sediment samples from the shelf slope (~1500 m depth) were collected from nine sites and three replicates/site using a multicorer (single deployment) with PVC cores of 11 cm diameter. The three cores collected from a single deployment were in rigor pseudoreplicates; but, the bias on the diversity estimates should be small, if any. This is because: (i) we combined the three cores to obtain a more robust estimate; and (ii) meiofauna (and nematodes) vary at spatial scales smaller than 1 m, therefore the three cores (separated by ~1 m) could effectively sample the variability in the distribution (and hence the diversity).

Samples were processed using mesh size of 45 µm as lowest limit of meiofauna. Meiofauna retained in sieves was washed and preserved in formalin (4%) or ethanol (70%). Nematodes were sorted out of the sediment by the flotation technique in high-density sucrose solution [44]. For CDZ and MT, the epilithic and epiphytic nematodes were detached from respective substrates by gentle washing with filtered tap water. Nematodes were mounted in permanent preparations using the general technique in [45]. The identification of nematodes to species level was made using the original descriptions of the species and faunal synopses [46–48]. However, many species could not be assigned with reasonable confidence to known species and thus they were named as sp. The nomenclature followed the online database World Record of Marine Species (https://www.marinespecies.org).

2.3. Biological Traits

We gathered information about five biological traits for all of the identified species. The data are presented in the Supplementary Table S2.

The life strategy, or c–p scale [49–51], an analog of K/r-strategists, was analyzed based on a scale from c–p = 1 (good colonizers) to c–p = 5 (good persisters). For some genera, c–p values were not available and information from other genera in the same family were used.

The trophic group was defined based on the Wieser's scheme that takes into account the morphology of the buccal cavity [52]: selective deposit feeders (i.e., detritivores), non-selective deposit feeders, epigrowth feeders, and omnivores/predators. The deposit feeders lack buccal armature, and the differentiation between selective and non-selective is based on the size of the buccal cavity. The epigrowth feeders have sclerotized structures as teeth and denticles. Omnivores/predators bear mandibles or spear-shape structures in a large buccal cavity.

The amphidial fovea shape has been recently proposed in [25] as a meaningful biological trait with eight groups. We have used here only seven of them since blister-like was included as circular type: circular, spiral, slit-like, pocket, elongated/rounded loop, indistinct, and longitudinal slit.

The cuticle morphology has been proposed also in [26] as: smooth, punctated/annulated with or without lateral differentiation, punctated/annulated with longitudinal structures for the whole body length, wide body annules and longitudinal ridges, desmens, and covered by ectosymbiotic bacteria.

The tail shape was classified into four types [53]: rounded/blunt, clavate/conico-cylindrical, conical, and elongated/filiform.

2.4. Data Analysis

We pooled the data from sampling units coming from the same site (except for MT) to obtain robust estimates of diversity and to minimize the effects of the sampling bias (e.g., different sizes of the sampler). A matrix of nematode species × sites was compiled using the abundance (i.e., counts) as values. The software EstimateS 9.1.0 [54] was used for the analysis of the species richness (SR) using curves of accumulation of species versus individuals and for the calculation of confidence intervals based on permutations. The non-parametric estimator Chao 1 was calculated for the estimation of the SR taking into account the fraction of non-seen species [55]. The comparison of SR at the same level of abundance (i.e., rarefaction technique) was made with the number of expected species ($ES_{(n)}$) in the software PRIMER 6.1.15 [56]. A correlation between SR and abundance was made using the Spearman rank correlation coefficient. The similarity patterns across the samples were explored with numerical ordinations by non-metric multidimensional scaling (NMDS) in the software PRIMER 6.1.15. Data were transformed as square root to downweigh the influence of the most abundant species. The Bray–Curtis similarity index was used as measure of resemblance. Statistical differences among habitat types in the assemblage composition were tested using a permutational analysis of variance (PERMANOVA) in the software PERMANOVA 1.0.5 [57]. Note that PERMANOVA tests for differences in the central position (i.e., centroid or multivariate mean); we used the same setting as that for NMDS and an unrestricted permutation of raw data. The β-diversity 1 was calculated as the mean distance of the sites to the centroid of the habitat groups (i.e., multivariate dispersion) using the routine PERMDISP in latter software. Thus, β-diversity 1 measured the variation among sites within the same habitat. The criterion for interpreting significant differences in β-diversity 1 was the lack of overlapping of the standard error associated to the mean. We also calculated the β-diversity 2 to measure the variation in the assemblage structure among habitats as the mean (±standard deviation) of Bray–Curtis dissimilarity values. We applied the SIMPER procedure in order to identify the species responsible of the compositional differences among habitats using the PRIMER 6.1.15. Formal pairwise comparisons among the nine habitats yield 36 lists of species (i.e., habitat 1 versus 2, 1 versus 3, etc.) contributing to the pairwise dissimilarity, which is an unmanageable number of lists. Therefore, we presented those species that contribute up to 50% of cumulative similarity within habitats. We considered that the selected species were typical of a particular habitat type since they occurred at most (or all) of their sites. The comparison in a table of the presence/absence of species indicates those species that most contribute to the β-diversity patterns across the nine habitats.

3. Results

3.1. Species Richness

A total of 24,736 nematodes were identified yielding 410 species belonging to 216 genera, 46 families, eight orders, and two classes. The accumulation curve of observed SR versus individuals had a non-asymptotic shape indicating that the number of species would increase with further sampling effort (Figure 2a). The observed SR for the Cuban archipelago (i.e., regional SR) was 410 ± 12 species. The non-parametric estimator Chao 1 indicated roughly the same number of species (415 ± 7 species).

The accumulation curves of SR per habitat did not show any asymptote. Even habitats heavily sampled such as UM and CDZ (>7000 identified individuals) had a slight tendency to increase (Figure 2b). The SM were also well sampled (>3000 individuals) but the steep shape of the curve indicated a potentially higher SR. DW habitats contained a high SR, but were still underestimated because of the almost linear shape of the curves. The two cave habitats were undersampled suggesting that SR estimates should be interpreted with caution.

The observed SR was different among habitats as indicated by the lack of overlap of 0.95 confidence intervals (Figure 3a). SM had the highest SR, followed by CDZ and DW habitats. A group of three marine habitats had intermediate SR of about 100 species (BS, UM, and AT). MT had lower SR, and the lowest richness occurred in the caves (both freshwater and anchihaline). The SR and abundance were positively correlated across the sites ($R_S = 0.64$, $p < 0.001$, $n = 60$).

Sample size strongly affects the estimates of SR; therefore, we computed the expected number of species (ES) for a same level of abundance. We removed freshwater and anchihaline caves from the ES calculation because they had too few individuals for an informative estimate (168 and 31 individuals, respectively). The ES was then computed based on the lowest abundance: 540 individuals in DW. $ES_{(540)}$ reaffirmed that SM had the largest SR, followed by DW (Figure 3b). Three habitats had similar values of $ES_{(540)}$: BS, AT, and CDZ. UM had a lower number of species and the lowest richness occurred in MT.

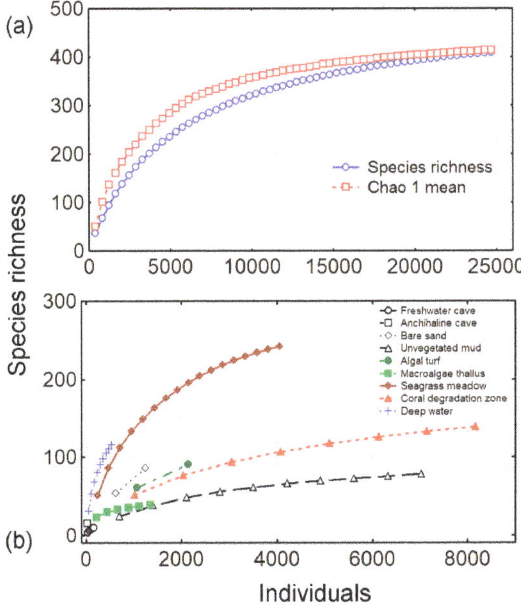

Figure 2. Accumulation curves of nematode species versus individuals. (**a**) For all the samples pooled using the observed species richness (SR) and the non-parametric estimator Chao 1; (**b**) the SR per habitat type.

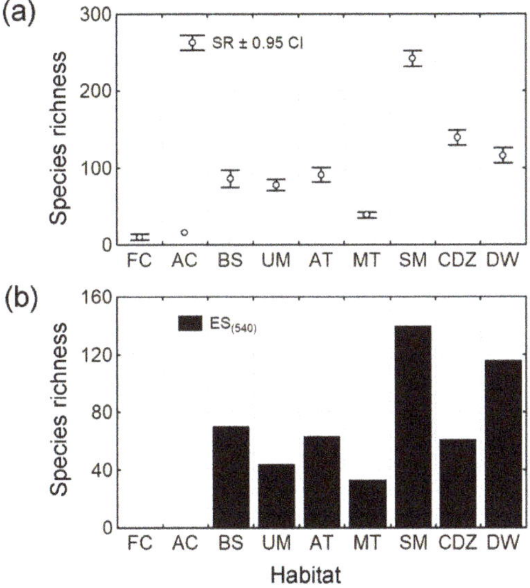

Figure 3. Species richness (SR) of nematode assemblages in 60 sites from nine aquatic habitats. (**a**) Observed SR ± 0.95 confidence intervals (CI); (**b**) Expected number of species (ES) in a subsample of 540 individuals. Note that ES (540) was not computed for two habitats due to their low abundance. Acronyms of the habitats: FC = freshwater cave, AC = anchihaline cave, BS = bare sand, UM = unvegetated mud, AT = algal turf, MT = macroalgae thallus, SM = seagrass meadow, CDZ = coral degradation zone, and DW = deep water.

3.2. β-Diversity

The multivariate structure of the assemblages was clearly different among habitats as indicated by a numerical ordination (Figure 4a) and PERMANOVA test (Pseudo-F = 6, *p* = 0.01, d.f.$_{(total)}$ = 59, d.f.$_{(effect)}$ = 8). Most of the samples were clustered by habitat type, which explained 45% of the total variance in the multivariate data. Some habitats had a different assemblage structure composition as indicated by separate groups of samples (e.g., DW, UM, and SM). Other habitats had remarkable similarity among sites but had some overlapping between them (e.g., CDZ, AT, and BS). FC showed a large variability in assemblage composition as indicated by the dispersal of the samples in the plot. This variation between habitats but also within habitats (i.e., among sites coded as a same habitat type) suggests further exploration of quantitative values of β-diversity.

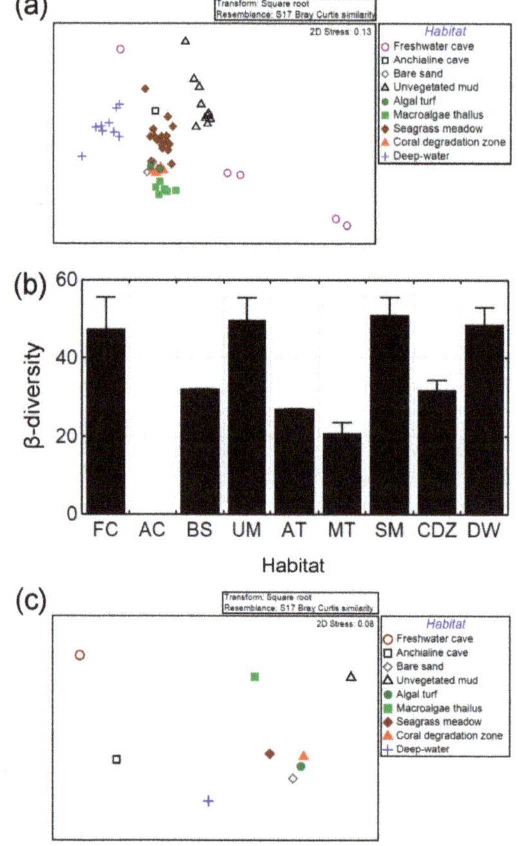

Figure 4. (**a**) Numerical ordination by non-metric multidimensional scaling (NMDS) of the samples coded by habitat type and based on square-root transformed abundance of nematodes. (**b**) β-diversity per habitat type represented as the mean (± standard deviation) of distance to centroid based on Bray–Curtis index. Note that β-diversity could not be computed in AC because only one site. (**c**) NMDS of habitats. Acronyms of the habitats: FC = freshwater cave, AC = anchihaline cave, BS = bare sand, UM = unvegetated mud, AT = algal turf, MT = macroalgae thallus, SM = seagrass meadow, CDZ = coral degradation zone, and DW = deep water.

The β-diversity 1 varied across the nine habitats (Figure 4b). There were four habitats with higher β-diversity (FC, UM, SM, and DW). The other five habitats had lower β-diversity.

The ordination of the nine habitats (all sites pooled) based on assemblage structure indicated that five habitats had a very different assemblage composition (Figure 4c): FC, AC, MT, UM, and DW. Four coastal marine habitats were relatively similar to each other: BS, AT, SM, and CDZ. The β-diversity 2 was very high (mean ± SD): 97% ± 14%.

There were several species typical of a particular habitat type (Table 1). FC had a unique set of species belonging to typical freshwater genera (*Aphanolaimus*, *Ironus*). Species in AC were marine and occurred in other habitats as well (e.g., SM and CDZ). UM had a unique set of species belonging to two known hypoxia-tolerant genera (*Sabatieria* and *Terschellingia*). Other species were widely distributed across the shallow coastal habitats: *Desmodora pontica*, *Euchromadora vulgaris*, and *Zalonema ditlevseni*. SM had a diverse set of species, some of them characterized by large body size such as *Cheironchus*, *Halichoanolaimus*, *Mesacanthion*, and *Viscosia*. Members of Comesomatidae (*Dorylaimopsis*, *Setosabatieria*), a family typically composed by detritivores, were characteristic of this habitat. Several genera of Chromadoridae were found typically on MT (*Chromadora*, *Chromadorella*, and *Euchromadora*). The set of species inhabiting DW was unique and included several genera typical of the deep-sea such as *Acantholaimus*, *Bolbolaimus*, *Cervonema*, and *Metadasynemella*.

Table 1. Nematode species typical of each habitat type based on SIMPER analysis. Species were chosen by the ordered contribution to the within-habitat similarity (measured by Bray–Curtis index) and only those that contribute up to 50% of cumulative similarity were listed. Acronyms of the habitats: FC = freshwater cave, AC = anchihaline cave, BS = bare sand, UM = unvegetated mud, AT = algal turf, MT = macroalgae thallus, SM = seagrass meadow, CDZ = coral degradation zone, and DW = deep water.

Species	FC	AC	BS	UM	AT	MT	SM	CDZ	DW
Average within-habitat similarity	25%		36%	26%	46%	68%	25%	53%	27%
Ironus ignavus	X								
Monhystrella sp.	X								
Aphanolaimus sp.	X								
Pomponema sp.		X							
Zalonema ditlevseni		X					X	X	
Paradesmodora immersa		X					X		
Desmodora pontica			X		X		X	X	
Tricoma sp.			X		X			X	
Enoploides bisulcus			X						
Innocuonema asymmetricum			X						
Viscosia abyssorum			X				X		
Sabatieria pulchra				X					
Terschellingia longicaudata				X			X		
Terschellingia communis				X					
Euchromadora vulgaris					X		X	X	
Epsilonema sp.					X				
Croconema cinctum					X			X	
Euchromadora gaulica					X			X	
Acanthopharynx denticulata					X				
Chromadora brevipapillata						X			
Paracanthonchus platypus						X			
Chromadorella paramucrodonta						X			
Chromadorella filiformis						X		X	
Euchromadora atypica						X			
Marylynnia sp.							X		
Halichoanolaimus chordiurus							X		
Cheironchus vorax							X		
Mesacanthion sp.							X		
Desmoscolex sp.							X		
Halichoanolaimus sp.							X		
Dorylaimopsis punctata							X		
Setosabatieria hilarula							X		
Daptonema sp.							X		
Draconema sp.								X	
Endeolophos fossiferus									X
Metadasynemella falciphalla									X

Table 1. Cont.

Species	FC	AC	BS	UM	AT	MT	SM	CDZ	DW
Average within-habitat similarity	25%		36%	26%	46%	68%	25%	53%	27%
Acantholaimus megamphis									X
Cervonema macramphis									X
Pselionema simile									X
Bolbolaimus sp.									X
Desmodorella tenuispiculum									X
Acantholaimus maks									X
Metadasynemella cassidiniensis									X

3.3. Biological Traits

The assemblage structure by life strategy (or colonizer/persister scale) varied between freshwater and marine habitats (Figure 5a). FC had higher contribution of extreme colonizer (c–p 1 = 16%) and persister species (c–p 4 = 59%) when compared with the marine habitats (c–p 1 = 0% and c–p 4 = 15%). Marine habitats also had differences among them, with UM having only intermediate colonizer species (c–p 2 = 57% and c–p 3 = 43%). The other seven marine habitats shared a similar life strategy structure with dominance of c–p 3 (52%) followed by c–p 2 (29%) and c–p 4 (18%).

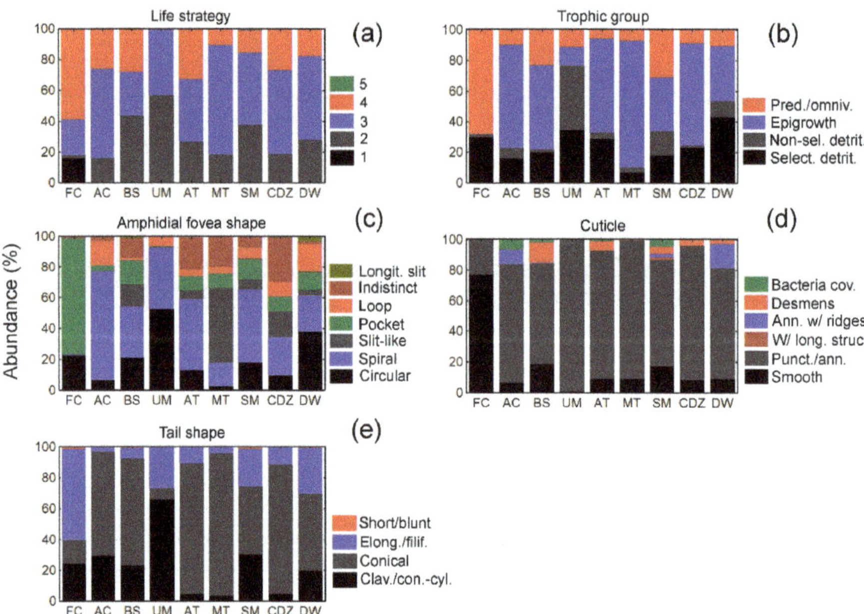

Figure 5. Biological traits of nematode assemblages across nine habitats. (**a**) Life strategy scale (scale from 1, colonizer to 5, persister) [43]. (**b**) Trophic groups based on buccal cavity morphology (detrit. = detritivore, pred. = predator, omniv. = omnivore) [46]. (**c**) Amphidial fovea shape (longit. = longitudinal) [34]. (**d**) Cuticle type (punct. = punctated, ann. = annulated, w/long. struct. = with longitudinal structures along the whole body length, ann. w/ridges = wide annules with ridges, cov. = covering) [34]. (**e**) Tail shape (clav. = clavate, con.-cyl. = conical-cylindrical, elong. = elongated, filif. = filiforme) [47]. Acronyms of the habitats: FC = freshwater cave, AC = anchihaline cave, BS = bare sand, UM = unvegetated mud, AT = algal turf, MT = macroalgae thallus, SM = seagrass meadow, CDZ = coral degradation zone, and DW = deep water.

The structure by trophic group was different among habitat types (Figure 5b). FC when compared with marine habitats had higher abundance of predators/omnivores (68% vs. 17%) and lower of epigrowth feeders (1% vs. 43%). UM had higher contribution of non-selective deposit feeders (42%) when compared with the other eight habitats (8%). Non-sedimentary habitats (i.e., AT, MT, and CDZ) had larger contribution of epigrowth feeders (72%) when compared with sedimentary habitats (28%).

The structure by amphidial fovea shape displayed differences among habitats (Figure 5c). FC had larger contribution of pocket fovea (76%) when compared with marine habitats (10%) and no species with spiral fovea.

The trait structure by cuticle type showed large differences in the contribution to assemblage structure (Figure 5d). Nematodes with smooth cuticle were dominant in freshwater (77%) when compared with marine habitats (10%). The marine habitats had a clear dominance of punctated/annulated cuticle (81%) when compared with freshwater (23%). UM had only nematodes with punctated/annulated cuticle type.

Tail shape assemblage structure varied among habitats (Figure 5e). FC had a larger contribution of elongated/filiforme tail shape (59%) when compared with marine habitats (20%). Nematodes with clavated/conical-cylindrical tail shape were more abundant in UM (66%) than in the other eight habitats (19%).

4. Discussion

The accumulation curves, for both observed SR and Chao 1, were close to an asymptote indicating that the regional richness (approximately 410 species) was reasonably well estimated. These figures should be interpreted as the minimum richness [31] and set a first SR estimate of aquatic nematodes for the Cuban archipelago. The relatively low number of unseen species (the difference between Chao 1 and observed SR being five species) could be due to a taxonomic bias. In other words, many species with one or two individuals (mostly juveniles or females) were removed from the matrix because of the uncertainty of the identification. Two solutions to this taxonomic issue are to increase the sampling effort looking for individuals in adult stages and to apply DNA barcoding for the species identification [58,59].

The accumulation curves per habitat showed a radically different scenario, with most of the curves lacking an asymptotic shape. In general, the increase of the sampled area clearly had a strong positive effect on the SR as indicated by the steep shape of the curves [60]. Even for well-sampled habitats such as coral degradation zones and unvegetated muds, the shape of the curves indicated the existence of undiscovered species. Seagrass meadow and deep-water habitats looked to be a promissory sources of undiscovered species since their curves were steep. The sample size of 300 individuals suggested in [61] as sufficient to describe the diversity in seagrass meadows at local scale seemed reasonable. However, at larger spatial scales, a substantially higher number of individuals (likely >4000) is needed to estimate the true species richness.

The low number of collected individuals in the deep-waters seemed to be the main limitation for a more accurate estimate of species richness. Low number of nematodes would be a bias due to the use of 45 μm mesh size that misses small-sized nematodes which are proportionally more abundant in the deep sea [62]. However, other studies in deep sea habitats of the Gulf of Mexico, using 45 μm mesh, have reported a consistently higher number of individuals than in our study [63]. The most plausible explanation for the low abundance of nematodes in the DW samples was the combination of: (i) oligotrophic environment in the water column causing a reduced input of organic carbon to sediments via sedimentation; and (ii) strong hydrodynamic regime given by the Cuban countercurrent flowing along the northwestern shelf slope [64].

Based on the Dynamic Equilibrium Model (DEM), Moens et al. [15] related two environmental axes (i.e., resource availability and physical disturbance) with habitat types and hence with SR and abundance of nematode assemblages. The model reasonably fitted well to our data of SR and abundance for the marine habitats (Figure 6). However, we could not directly measure abiotic variables related to these drivers (e.g., biopolymeric carbon, current speed). This issue limits the strength of our evidence to test the DEM model. The depth can be a good correlate of food availability in sediments, splitting coastal and deep waters habitats [62]. However, in shallow coastal habitats depth is a poor predictor of food availability or hydrodynamic regime because other local factors (e.g., turbidity, shoreline topography) may also have a significant influence on productivity and hydrodynamics.

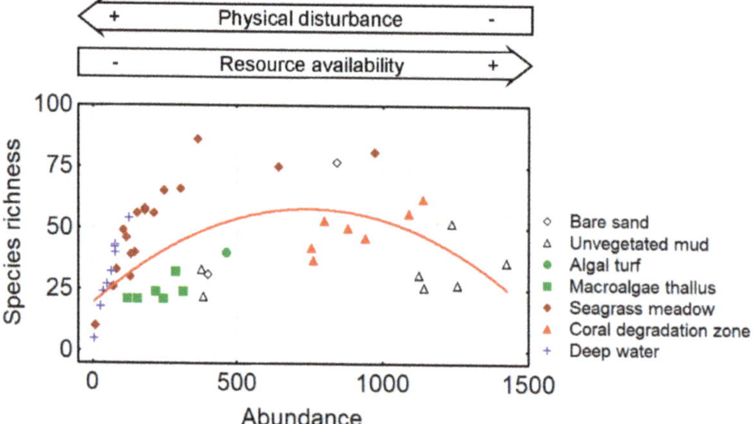

Figure 6. The relationships between species richness and abundance of nematodes across nine aquatic habitats. Gradients of physical disturbance and resource availability are superposed on qualitative bases. The curve (red line) is the best fitted line by a polynomial function. Based on the conceptual model in [15]. Four sites were removed because outlier values: two defaunated UM from the extremely polluted Havana Bay, one SM, and one CDZ represented aggregated samples with artificially high values of abundance.

A word of caution in the interpretation of values of abundance is needed because of the lack of standardization of counts into density/concentration values. The disparate structure of the sampled habitats, e.g., coral rubble, algal turf, macroalgae thallus, and sediments, prevented such standardization. Therefore, the values of abundance obtained in our study may be indicative of relative differences between habitats, but may not constitute a reference framework for the absolute abundance in the habitats.

The unimodal curve of SR versus abundance suggested that the most specious habitats were those subjected to intermediate levels of resource availability (e.g., seagrass meadow and coral degradation zone). The sampled deep-waters were located on the shelf slope (~1500 m depth) and likely subjected to oligotrophic conditions in sediments and intense hydrodynamic regime. This habitat harbored a diverse but relatively scarce nematofauna with genera typical from the deep sea and reported from other oceanic basins [65]. Macroalgae thallus were also subjected to an intense hydrodynamic regime but larger resource availability is likely as biofilm and particulate organic matter accumulate on the thalli.

Coastal habitats (seagrass meadow, bare sand, algal turf, and coral degradation zone) were subjected to a moderately high level of physical disturbance by wave and current actions. The suite of species typical of these habitats had adaptations to physical disturbance such as ornamented cuticle, conical tails, and small body size. The most specious habitat was seagrass meadow, likely reflecting a large resource availability (e.g., interstitial space and food) and good protection from physical disturbance. The relatively larger body size of nematodes in seagrass meadows [66] indicated more available energy that supported in turn higher biomass [22]. Seagrass meadows are highly productive habitats, not only because of the seagrass primary production, but also because of the associated epiphytic microalgae [67]. The taxonomic and functional structure of nematode assemblages have been positively correlated with food availability in seagrass meadows [68–70]. The seagrass vegetation also enhances the nematode assemblages by increasing refuge from predators [71] and reducing the resuspension by currents [72]. The other coastal habitats with intermediate levels of richness (bare sand, algal turf, and coral degradation zone) likely were influenced by the biofilm produced by microphytobenthos as a main source of food for nematodes. Recent characterization of the biofilm growing on the *Acropora palmata* coral rubble indicated an associated rich microbial community of algae, sponges, virus, bacteria, and other microorganisms [73] that potentially provides food for epigrowth feeder nematodes.

The unvegetated muds were located in semi-enclosed bays with organically enriched sediments because of the augmented primary production by phytoplankton and microphytobenthos due to human-driven eutrophication. The accumulation and oxidation of organic matter in the bottoms caused seasonal hypoxia enhanced by a reduced hydrodynamic regime by weak tidal currents and salinity-driven stratification of the water column [74,75]. In these harsh conditions, only tolerant and opportunistic species (e.g., *Terschellingia longicaudata* and *Sabatieria pulchra*) were able to survive.

We removed the caves from the DEM model because extreme oligotrophy and physical isolation likely constituted the major drivers of the nematode assemblages. Several studies have indicated that food availability is the most important driver for nematode assemblages in caves (e.g., [76], and references therein). Stagnation of waters usually create hypoxia in the caves' sediments adding additional stress. Isolation of the cave systems also limits the immigration of species promoting a low SR as well [76].

We analyzed the β-diversity at two different scales: β-diversity 1 measured the variation within habitats (i.e., among sites in a same habitat type), and β-diversity 2 measured the variation among habitats. Differences in β-diversity 1 likely were due to the small-scale habitat heterogeneity that promotes a larger number of different niches, a finer partition and specialization in the use of resources by the species (reducing the interspecific competition), and thus more distinctive assemblages [77]. Limitations to the dispersal imposed by geographical distances between sites and/or physical barriers also promoted the β-diversity 1, evidenced by the higher value in freshwater caves.

The β-diversity 2 was very high, indicating that habitat types harbored distinctive nematode assemblages, and highlighting its important role as an assembly factor. This pattern emerged despite the considerable small-scale variability displayed by nematode assemblages as a response of microhabitat heterogeneity and/or species interactions [15]. In addition to the combined effects of resource availability and physical disturbance (i.e., habitat filtering sensu latu), the connectivity among habitats was likely very important as an assembly process given by the dispersal of organisms [5]. The uniqueness of nematode assemblages in some habitats (e.g., caves, macroalgae thallus, and deep water) was likely caused by the limited dispersal capacity of nematodes [78]. Actually, the largest dissimilarity in nematode assemblage structure occurred in freshwater caves stressing the important role of dispersal and vicariance on the stygobiont meiofauna [79]. However, habitats that usually were located close to each other in the shallow waters (i.e., the coral reef complex: seagrass meadow, bare sand, coral degradation zone, and algal turf) had higher similarity in the assemblage structure when compared with other habitats more separated geographically (e.g., unvegetated bottoms or deep waters). This connectivity of coastal habitats, also termed as a mosaic, has been highlighted in [80] and it is relevant for ecosystem functioning and fisheries.

Macroalgae thalli likely acted as keystone structures providing resources (mainly biofilm) and substrate to nematodes. These keystone structures are linked to habitat heterogeneity, which in turn may promote the species richness and β-diversity [81]. However, macroalgae thalli were subjected to an intense hydrodynamic regime that likely limited the amount of epiphytic nematodes and overcame the role of the fractal complexity of the thalli [82]. Latter authors concluded that neither complexity nor deposited sediment influenced the epiphytic nematode assemblages. Apparently, the most similar habitat to the macroalgae thallus was the algal turf, but the latter had highly similar structure to the other coastal habitats (e.g., bare sand and seagrass meadow) likely reflecting the role of the passive dispersal of nematodes by bottom currents and the spatial nearness among these habitats.

The link between functional diversity and biological traits is complex and depends on the particular ecosystem and taxon under study [83]. For aquatic nematodes, only the life strategy (or c–p scale) takes into account physiological and developmental features. The other four traits analyzed here were based purely on morphological features (i.e., buccal cavity, cuticle, amphidial fovea, and tail). Life strategy, trophic group, and tail shape percentages meaningfully distributed across the habitats likely reflecting ecological adaptations of nematode species to the environment.

Freshwater caves harbored more nematodes with extreme c–p strategies (i.e., extreme colonizers and extreme persisters) due to the impoverished sediments that demanded fast colonization on ephemeral food patches or very low physiological rate to cope with the oligotrophic conditions. Dominance of predators/omnivores likely also reflected the oligotrophy of the cave sediments and the necessity to feed on detritus and also on other meiofauna [84]. In other cave systems with some accumulation of detritus and guano, the most dominant groups were bacterivores, followed by omnivores and predators [85]. The anchialine cave had a trophic group composition dominated by epigrowth feeders followed by selective detritivores. This likely reflected an important role of the immigration of nematode species from adjacent coral reef environments such as *Paradesmodora immersa*, *Pomponema* sp., and *Zalonema ditlevseni*. Other studies in submarine caves have reported different findings with the dominance of either non-selective deposit feeders or omnivores/predators [85,86].

Trophic groups in the marine habitats gave further indication about the food sources used by the nematodes, although many species behave as opportunistic feeders which may change feeding strategies in response to available food [87]. Sedimentary environments where detritus constitutes the main source of food (unvegetated mud and deep-water) had dominance of deposit feeding nematodes. Dominance of deposit feeders and pollution-tolerant genera (e.g., *Sabatieria*, *Terschellingia*) in unvegetated muds suggested that organic accumulation further fostered hypoxia in this habitat type [40,74]. The other coastal habitats supported a larger proportion of epigrowth feeders able to feed on biofilm growing on sand grains, macroalgae thallus or coral rubble.

Amphidial fovea shape varied without a clear pattern across the habitats. Semprucci et al. [26] have indicated some relationships between fovea shape and sediment type and hydrodynamic stress. Nematodes may be attracted or repelled by compounds released by cyanobacterial biofilms [88], but whether particular types of amphidial fovea play an adaptive role in the behavior of aquatic nematodes is yet to be tested. Our study covered a large range of environmental conditions across the nine habitats, so this lack of pattern in amphidial fovea composition likely pointed to weak, if any, ecological relationships. Further studies about the usefulness of this trait are needed in order to refine the classification by fovea type or discard the trait as a predictor of ecological conditions.

Cuticle showed a dominance of punctated/annulated types likely reflecting a phylogenetic signature instead of ecological adaptations to the environment. Enoplids and dorylaimids, mostly with smooth cuticle, dominated in freshwater caves, although one species (*Ironus ignavus*) contributed to most of the abundance as usually occurs in freshwater systems [89]. Six orders of chromadorids dominated in the studied marine habitats: Chromadorida, Desmodorida, Monhysterida, Plectida, Araeolaimida, and Desmoscolecida. Most of these orders included nematodes with punctated/annulated cuticle. An important fraction of nematodes from deep water sediments had a cuticle with wide body annules and longitudinal ridges that is explained by the relative dominance of the family Ceramonematidae. This fact pointed to a phylogenetic signature given by a particular family inhabiting deep-waters and bearing a cuticle with wide body annules and longitudinal ridges, but without obvious ecological link between habitat setting and cuticle type.

Tail shape may bring ecological information useful for the interpretation of trait composition patterns [53]. The elongated/filiforme tail shape was well-represented in fine sediments (freshwater cave, unvegetated mud, and deep water). These nematodes act as burrowing species pushing the sediment particles to create their space within the sediment; in these conditions, elongated/filiform tail may enhance locomotion. For the other habitats where particle size is larger (bare sand) or substrate is non-sedimentary, the interstitial strategy should be more important and conical tails seem to have higher adaptive value.

In summary, we reported a regional richness of 410 ± 12 species of nematodes for the Cuban archipelago; however, some habitats need to be sampled more intensively (e.g., caves and deep waters). Our data supported the dynamic equilibrium model with habitats reasonably ordered across gradients of resource availability and physical disturbance according to their SR and abundance. Seagrass meadow was the most specious habitat, and freshwater and anchihaline caves the least diverse. Differences in β-diversity likely were due to habitat heterogeneity (e.g., in seagrass meadows) and limitations for nematode dispersal (e.g., in caves). The nematode assemblage composition was unique in some habitats reflecting the effects of habitat filtering (e.g., deep waters, macroalgae thallus). However, coastal shallow habitats shared many species indicating a high connectivity due to geographic proximity and dispersal by bottom currents. The biological traits "life strategy", "trophic group", and "tail shape" reflected ecological adaptations to environmental setting. Instead, "amphidial fovea" and "cuticle" likely reflected phylogenetic signatures from families/genera living in different habitats. Habitat type played an influential role in the diversity patterns of aquatic nematodes from taxonomic and functional points of view.

Supplementary Materials: The following are available online at http://www.mdpi.com/1424-2818/11/9/166/s1, **Table S1**: List of the sites and associated information, **Table S2**: List of biological traits of nematodes.

Author Contributions: Conceptualization, M.A; methodology, M.A., and J.A.P.-G; formal analysis, M.A.; data curation, J.A.P.-G., D.M.-P., and P.R.-G.; writing—original draft preparation, M.A.; writing—review and editing, M.A., J.A.P.-G., D.M.-P., and P.R.-G.; funding acquisition, M.A.

Funding: This research was partially funded by THE OCEAN FOUNDATION, grant PROYECTO TRES GOLFOS; CARLSBERG FOUNDATION, grants 2013_01_0779, and 2013_01_0501; THE GULF OF MEXICO RESEARCH INITIATIVE through its consortia The Center for the Integrated Modeling and Analysis of the Gulf Ecosystem (C-IMAGE).

Acknowledgments: We deeply appreciate the contribution of past and current students that contributed with the identification of nematodes during almost 10 years: Adriana Parages, Alejandro Pérez, Ariadna Rojas, Arturo del Pino, Claudia Hernández, Gabriela Alvarez, Susset González, Yarima Díaz, and Yimay Sosa. We thank the staff of the Centro de Investigaciones Marinas, Universidad de La Habana, for the support during the expeditions and field trips. Acknowledgements to Katrine Worsaae, Alejandro Martínez-García and Brett C. González for the cave sampling and support of field work. We thank Adrián Martínez for the help with the map of the study sites and Fernando Bretos, Amy Apprill, and Steven A. Murawski for their help in the field working and funding. We thank the comments/criticisms by two anonymous reviewers which improved an earlier version of the manuscript.

Conflicts of Interest: The authors declare no conflict of interest. The funders had no role in the design of the study; in the collection, analyses, or interpretation of data; in the writing of the manuscript, or in the decision to publish the results.

References

1. Begon, M.; Townsend, C.R.; Harper, J.L. *Ecology: From Individuals to Ecosystems*; Blackwell Publishing: Hoboken, NJ, USA, 2006.
2. Southwood, T.R.E. Habitat, the templet for ecological studies? *J. Anim. Ecol.* **1977**, *46*, 337–365. [CrossRef]
3. Southwood, T.R.E. Tactics, strategies and templets. *Oikos* **1988**, *52*, 3–18. [CrossRef]
4. Ferraro, S.P. Ecological periodic tables: In principle and practice. *Oikos* **2013**, *122*, 1541–1553. [CrossRef]
5. Vellend, M. Conceptual synthesis in community ecology. *Q. Rev. Biol.* **2010**, *85*, 183–206. [CrossRef] [PubMed]
6. Kraft, N.J.B.; Adler, P.B.; Godoy, O.; James, E.C.; Fuller, S.; Levine, J.M. Community assembly, coexistence and the environmental filtering metaphor. *Funct. Ecol.* **2015**, *29*, 592–599. [CrossRef]
7. Townsend, C.R.; Hildrew, A.G. Species traits in relation to a habitat templet for river systems. *Freshw. Biol.* **1994**, *31*, 265–275. [CrossRef]
8. Hewitt, J.E.; Thrush, S.F.; Dayton, P.D. Habitat variation, species diversity and ecological functioning in a marine system. *J. Exp. Mar. Biol. Ecol.* **2008**, *366*, 116–122. [CrossRef]
9. Ferraro, S.P.; Cole, F.A. Ecological periodic tables for benthic macrofaunal usage of estuarine habitats in the US Pacific Northwest. *Estuar. Coast. Shelf Sci.* **2011**, *94*, 36–47. [CrossRef]
10. Ferraro, S.P.; Cole, F.A. Ecological periodic tables for benthic macrofaunal usage of estuarine habitats: Insights from a case study in Tillamook Bay, Oregon, USA. *Estuar. Coast. Shelf Sci.* **2012**, *102–103*, 70–83. [CrossRef]
11. Anderson, M.J.; Crist, T.O.; Chase, J.M.; Vellend, M.; Inouye, B.D.; Freestone, A.L.; Sanders, N.J.; Cornell, H.V.; Comita, L.S.; Davies, K.F.; et al. Navigating the multiple meanings of b diversity: A roadmap for the practicing ecologist. *Ecol. Lett.* **2011**, *14*, 19–28. [CrossRef]

12. Heip, C.; Vincx, M.; Vranken, G. The ecology of marine nematodes. *Oceanogr. Mar. Biol. Annu. Rev.* **1985**, *23*, 399–489.

13. Abebe, E.; Andrássy, I.; Traunspurger, W. *Freshwater Nematodes: Ecology and Taxonomy*; Abebe, E., Andrássy, I., Traunspurger, W., Eds.; CABI Publishing: Oxfordshire, UK, 2006; p. 253.

14. Appeltans, W.; Ahyong, S.T.; Anderson, G.; Angel, M.V.; Artois, T.; Bailly, N.; Bamber, R.; Barber, A.; Bartsch, I.; Berta, A.; et al. The magnitude of global marine species diversity. *Curr. Biol.* **2012**, *22*, 1–14. [CrossRef]

15. Moens, T.; Braeckman, U.; Derycke, S.; Fonseca, G.; Gallucci, F.; Gingold, R.; Guilini, K.; Ingels, J.; Leduc, D.; Vanaverbeke, J.; et al. Ecology of free-living marine nematodes. *Hand. Zool.* **2013**, *2*, 109–152.

16. Danovaro, R.; Carugati, L.; Corinaldesi, C.; Gambi, C.; Guilini, K.; Pusceddu, A.; Vanreusel, A. Multiple spatial scale analyses provide new clues on patterns and drivers of deep-sea nematode diversity. *Deep Sea Res. II* **2013**, *92*, 97–106. [CrossRef]

17. Rosli, N.; Leduc, D.; Rowden, A.A.; Probert, P.K.; Clark, M.R. Regional and sediment depth differences in nematode community structure greater than between habitats on the New Zealand margin: Implications for vulnerability to anthropogenic disturbance. *Prog. Oceanogr.* **2018**, *160*, 26–52. [CrossRef]

18. Pusceddu, A.; Gambi, C.; Zeppilli, D.; Bianchelli, S.; Danovaro, R. Organic matter composition, metazoan meiofauna and nematode biodiversity in Mediterranean deep-sea sediments. *Deep Sea Res. II* **2009**, *56*, 755–762. [CrossRef]

19. McGill, B.J.; Enquist, B.J.; Weiher, E.; Westoby, M. Rebuilding community ecology from functional traits. *Trends Ecol. Evol.* **2006**, *21*, 178–185. [CrossRef] [PubMed]

20. Huston, M. *Biological Diversity: The Coexistence of Species on Changing Landscapes*; Cambridge University Press: Cambridge, UK, 1994; p. 681.

21. Connell, J.H. Diversity in tropical rain forests and coral reefs. *Science* **1978**, *199*, 1302–1310. [CrossRef]

22. Wright, D.H. Species-energy theory: An extension of species-area theory. *Oikos* **1983**, *41*, 496–506. [CrossRef]

23. Raes, M.; De Troch, M.; Ndaro, S.G.M.; Muthumbi, A.; Guilini, K.; Vanreusel, A. The structuring role of microhabitat type in coral degradation zones: A case study with marine nematodes from Kenya and Zanzibar. *Coral Reefs* **2007**, *26*, 113–126. [CrossRef]

24. Raes, M.; Decraemer, W.; Vanreusel, A. Walking with worms: Coral-associated epifaunal nematodes. *J. Biogeogr.* **2008**, *35*, 2207–2222. [CrossRef]

25. Semprucci, F.; Colantoni, P.; Sbrocca, C.; Baldelli, G.; Balsamo, M. Spatial patterns of distribution of meiofaunal and nematode assemblages in the Huvadhoo lagoon (Maldives, Indian Ocean). *J. Mar. Biol. Assoc. UK* **2014**, *94*, 1377–1385. [CrossRef]

26. Semprucci, F.; Cesaroni, L.; Guidi, L.; Balsamo, M. Do the morphological and functional traits of free-living marine nematodes mirror taxonomical diversity? *Mar. Environ. Res.* **2018**, *135*, 114–122. [CrossRef] [PubMed]

27. Santos, T.M.T.; Venekey, V. Meiofauna and free-living nematodes in volcanic sands of a remote South Atlantic, oceanic island (Trindade, Brazil). *J. Mar. Biol. Assoc. UK* **2017**. [CrossRef]

28. Prasath, D.; Balasubramaniam, J.; Jayarat, K.A. Diversity and distribution of the free-living marine nematodes in the mangrove sediments of the Andaman Islands. *Indian J. Geo Mar. Sci.* **2018**, *47*, 2217–2224.

29. Thai, T.T.; Lam, N.L.Q.; Yen, N.T.M.; Vanreusel, A.; Quang, N.X. Biodiversity and distribution patterns of free-living nematode communities in Ba Lai River, Ben Tree Province. *Vietnam J. Sci. Technol.* **2018**, *56*, 224–235. [CrossRef]

30. Colwell, R.K.; Coddington, J.A. Estimating terrestrial biodiversity through extrapolation. *Philos. Trans. R. Soc. Lond. B Biol. Sci.* **1994**, *345*, 101–118. [PubMed]

31. Gotelli, N.J.; Colwell, R.K. Estimating species richness. In *Biological Diversity. Frontiers in Measurement and Assessment*; Magurran, A.E., McGill, B.J., Eds.; Oxford University Press: Oxford, UK, 2011; pp. 39–54.

32. Violle, C.; Navas, M.L.; Vile, D.; Kazakou, E.; Fortunel, C.; Hummel, I.; Garnier, E. Let the concept of trait be functional! *Oikos* **2007**, *116*, 882–892. [CrossRef]

33. Kiørboe, T.; Visser, A.; Andersen, K.H. A trait-based approach to ocean ecology. *ICES J. Mar. Sci.* **2018**, *75*, 1849–1863. [CrossRef]

34. Van der Plas, F. Biodiversity and ecosystem functioning in naturally assembled communities. *Biol. Rev.* **2019**, *94*, 1220–1245. [CrossRef]

35. Bremner, J.; Rogers, S.I.; Frid, C.L.J. Assessing functional diversity in marine benthic ecosystems: A comparison of approaches. *Mar. Ecol. Prog. Ser.* **2003**, *254*, 11–25. [CrossRef]

36. Bremner, J.; Rogers, S.I.; Frid, C.L.J. Methods for describing ecological functioning of marine benthic assemblages using biological traits analysis (BTA). *Ecol. Ind.* **2006**, *6*, 609–622. [CrossRef]

37. Schratzberger, M.; Warr, K.; Rogers, S.I. Functional diversity of nematode communities in the southwestern North Sea. *Mar. Environ. Res.* **2007**, *63*, 368–389. [CrossRef] [PubMed]

38. Alves, A.S.; Veríssimo, H.; Costa, M.J.; Marques, J.C. Taxonomic resolution and Biological Traits Analysis (BTA) approaches in estuarine free-living nematodes. *Estuar. Coast. Shelf Sci.* **2014**, *138*, 69–78. [CrossRef]

39. Liu, X.; Liu, Q.; Zhang, Y.; Hua, E.; Zhang, Z. Effects of Yellow Sea Cold Water Mass on marine nematodes based on biological trait analysis. *Mar. Environ. Res.* **2018**, *141*, 167–185. [CrossRef] [PubMed]

40. Armenteros, M.; Ruiz-Abierno, A.; Fernández-Garcés, R.; Pérez-García, J.A.; Díaz-Asencio, L.; Vincx, M.; Decraemer, W. Biodiversity patterns of free-living marine nematodes in a tropical bay: Cienfuegos, Caribbean Sea. *Estuar. Coast. Shelf Sci.* **2009**, *85*, 179–189. [CrossRef]

41. Pérez-García, J.A.; Marzo-Pérez, D.; Armenteros, M. Spatial scale influences diversity patterns of free-living nematode assemblages in coral degradation zones from the Caribbean Sea. *Mar. Biodivers.* **2019**, *49*, 1831–1842. [CrossRef]

42. Andrássy, I. Nematoden aus strand- und höhlenbiotopen von Kuba. *Acta Zool. Acad. Sci. Hung.* **1973**, *19*, 233–270.

43. López-Cánovas, C.I.; Pastor De Ward, C. Lista de los nemátodos de la clase Adenophorea (Subclase Chromadoria y Enoplia) de los pastos marinos del archipiélago Sabana-Camagüey, Cuba. *Poeyana* **2006**, *494*, 38–42.

44. Armenteros, M.; Pérez-García, J.A.; Pérez-Angulo, A.; Williams, J.P. Efficiency of extraction of meiofauna from sandy and muddy marine sediments. *Rev. Investig. Marian. Univ. Habana* **2008**, *29*, 113–118.

45. De Grisse, A.T. Redescription ou modification de quelques techniques utilises dans l' étude des nématodes phytoparasitaires. *Meded. Rijksfac. Landbouwwet. Gent* **1969**, *34*, 351–369.

46. Platt, H.M.; Warwick, R.M. *Free-Living Marine Nematodes. Part I. British Enoplids*; The Linnean Society of London and The Estuarine and Brackish-Water Sciences Association: Cambridge, UK, 1983; Volume 28.

47. Platt, H.M.; Warwick, R.M. *Free-Living Marine Nematodes. Part II. British Chromadorids*; The Linnean Society of London and The Estuarine and Brackish-water Sciences Association: Leiden, The Netherlands, 1988; Volume 38.

48. Warwick, R.M.; Platt, H.M.; Somerfield, P.J. *Free-Living Marine Nematodes. Part III. Monhysterids*; The Linnean Society of London and The Estuarine and Coastal Sciences Association: Shrewsbury, UK, 1998; Volume 53.

49. Bongers, T.; Alkemade, R.; Yeates, G.W. Interpretation of disturbance-induced maturity decrease in marine nematode assemblages by means of the Maturity Index. *Mar. Ecol. Prog. Ser.* **1991**, *76*, 135–142. [CrossRef]

50. Bongers, T.; Bongers, M. Functional diversity of nematodes. *Appl. Soil Ecol.* **1998**, *10*, 239–251. [CrossRef]

51. Bongers, T.; De Goede, R.G.M.; Korthals, G.W.; Yeates, G.W. Proposed changes of c–p classification for nematodes. *Russ. J. Nematol.* **1995**, *3*, 61–62.

52. Wieser, W. Die Bezichung swischen Mundhöhlengestalt, Ernährungsweise und Vorkommen bei freilebenden marinen Nematoden. *Ark. Zool.* **1953**, *4*, 439–484.

53. Thistle, D.; Lambshead, P.J.D.; Sherman, K.M. Nematode tail-shape groups respond to environmental differences in the deep sea. *Vie Milieu* **1995**, *45*, 107–115.

54. Colwell, R.K. EstimateS: Statistical Estimation of Species Richness and Shared Species from Samples. Version 9. Available online: http://purl.oclc.org/estimates (accessed on 26 December 2018).

55. Chao, A.; Chazdon, R.L.; Colwell, R.K.; Shen, T.J. Abundance-based similarity indices and their estimation when there are unseen species in samples. *Biometrics* **2006**, *62*, 361–371. [CrossRef] [PubMed]

56. Clarke, K.R.; Gorley, R.N. *Primer V6: User Manual/Tutorial*; Primer-E, Ltd: Plymouth, UK, 2006.

57. Anderson, M.J.; Gorley, R.N.; Clarke, K.R. *PERMANOVA+ for PRIMER: Guide to Software and Statistical Methods*; Primer-E, Ltd: Plymouth, UK, 2008.

58. Armenteros, M.; Rojas-Corzo, A.; Ruiz-Abierno, A.; Derycke, S.; Backeljau, T.; Decraemer, W. Systematics and DNA barcoding of free-living marine nematodes with emphasis on tropical desmodorids using nuclear SSU rDNA and mitochondrial COI sequences. *Nematology* **2014**, *16*, 979–989. [CrossRef]

59. Macheriotou, L.; Guilini, K.; Bezerra, T.N.C.; Tytgat, B.; Nguyen, D.T.; Nguyen, T.X.P.; Noppe, F.; Armenteros, M.; Boufahja, F.; Rigaux, A.; et al. Metabarcoding free-living marine nematodes using curated 18S and CO1 reference sequence databases for species-level taxonomic assignments. *Ecol. Evol.* **2019**, *9*, 1211–1226. [CrossRef] [PubMed]

60. Steinmann, K.; Eggenberg, S.; Wohlgemuth, T.; Linder, H.P.; Zimmermann, N.E. Niches and noise—Disentangling habitat diversity and area effect on species diversity. *Ecol. Complex.* **2011**, *8*, 313–319. [CrossRef]
61. Liao, J.X.; Yeh, H.M.; Mok, H.K. Meiofaunal communities in a tropical seagrass bed and adjacent unvegetated sediments with note on sufficient sample size for determining local diversity indices. *Zool. Stud.* **2015**, *54*, 14. [CrossRef]
62. Udalov, A.A.; Azovsky, A.I.; Mokievsky, V.O. Depth-related pattern in nematode size: What does the depth itself really mean? *Prog. Oceanogr.* **2005**, *67*, 1–23. [CrossRef]
63. Sharma, J.; Baguley, J.; Montagna, P.A.; Rowe, G.T. Assessment of Longitudinal Gradients in Nematode Communities in the Deep Northern Gulf of Mexico and Concordance with Benthic Taxa. *Int. J. Oceanogr.* **2012**, *2012*, 15. [CrossRef]
64. Brooks, G.R.; Larson, R.A.; Schwing, P.T.; Diercks, A.R.; Armenteros, M.; Diaz-Asencio, M.; Martínez-Suárez, A.; Sanchez-Cabeza, J.A.; Ruiz-Fernandez, A.C.; Herguera, J.C.; et al. Gulf of Mexico (GoM) bottom sediments and depositional processes: A baseline for future oil spills. In *Scenarios and Responses to Future Deep Oil Spills*; Murawski, S.A., Ainsworth, C.H., Gilbert, S., Hollander, D.J., Paris, C.B., Schlüter, M., Wetzel, D.L., Eds.; Springer Nature: Basel, Switzerland, 2020; pp. 75–95.
65. Vanreusel, A.; Fonseca, G.; Danovaro, R.; da Silva, M.C.; Esteves, A.M.; Ferrero, T.; Gad, G.; Galtsova, V.; Gambi, M.C.; Fonsêca-Genevois, V.G.; et al. The contribution of deep-sea macrohabitat heterogeneity to global nematode diversity. *Mar. Ecol.* **2010**, *31*, 6–20. [CrossRef]
66. Armenteros, M.; Ruiz-Abierno, A. Body size distribution of free-living marine nematodes from a Caribbean coral reef. *Nematology* **2015**, *17*, 1153–1164. [CrossRef]
67. Hemminga, M.A.; Duarte, C.M. *Seagrass Ecology*; Cambridge University Press: Cambridge, UK, 1999; p. 299.
68. Danovaro, R.; Gambi, C. Biodiversity and trophic structure of nematode assemblages in seagrass systems: Evidence for a coupling with changes in food availability. *Mar. Biol.* **2002**, *141*, 667–677.
69. Fisher, R. Spatial and temporal variations in nematode assemblages in tropical seagrass sediments. *Hydrobiologia* **2003**, *493*, 43–63. [CrossRef]
70. Fisher, R.; Sheaves, M.J. Community structure and spatial variability of marine nematodes in tropical Australian pioneer seagrass meadows. *Hydrobiologia* **2003**, *495*, 143–158. [CrossRef]
71. Canion, C.R.; Heck, K.L., Jr. Effect of habitat complexity on predation success: Re-evaluating the current paradigm in seagrass beds. *Mar. Ecol. Prog. Ser.* **2009**, *393*, 37–46. [CrossRef]
72. Walters, K.; Bell, S.S. Diel patterns of active vertical migration in seagrass meiofauna. *Mar. Ecol. Prog. Ser.* **1986**, *34*, 95–103. [CrossRef]
73. Sánchez-Quinto, A.; Falcón, L.I. Metagenome of *Acropora palmata* coral rubble: Potential metabolic pathways and diversity in the reef ecosystem. *PLoS ONE* **2019**, *14*, e0220117. [CrossRef] [PubMed]
74. Armenteros, M.; Pérez-Angulo, A.; Regadera, R.; Beltrán, J.; Vincx, M.; Decraemer, W. Effects of heavy and chronic pollution on macro- and meiobenthos of Havana Bay, Cuba. *Rev. Investig. Marina. Univ. Habana* **2009**, *30*, 203–214.
75. Díaz-Asencio, L.; Helguera, Y.; Fernández-Garcés, R.; Gómez-Batista, M.; Rosell, G.; Hernández, Y.; Pulido, A.; Armenteros, M. Two-year temporal response of benthic macrofauna and sediments to hypoxia in a tropical semi-enclosed bay (Cienfuegos, Cuba). *Rev. Biol. Trop.* **2016**, *64*, 177–188. [CrossRef] [PubMed]
76. Du Preez, G.; Majdi, N.; Swart, A.; Traunspurger, W.; Fourie, H. Nematodes in caves: A historical perspective on their occurrence, distribution and ecological relevance. *Nematology* **2017**, *19*, 627–644. [CrossRef]
77. Hewitt, J.E.; Thrush, S.F.; Halliday, J.; Duffy, C. The importance of small-scale habitat structure for maintaining beta diversity. *Ecology* **2005**, *86*, 1619–1626. [CrossRef]
78. Derycke, S.; Backeljau, T.; Moens, T. Dispersal and gene flow in free-living marine nematodes. *Front. Zool.* **2013**, *10*, 12. [CrossRef] [PubMed]
79. Culver, D.C.; Pipan, T.; Schneider, K. Vicariance, dispersal and scale in the aquatic subterranean fauna of karst regions. *Freshw. Biol.* **2009**, *54*, 918–929. [CrossRef]
80. Kritzer, J.P.; Delucia, M.B.; Greene, E.; Shumway, C.; Topolski, M.F.; Thomas-Blate, J.; Chiarella, L.A.; Davy, K.B.; Smith, K.P. The importance of benthic habitats for coastal fisheries. *BioScience* **2016**, *66*, 274–284. [CrossRef]
81. Tews, J.; Brose, U.; Grimm, V.; Tielbörger, K.; Wichmann, M.C.; Schwager, M.; Jeltsch, F. Animal species diversity driven by habitat heterogeneity/diversity: The importance of keystone structures. *J. Biogeogr.* **2004**, *31*, 79–92. [CrossRef]

82. Pérez-García, J.A.; Ruiz-Abierno, A.; Armenteros, M. Does morphology of host marine macroalgae drive the ecological structure of epiphytic meiofauna? *J. Mar. Biol. Oceanogr.* **2015**, *4*, 7. [CrossRef]

83. Petchey, O.L.; Gaston, K.J. Functional diversity: Back to basics and looking forward. *Ecol. Lett.* **2006**, *9*, 741–758. [CrossRef]

84. Pérez-García, J.A.; Díaz-Delgado, Y.; García-Machado, E.; Martínez-García, A.; Gonzalez, B.C.; Worsaae, K.; Armenteros, M. Nematode diversity of freshwater and anchialine caves of Western Cuba. *Proc. Biol. Soc. Wash.* **2018**, *131*, 144–155. [CrossRef]

85. Zhou, H.; Zhang, Z.N. Nematode assemblages from submarine caves in Hong Kong. *J. Nat. Hist.* **2008**, *42*, 781–795. [CrossRef]

86. Ape, F.; Arigó, C.; Gristina, M.; Genovese, L.; Di Franco, A.; Di Lorenzo, M.; Baiata, P.; Aglieri, G.; Milisenda, G.; Mirto, S. Meiofaunal diversity and nematode assemblages in two submarine caves of a Mediterranean marine protected area. *Mediterr. Mar. Sci.* **2016**, *17*, 202–215. [CrossRef]

87. Moens, T.; Vincx, M. Observations on the feeding ecology of estuarine nematodes. *J. Mar. Biol. Assoc. UK* **1997**, *77*, 211–227. [CrossRef]

88. Höckelmann, C.; Moens, T.; Jüttner, F. Odor compounds from cyanobacterial biofilms acting as attractants and repellents for free-living nematodes. *Limnol. Oceanogr.* **2004**, *49*, 1809–1819. [CrossRef]

89. Traunspurger, W. Ecology of freshwater nematodes. *Hand. Zool.* **2013**, *2*, 153–169.

 *diversity*

Article

The Effect of a Dam Construction on Subtidal Nematode Communities in the Ba Lai Estuary, Vietnam

Nguyen Thi My Yen [1], Ann Vanreusel [2], Lidia Lins [2], Tran Thanh Thai [1], Tania Nara Bezerra [2] and Ngo Xuan Quang [1,3,*]

[1] Department of Environmental Management and Technology, Institute of Tropical Biology, Vietnam Academy of Science and Technology, 85, Tran Quoc Toan, Dist.3, Ho Chi Minh City 72415, Vietnam; myyenitb@gmail.com (N.T.M.Y.); thanhthai.bentrect@gmail.com (T.T.T.)

[2] Marine Biology Research Group, Biology Department, Ghent University, Krijgslaan 281, S8, B-9000 Ghent, Belgium; Ann.Vanreusel@ugent.be (A.V.); Lidia.LinsPereira@ugent.be (L.L.); Tania.CampinasBezerra@ugent.be (T.N.B.)

[3] Vietnam Academy of Science and Technology, Graduate University of Science and Technology, 18, Hoang Quoc Viet, Cau Giay, Hanoi 122300, Vietnam

* Correspondence: ngoxuanq@gmail.com; Tel.: +84-283-932-6296

Received: 28 February 2020; Accepted: 31 March 2020; Published: 2 April 2020

Abstract: Nematode communities and relevant environmental variables were investigated to assess how the presence of a dam affects the Ba Lai estuary benthic ecosystem, in comparison to the adjacent dam-free estuary Ham Luong. Both estuaries are part of the Mekong delta system in Vietnam. This study has shown that the dam's construction had an effect on the biochemical components of the Ba Lai estuary, as observed by the local increase in total suspended solids and heavy metal concentrations (Hg and Pb) and by a significant oxygen depletion compared to the natural river of Ham Luong. The nematode communities were also different between the two estuaries in terms of density, genus richness, Shannon–Wiener diversity, and dominant genera. The Ba Lai estuary exhibited lower nematode densities but a higher diversity, while the genus composition only slightly differed between estuaries. The results indicate that the present nematode communities may be well adapted to the natural organic load, to the heavy metal accumulation and to the oxygen stress in both estuaries, but the dam presence may potentially continue to drive the Ba Lai's ecosystem to its tipping point.

Keywords: dam impact; estuary; heavy metals; free-living nematodes; density; diversity

---

1. Introduction

The Mekong estuarine system in Vietnam is an ecologically important habitat, supporting, at the same time, many different socio-economical activities in agriculture, fisheries and aquaculture. The rivers within this system carry a lot of alluvium to form the lower Mekong River Delta, resulting in a high diversity of bio-resources along the southern coastal area of the East Sea [1]. As in every estuary, the Mekong supports a high diversity of both freshwater and marine species, including fish and crustaceans, which use the present habitats for feeding, as refuge, as a migration route, and as a nursery during the different stages of their life cycle [2].

In 2002, the first Mekong dam was built across the Ba Lai estuary, in the Ben Tre province [3]. The barrage aimed to improve the province's agricultural production and its economic development, by transforming rice fields into polyculture. This action was supposed to help residents in minimizing the damage caused by drought and inland salinization on farm production in the dry season, in preventing flooding during the rainy season, and in irrigating crop fields with fresh water [3]. However, results from previous studies at other locations worldwide have shown that dams actually change the physicochemical characteristics of the environment, disturbing the aquatic and riverine ecosystems, as well as the socio-economical structure of local communities [4,5]. Dams create water reservoirs upstream from the dam, transform lotic systems into lentic environments, slow down the water flow and cause reduced dissolved oxygen concentrations and a higher deposition of fine sediments in the reservoirs [6–9]. Consequently, these changes influence the aquatic biodiversity and the local productivity, as well as food web interactions [10,11]. Numerous studies recorded a reduction in abundance and diversity of aquatic communities (e.g., fish, algal, and benthic invertebrates), due to changes in the environment following the construction of a dam. [4,10,12–14].

As observed in other dams worldwide, the construction of the Ba Lai dam has in many aspects caused negative effects on the estuarine ecosystem and even on its surrounding environment, affecting the local residents' livelihood. The presence of the Ba Lai dam has changed the salinity gradient of the natural estuary, and converted brackish habitats into a freshwater ecosystem and the sea-estuary dynamics. The greatest impact of the dam is the transformation of forest land into farming areas in the Ba Tri district, as the natural flow has been artificially controlled [2]. Not only has the production of fish populations declined compared to the original estuary [7], but the reduction of water circulation also resulted in the accumulation of organic pollutants and waste discharge [2,15]. Since the dam's operation, local communities were facing gradual depletion of natural aquatic resources, while landslides have been threatening their land and houses [7,15]. Also, the Ba Lai estuary itself has started to decrease its water volume, since the reduction in flow can no longer carry away all the alluvium [16]. This process especially changes the benthic environment and its associated biota [5,10], of which free-living nematodes represent one of the most abundant invertebrate groups.

Nematodes provide many advantages in biomonitoring studies, including their high density and species richness, their ubiquity, and their different feeding strategies and life mode [17–21]. They lack a pelagic larval stage and respond promptly to environmental changes on the seafloor, since their entire life cycle is associated with benthic compartment. Different attributes of nematode communities, especially those related to their biodiversity, such as density and diversity indices, are used to assess the ecological status and disturbance impacts in estuaries [22–29]. However, there have been only a few studies that analyzed the response of nematode communities to the effects of dam construction [27,30–33]. Depending on the area, different responses were found, among which were significant reductions in diversity in the river Elbe, Germany [31] or reduced abundances of nematodes in the river Murray, Australia [32].

Ngo et al. (2016) studied the intertidal nematodes assemblages of the eight Mekong estuaries in Vietnam and authors observed that the community structure near the Ba Lai dam strongly deviated from what was expected based on the sediment characteristics [25]. The mesohaline downstream station (PSU of 22.9), which was located close to the barrier dam, was characterized by the dominance of the rapid colonizer genus *Diplolaimella,* and by the absence of the typical dominant genera in silty sediments, such as *Parodontophora, Halalaimus, Thalassomonhystera,* and *Terschellingia,* which resulted in a very low maturity index, an index generally used to assess levels of disturbance. At this location, unusually high ammonium concentrations where also observed, pointing to a source of disturbance.

Building upon the observations by Ngo et al. (2016) [25] for the Ba Lai estuary, the present study evaluates the ecological impacts of the dam by characterizing free-living nematode assemblages and their link to the environmental characteristics. As there are no historical data available from the time before the dam construction, we also investigated an adjacent dam-free estuary, Ham Luong, as a reference for our comparisons. We tested the following hypotheses: 1) The presence of a dam affects the environmental conditions from both the downstream and upstream part of an estuary compared to a natural estuary (i.e., dam-free). 2) The environmental differences between estuaries and estuarine sections (upstream versus downstream) resulted in different nematode communities in terms of diversity and composition.

2. Materials and Methods

2.1. Study Area and Sampling Location

The Ba Lai estuary is on average 59 km long and 3–4 m deep, with a water flow volume of 50–60 m^3/s in the dry season, while it is five times higher in the rainy season [2]. The Ham Luong estuary is 72 km long and with a water flow around 800–850 m^3/s during the dry season, and approximately 3300–3400 m^3/s in the rainy season [2]. Both estuaries flow from the My Tho River to the East Sea and they are located in the Ben Tre province.

Sampling was carried out in the Ba Lai (BL) and Ham Luong (HL) estuaries, during the dry season in March 2017. In each estuary, 6 subtidal stations were identified from the mouth to the upstream. Within the Ba Lai estuary, BL1, BL2, and BL3 (BL1-BL3) are the downstream stations, while BL4, BL5 and BL6 are located upstream of the dam. In the Ham Luong estuary, HL1, HL2, HL3, HL4, HL5, and HL6 are reference stations situated at similar positions along the original estuarine gradient as in the Ba Lai. The sampling area is shown in Figure 1.

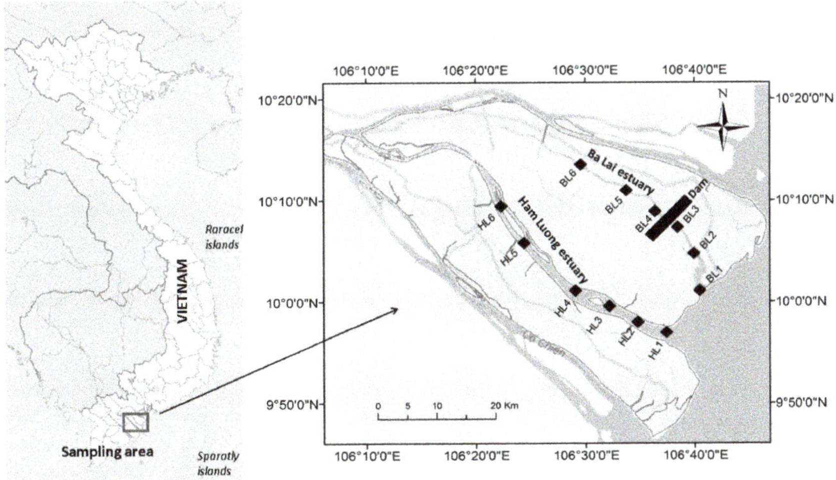

Figure 1. Sampling locations in the Ba Lai and Ham Luong estuary, Ben Tre province, Vietnam.

2.2. Sampling and Environmental Variable Analysis

One sample per station was collected for all environmental variables measured in the water column and in the sediment.

In the overlying water, different physicochemical parameters, including salinity, pH, dissolved oxygen (DO), total dissolved solids (TDS) and total suspended solids (TSS) were measured. Salinity was measured in situ using a Multiparameter Water Quality Meter Model WQC22A. Based on the measured salinity, stations were classified into different estuarine zones from polyhaline (18–30 PSU), mesohaline (5–18 PSU), oligohaline (0.5–5 PSU), to freshwater (< 0.5 PSU), following Montagna et al. (2013) [34]. The pH was measured with a pH-62K APEL equipment, DO was determined by DO-802 APEL instruments, TSS was measured using SMEWW 2540, and TDS was measured using a Water Quality Checker WQC-22A. The methods followed the international guidance sampling technique with ISO 5667-1: 2006 and ISO 5667-3: 2018 [35].

Sediment samples were collected using a Ponar type grab deployed from a small boat. The undisturbed sediment was subsampled using different techniques for different purposes (Application of ISO 5667-1: 2006, ISO 5667-12: 2017 and ISO 5667-15: 2009) [36].

Sediment samples for grain size analysis were collected by means of a cut off syringe of 3 cm diameter and 10 cm deep. The granulometry was analyzed by a Coulter type Mastersizer APA2000, equipped with Hydro 2000G model AWA2000-Malvern. The sediment fractions were classified as sand (> 63 μm), silt (4–63 μm), and clay (< 4 μm) [37].

Samples for the analysis of nutrients (including total organic carbon (TOC), total phosphorus (TP), total nitrogen (TN), ammonium (NH_4^+), and nitrate (NO_3^-)) were collected with a core of 6 cm inner diameter pushed in the sediment up to 10 cm and preserved at 4 °C until arrival at the laboratory, where they were frozen at −20 °C until further analysis. TOC, TN, TP content in the sediment were analyzed with an Element Analyzer Flash 2000 after lyophilization, homogenization and acidification with 1% HCl. The sediment was thawed and processed for measurements of nitrate (NO_3^-) and ammonia (NH_4^+) concentration, using an automatic chain (SANplus Segmented Flow Analyser, SKALAR).

Heavy metal sediment samples were collected, kept in glass bottles and transported to the laboratory (3050B method). After centrifugation followed by decantation, the sediment fraction was collected and kept frozen [38]. Amounts of 0.5 g of the wet samples (exact to 0.1 mg) were weighted into Teflon-closed vessels, followed by the addition of 6 mL sub-boiling HNO_3 (d = 1.42 g/mL) and 2 mL distilled, sub-boiling HCl (d = 1.19 g/mL). The vessels were then sonicated for 10–15 min, then simmered for at least 12 h at 110 °C. During the simmering, the vessels were sonicated for 5 min after every 4-hour simmering. After sonication, they were ramped to 160 °C and kept there for 4 h. After cooling down to room temperature, the samples were quantitatively transferred into 25 mL volumetric flasks and filled to the mark with deionized water. The bulk solutions were then transferred to 50 mL PE tubes and centrifuged at 3000 rpm for 15 min. The supernatant was divided into two separated 15 mL test tubes: one tube for archive and one tube for CV-AAS/F-AAS/ICP-MS analysis. The solutions were directly measured using the ICP-MS method for Cr, Cu, As, Se, Cd, Pb targets; as 1 g (to the nearest milligram) for Hg target; and diluted 200 times prior to F-AAS measurement for Fe target. Cr, Cu, As, Se, Cd, Pb were analyzed by inductively coupled plasma mass spectrometry (ICPMS). Fe was analyzed by flame atomic absorption spectrometry (F-AAS). Hg was analyzed by cold vapor atomic absorption spectrometry (CV-AAS).

For the Sulfide (H_2S) analysis, the top 2 cm of undisturbed sediment samples were immediately isolated after the grab returned to the boat and stored in 50 mL capped vials (polypropylene screw cap, conical bottom tubes, ISOLAB) on dry ice (Ion-selective electrode method). Once returned to the laboratory on shore, samples were transferred to a −18 °C freezer until further analysis. Total free H_2S concentrations were determined following the method of Brown et al. (2011) [39]. Briefly, the sediment sample in the 50 mL plastic vial was mildly defrosted at 4 °C, then centrifuged at 3000 rpm for 5 min. After the water layer was discarded, the sediment was homogenized with a stainless steel spatula. A 10 mL portion of the sample was transferred into another graduated plastic vial containing 10 mL of SAOB and further vortexed. The mixture was measured as quick as possible to avoid sulfide conversion.

In order to analyze the Methane (CH_4) concentration, about 10 mL of the top 2 cm of undisturbed sediment was immediately stored after sampling into a tared 40 mL serum vial containing 5 mL of 0.1 N NaOH to terminate further bacterial activity (GC-FID method). The vial was quickly sealed with a silicone stopper to minimize potential loss of methane and placed on dry ice. Once returned to the laboratory, samples were transferred to a −18 °C freezer until further analysis.

2.3. Sampling and Analysis of Nematodes

Triplicate samples per station were collected for nematode analysis. Sediment samples were taken from the boat with a Ponar grab. The grab was subsampled with a PVC core (30 cm long, 3.5 diameter) up to 10 cm depth in the sediment (10 cm^2 surface area) and placed in a sample bottle (about 300 mL). Samples were fixed and preserved in the field with 7% hot neutralized formalin (60–70 °C) and gently stirred.

In the laboratory, samples were washed over a 1 mm sieve to remove any big fractions of stone, debris and sands. Nematodes were then separated and collected by flotation technique using

Ludox-TM50 at a specific gravity of 1.18 g/cm^3, using a 38 µm sieve [40]. The procedure was repeated three times to make sure all organisms were extracted from the sample. In order to facilitate the nematode counting, samples were stained with a 1% solution of Rose Bengal. A quantitative analysis of nematodes was done using a Stereo microscope. From each sample, 200 individuals were randomly picked out and gradually transferred to pure glycerin and then mounted on permanent slides for taxonomic analysis, following the method of De Grisse (1969) [41]. For those samples with less than 200 individuals, all nematodes were picked out.

Nematode specimens on permanent slides were identified to the genus level using a Leica light microscope. For nematode identification, we used the specialized literatures [42–47], including free-living nematodes reported for Vietnam [48] and the Nemys database [49].

2.4. Data Processing and Statistical Analysis

Genus richness and Shannon–Wiener [50] diversity indices were used for biodiversity analysis.

Univariate data analyses were applied on the environmental variables, nematode densities (individuals per 10 cm^2) and diversity indices. Based on the dam presence and the measured salinity zones (mesohaline zone and oligohaline-freshwater zone), we identified two factors for further statistical analysis. The "estuary" factor, which includes 2 levels: the two estuaries Ba Lai and Ham Luong (dammed versus reference estuary); The second factor was indicated as "estuarine-section" and consists of 2 levels, upstream and downstream. The combination of the two factors resulted in 4 groups referred as: "dammed downstream", "dammed upstream", "reference downstream", "reference upstream". We chose to pool the stations per three according to their up-or downstream position (based on the dam presence and the measured salinity zones), to reduce the effect of local patchiness and to address the main question related to the dam effect.

The Shapiro–Wilk test was used to check for normal distributions and Levene's test to evaluate the homogeneity of variances ($p > 0.05$). The data were transformed by either square-root or log transformation if assumptions were not met.

If assumptions were fulfilled, a two-factor ANOVA (analysis of variance) was performed in RStudio [51]. When the assumptions were not fulfilled, the analyses were replaced by a non-parametric permutational PERMANOVA on software PRIMER 6, in order to identify significant differences between estuarine-sections and estuaries [52]. When significant differences were found ($p < 0.05$), a post hoc test (Tukey HSD) was applied for pairwise comparisons between estuarine-sections.

The obtained p-values were corrected with the Benjamini and Hochberg (1995) [53] correction method.

Multivariate analyses were performed using the software PRIMER 6 with the PERMANOVA add-on package, and the same the design was used as the one described in the univariate analyses section [52].

Significant differences between groups in nematode community composition datasets were tested with PERMANOVA. The test was based on square-root transformed data for community composition. Significant values were considered when $p < 0.05$. After the PERMANOVA tests, PERMDISP routines were performed to test for the homogeneity of multivariate dispersions. Subsequently, pairwise comparison tests were performed, to identify which pairs of estuarine sections were significantly different from each other. Then, a SIMPER analysis (SIMilarity PERcentages) was performed to identify the taxa responsible for dissimilarities between groups.

DistLM (distance-based linear model) analyses were conducted for environmental variables with correlations lower than 0.9, in order to identify which environmental factors were significantly associated with the variability observed in the nematode multivariate community structure. The DistLM model was performed using a step-wise selection procedure and adjusted R^2 as selection criteria. Results from the DistLM were visualized using dbRDA (distance-based redundancy analysis) plots.

Correlation analyses between environmental and nematode variables were computed in R-Studio, either using the Pearson or Spearman rank method (when the data were not normally distributed).

3. Results

3.1. Environmental Characteristics of Ba Lai and Ham Luong Estuaries

3.1.1. Water Environmental Characteristics

In both Ba Lai and Ham Luong estuaries, the values for salinity, TDS, pH and DO decreased in general from the downstream to the upstream sections, whereas the opposite trend was found for TSS (Figure 2). Salinity and TDS showed higher variability between the two sections of the Ba Lai, in comparison to the variation found between the two sections within Ham Luong.

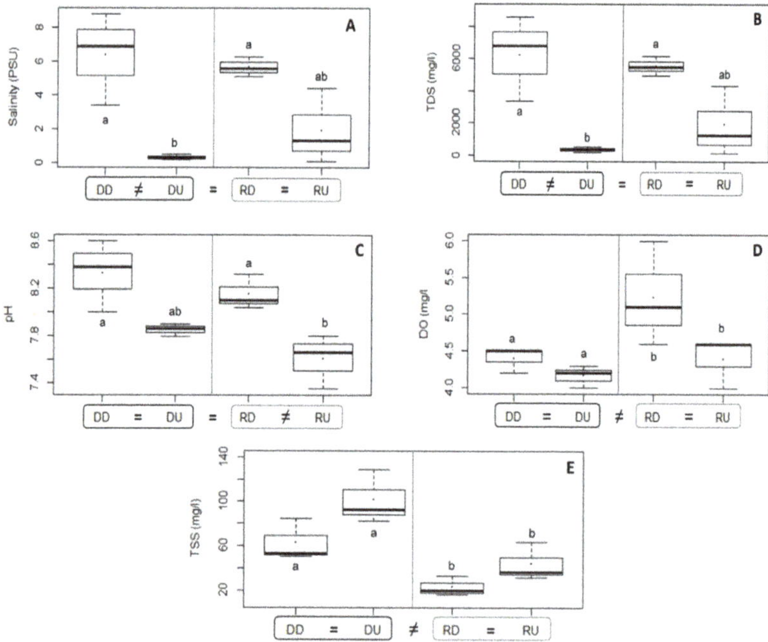

Figure 2. Box and whisker plots and mean values (.) for salinity (**A**) and TDS (**B**), pH (**C**), DO (**D**) and TSS (**E**) in the water column of the estuarine sections. ▢ point to estuary level. Abbreviations: "DD" = dammed downstream, "DU" = dammed upstream, "RD" = reference downstream, "RU" = reference upstream. Letters a and b pointed to significant differences. "≠" sign indicated the significant differences between 2 estuaries or/and estuarine sections within an estuary, while "=" sign meant that no significant difference was found.

Based on the average salinity, we classified the DD part as mesohaline (6.37 ± 2.74 PSU) and the DU section as freshwater (0.33 ± 0.15 PSU), whereas the two sections of the Ham Luong estuary were characterized as mesohaline and oligohaline (5.67 ± 0.60 and 1.93 ± 2.22 PSU, respectively) (Figure 2A). The TDS showed a concentration of 6220 ± 2628.52 mg/L in DD and 5536.67 ± 611.34 mg/L in RD. These values were higher than that at both upstream sections, with 355.33 ± 147.02 mg/L in DU and 1916 ± 2181.28 mg/L in RU (Figure 2B). The pH in all sections was higher than 7, pointing to an alkaline environment. The level of pH only slightly fluctuated with higher values in the downstream parts, compared to the upstream sides in both estuaries (Figure 2C). The Ba Lai estuary had, in general, lower DO levels than Ham Luong. Both downstream sides of the two estuaries had greater DO concentrations than the upstream parts, 4.40 ± 0.17 mg/L in DD and 5.23 ± 0.71 mg/L in RD, compared to values of 4.17 ± 0.15 in DU and 4.40 ± 0.35 in RU (Figure 2D). From Figure 2 (A, B, D), it can be seen that the dammed upstream (DU) section of Ba Lai was characterized by the lowest concentrations of salinity, TDS and DO, compared to all other sections.

The TSS concentration was higher in Ba Lai than in Ham Luong and it increased from downstream to upstream in both estuaries, with 62.63 ± 18.38 mg/L in DD and 22.93 ± 8.78 mg/L in RD, compared to 101.30 ± 24.49 mg/L in DU and 43.73 ± 17 mg/L in RU (Figure 2E).

The two-way ANOVA and permutational PERMANOVA analyses showed significant differences between the two estuaries, in terms of DO and TSS. The estuarine sections were also different in terms of salinity, TDS; pH, and DO. The significant p values were presented in Table 1. No significant interaction effect was found for any of these variables.

Table 1. The significant p values generated from ANVOVA / permutational PERMANOVA and post hoc comparison analyses for environmental variables (*pe:* the significant estuary effects, *ps:* estuarine-section effect. "≠" sign indicated for the significant differences).

		P Value	
Variable	*pe*	*ps*	**Post Hoc Test**
Salinity		0.028	DD ≠ DU: 0.034, DU ≠ RD: 0.035
pH		0.035	DD ≠ RU: 0.034, RD ≠ RU: 0.048
DO	0.037	0.026	
TSS	0.034		
TDS		0.028	DD ≠ DU: 0.034, DU ≠ RD: 0.035
Pb	0.035		
Hg	0.035	0.034	DD ≠ DU: 0.049, DU ≠ RD: 0.044

3.1.2. Sediment Environmental Characteristics

The sedimentary variables such as grain size (sand, silt, clay) and nutrient concentrations (TOC, TP, TN, NH_4^+ and NO_3^-) are presented in Figure A1. They did not show any statistically significant difference between estuaries and estuarine-sections, for either the two-way ANOVA or the permutational PERMANOVA.

The concentrations of heavy metals (Figure 3) in the Ba Lai estuary were higher than those found in Ham Luong. In addition, the values were all highest in DU compared to other estuarine section. The same trend was found in the Ham Luong estuary, although with smaller variations, except for Pb, As and Cd, which showed slightly lower average values than in the RU section. A two-way ANOVA analysis showed significant differences between the two estuaries only for Pb and Hg. The "estuarine-section" factor had a significant effect on the concentration of Hg and the Tukey HSD pairwise comparison showed differences between DD and DU, and between DU and RD. The significant p values are presented in Table 1.

The H_2S concentration was higher in Ba Lai than in Ham Luong, and showing a higher variation within the two estuarine sections of the dammed estuary (Figure 3I). The DU showed higher values than the DD part, with values of 3.60 ± 3.75 and 4.43 ± 3.94 ug/g, respectively, while the concentrations for both sections in Ham Luong varied less, with 0.55 ± 0.62 ug/g and 0.74 ± 0.98 ug/g in RD and RU, respectively (Figure 3I). At station level, the highest concentration of H_2S was observed in BL4 (8.58 ug/g) and BL3 (7.67 ug/g), which are the two nearest upstream and downstream stations to the Ba Lai dam.

The CH_4 concentration was also higher in the Ba Lai than in Ham Luong. In particular, at DD, the CH_4 concentration was 2028.78 ug/g, which is almost 10 times higher than in DU (212.76 ug/g), while Ham Luong RU (1428.28 ug/g) contained much higher values of CH_4 than RD (23.98 ug/g) (Figure 3J). Nevertheless, within the Ba Lai estuary, again a high variation was observed between stations within each estuarine section, with the highest concentration of CH_4 found in station BL3 near the dam (5975.56 ug/g). Additionally, BL4 showed elevated levels of CH_4 (418.33 ug/g), however an order of magnitude smaller than in BL3. In the Ham Luong estuary, the highest average values were present in the stations HL6 and HL5 of the upstream part.

Due to the high variation within some estuarine sections, no statistically significant difference was found for both H_2S and CH_4 concentrations, either for the 2-way ANOVA or for the permutational PERMANOVA analysis.

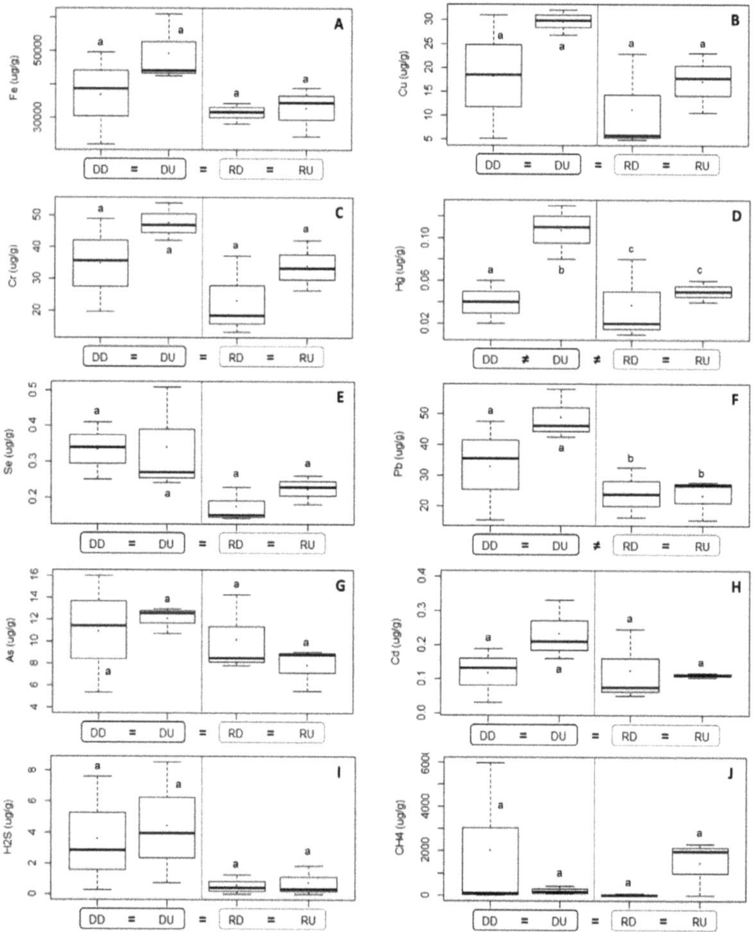

Figure 3. Box and whisker plots and mean values (.) for heavy metals including Fe (**A**), Cu (**B**), Cr (**C**), Hg (**D**), Se (**E**), Pb (**F**), As (**G**), Cd (**H**), H_2S (**I**) and CH_4 (**J**) concentration in sediment of the estuarine sections. ▢ point to estuary level. Abbreviations: "DD" = dammed downstream, "DU" = dammed upstream, "RD" = reference downstream, "RU" = reference upstream. Letters a, b and c refer to significant differences. "≠" sign indicated for the significant differences between 2 estuaries or/and estuarine sections within an estuary, while "=" sign meant that no significant difference was found.

3.2. Nematode Assemblages in Ba Lai and Ham Luong Estuaries

3.2.1. Density of Nematode Communities in Ba Lai and Ham Luong Estuaries

Nematode densities in DU were 111 ± 76 ind./10 cm^2, being twice as low as in DD, with 218 ± 138 ind./10 cm^2. The opposite pattern was found for Ham Luong, with RU showing 248 ± 195 ind./10 cm^2, almost twice as high as the densities at RD (139 ± 109 ind./10 cm^2) (Figure 4). A

PERMANOVA analysis showed a significant interaction effect on the nematode densities for the estuary and estuarine section (Table 2).

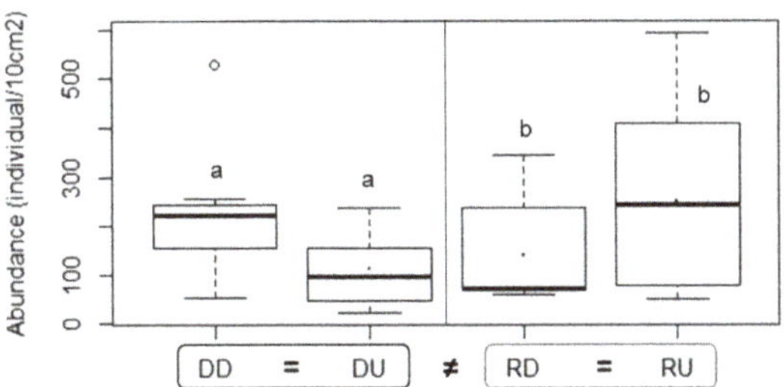

Figure 4. Box and whisker plots and mean values (.) for nematode densities of the estuarine sections. ⬭ point to estuary level. Abbreviations: "DD" = dammed downstream, "DU" = dammed upstream, "RD" = reference downstream, "RU" = reference upstream. Letters a and b pointed to significant differences. "≠" sign indicated for the significant differences between 2 estuaries, while "=" sign meant that no significant difference was found.

Table 2. The significant *p* values generated from ANVOVA / permutational PERMANOVA and pairwise comparison analyses for characteristics of nematode communities. (*pe&s:* the significant interaction effect of both factors estuary and estuarine-section, *pe:* the significant estuary effects, *ps:* estuarine-section effect. "≠" sign indicated for the significant differences).

Variable	*p* Value			
	pe&s	*pe*	*ps*	Post hoc Test
Density	0.0466		0.028	DD ≠ DU: 0.034, DU ≠ RD: 0.035
Genus richness	0.0466		0.035	DD ≠ DU: 0.0001, DD ≠ RD: 0.00026, DD ≠ RU: 0.000003
H′		0.0015	0.026	DD ≠ DU: 0.014, DD ≠ RD: 0.0065, DD ≠ RU: 0.0025
Genus composition (square-root transformation)		0.0003	0.0001	DD ≠ DU: 0.0001, DD ≠ RD: 0.0002, DD ≠ RU: 0.0001, DU ≠ RD: 0.0003, DU ≠ RU: 0.0039, RD ≠ RU: 0.0001

3.2.2. Diversity of Nematode Communities in Ba Lai and Ham Luong Estuaries

The genus richness was highly variable in Ba Lai, with DD containing twice as many genera per sample (35) as DU (14). In Ham Luong, richness was less fluctuating between the two sections, with 15 and 12 genera encountered in RD and RU samples, respectively (Figure 5A). The two-way ANOVA based on richness, followed by a pairwise comparison analysis, showed a significant interaction effect, with significant differences between the pairs DD&DU, DD&RD and DD&RU.

The Shannon–Wiener diversity index (H′) was higher in Ba Lai compared to Ham Luong, while H′ values were lower in upstream sections of both estuaries. The H′ value at DD was 2.62, being significantly higher than in DU (H′ = 1.97) (Figure 5B), while the two sections in Ham Luong showed H′s values of 1.79 and 1.56 in RD and RU respectively. Permutational 2-factor PERMANOVA analysis showed a significant difference between the two estuaries and estuarine section (Table 2, Figure 5B). The pairwise comparison showed differences between DD and DU, between DD and RD, and between DD and RU (Table 2, Figure 5B). No significant interaction effect was found.

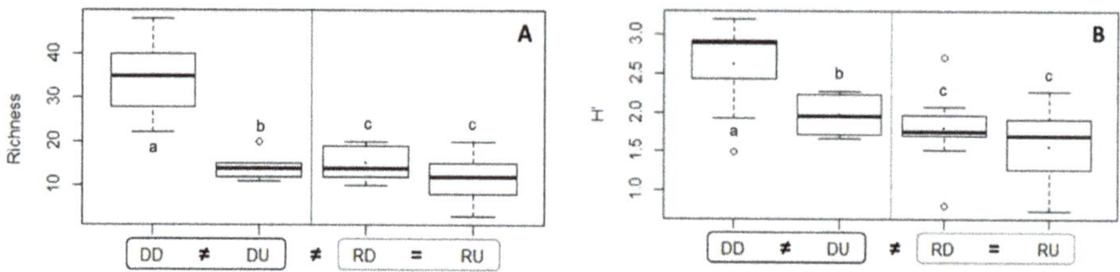

Figure 5. Box and whisker plots and mean values (.) for the richness (**A**) and Shannon–Wiener index (**B**) of nematode community of estuarine sections. ☐ point to estuary level. Abbreviations: "DD" = dammed downstream, "DU" = dammed upstream, "RD" = reference downstream, "RU" = reference upstream. Letters a, b and c pointed to significant differences. "≠" sign indicated for the significant differences between 2 estuaries or/and estuarine sections within an estuary, while "=" sign meant that no significant difference was found.

3.2.3. Nematode Community Composition in Ba Lai and Ham Luong Estuaries

The nematode communities consisted of taxa belonging to both classes Enoplea and Chromadorea. In total, 144 genera belonging to 51 families and representing 11 orders were identified. Orders included Enoplida, Triplonchida, Dorylaimida, Mononchida, Chromadorida, Desmodora, Desmoscolecida, Araeolaimida, Monhysterida, Plectida, and Rhabditida. The orders Desmoscolecida and Rhabditida were absent in Ham Luong, while all occurred in Ba Lai. The total number of nematode genera observed in the Ba Lai estuary was two times higher than those found in the Ham Luong estuary, representing 129 (47 families, 11 orders) and 57 genera (29 families, 9 orders), respectively. The total number of genera encountered in DD, DU, RD, RU were 125, 33, 45 and 77 genera, respectively.

A PERMANOVA analysis, based on the relative abundance of nematode genus nematode composition, showed a significant effect for both factors "estuary" (68.29%) and "estuarine-section". Pairwise comparisons showed that all estuarine sections were significantly different. A SIMPER analysis resulted in a percentage of dissimilarity between groups, ranging from 61.45% to 71.76%. The most important genera are responsible for more than 50% of the differences between groups (i.e., between the two estuaries and between each pair of all estuarine sections), and their percentage contribute to the dissimilarities are shown in Table A1. *Parodontophora* and *Theristus* were the two major genera responsible for the differences in community composition in terms of abundances between the two estuaries and among estuarine sections.

In general, *Parodontophora*, *Theristus*, *Daptonema*, *Terschellingia*, *Sphaerotheristus*, and *Viscosia* comprised the most abundant genera (represented each by more than 4% of the total relative abundance), and together they contributed 67.48% of the total relative abundance of both estuaries. *Parodontophora*, *Theristus* and *Daptonema* were the most common genera, representing 32.70%, 12.67%, and 8.36% of the total community, respectively.

The eight most abundant genera in each estuarine section are shown in Figure 6. *Parodontophora* was the dominant genus and contributed more than 26% to the total community in DD, RD and RU, whereas in DU, *Theristus* was the most abundant genus (26.78%), with a relative abundance which was twice as high that of *Parodontophora* (13.96%). *Theristus* and *Terschellingia* were present in higher percentages in the upstream parts, especially in the DU section of Ba Lai. *Parodontophora*, nevertheless, was found to be more abundant in the downstream sections. Especially within Ba Lai, in DD *Parodontophora* was twice as abundant as in DU.

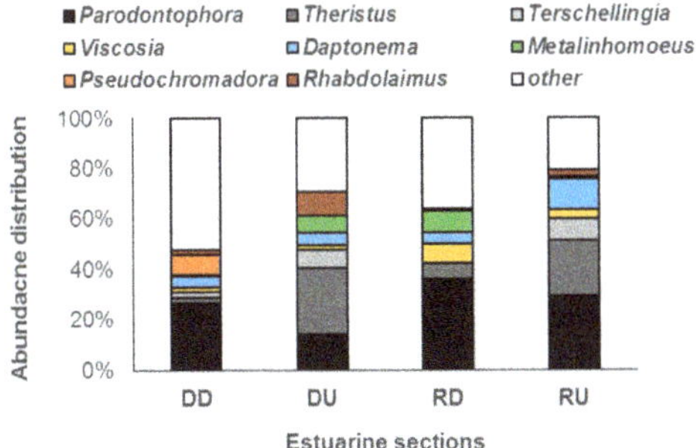

Figure 6. The percentage of the eight most abundant genera in each "estuarine section" (DD, DU, RD and RU), as identified by the SIMPER analysis. Abbreviations: "DD" = dammed downstream, "DU" = dammed upstream, "RD" = reference downstream, "RU" = reference upstream.

3.3. Correlation Between Nematode Communities and Environmental Variables

Nematode generic richness was positively correlated with pH ($p = 0.005$, $r = 0.753$), and negatively correlated with silt proportion in the sediment ($p = 0.006$, $r = -0736$). The H′ index showed positive correlations with both pH ($p = 0.01$, $r = 707$) and NO_3^- ($p = 0.012$, $r = 695$).

Furthermore, the DISTLM analysis showed that Hg ($p = 0.0204$), NO_3^- ($p = 0.019$) and Fe ($p = 0.0348$) best explained the observed nematode distribution patterns of genus composition, explaining 49% of the variation. The dbRDA analysis (with axis 1 explaining 32% of the variation and axis 2 explaining 19.6%) illustrated the lack of clear differences between the four estuarine sections (Figure 7).

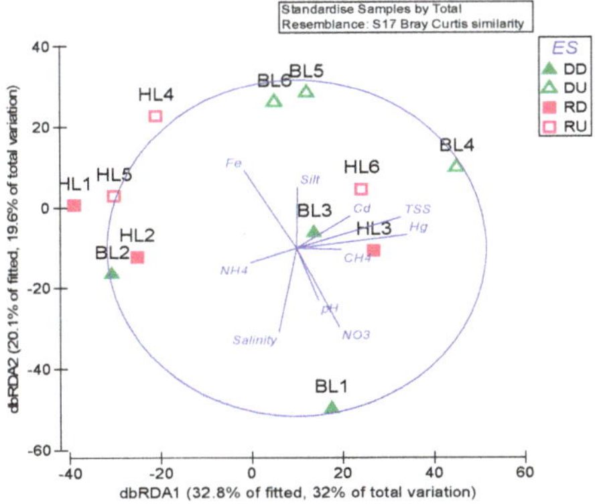

Figure 7. The dbRDA plot on the correlation between nematode generic composition on abundance with environment variables for all stations of estuarine sections. Abbreviations: (Es: estuarine-section), "DD" = dammed downstream, "DU" = dammed upstream, "RD" = reference downstream, "RU" = reference upstream.

4. Discussion

4.1. Differences in Environmental Variables Related to Dam Effects

Several environmental variables measured in the overlying water of the Ba Lai estuary showed a different trend than the ones observed for the reference estuary Ham Luong, such as the significantly lower DO values and higher TSS concentrations. The level of DO in the dammed estuary was also lower than in other dam-free branches of the Mekong estuarine system [25], whereas the level of TSS in Ba Lai, especially in the upstream part, was 1.5 times higher than that of the Mekong River system in the dry season of 2017 (61.4 mg/L), and more than twice as high as in the Bassac River (47 mg/L)—one of the two main tributaries of the Mekong River delta [1]. TSS includes all particles which are suspended in water, primarily composed of micro-algae, and organic and mineral particles linked to river input by land erosion, as well as the re-suspension of sediment by the vertical mixing of the water column. In the Mekong River delta, TSS is influenced by both natural and anthropogenic activities, including the release of municipal and domestic waste water, synthetic and organic chemicals from industrial waste, agriculture, aquaculture, construction, and soil erosion [1]. Therefore, the increase in suspended solid loads tends to coincide with a decrease in dissolved oxygen, which could lead to hypoxic stress, which in turn can result in lower abundances, diversity, and survival rates (less viable embryos and larvae) during the development in fish species, as observed for the *Notropis girardi* in the Arkansas River [13]. Additionally, TSS can also capture heavy metals and nutrients in the water column and might contain high concentrations of contaminants. Once deposited on the bottom, it may disturb the benthic environment and its biota [13]. Deposited sediment coming from upstream has been reported to be associated with contaminants, including heavy metals and other toxic compounds, such as pesticides, which can bioaccumulate in the river bed [54].

The heavy metal contamination also reached higher levels in the Ba Lai estuary, especially at the upstream side in terms of Pb and Hg concentration. The level of Pb in the dammed upstream section (48.84 µg/g) was higher than the lowest effect level (concentrations of Pb from about 31 µg/g can already cause negative effects on the benthic life) as reported by Burton (2002) [55]. The concentration of Hg in the dammed upstream area was also higher than in the natural biosphere reserve part of the protected Can Gio mangrove forests in the southern part of Vietnam. Here, the amount of Hg varied between 0.040 and 0.048 µg/g [56]. It has been previously reported that heavy metals are adsorbed by suspended particulates and in this way they can settle down in the sediment and be remobilized into the food chain [9]. Therefore, it seems plausible that the barrier created by the dam enhanced the accumulation of contaminants especially in the upstream area.

In summary, despite the fact that dam effects are less prominent in terms of siltation and changes in biochemical processes, the presence of the dam is clearly associated with lower DO and higher TSS, Hg and Pb concentrations in the Ba Lai estuary.

4.2. Differences in Nematode Communities Explained by Differences in Environmental Conditions related to Dam Effects

The subtidal nematode assemblages in Ba Lai and Ham Luong estuaries were significantly different between both estuaries and estuarine sections. There were interaction effects on the nematode density and genus richness. The dammed estuary showed lower densities and lower relative abundances of dominant genera, such as *Parodontophora*, *Theristus* and *Daptonema*, but a higher generic richness and H' index, especially downstream, compared to the reference estuary.

The nematode densities of both the Ba Lai and Ham Luong estuaries were lower than the counts from intertidal areas of the natural Mekong estuaries, which had densities ranging between 88 and 4580 ind./10 cm^2 [57–59]. They were also lower than those in subtidal areas from other estuaries worldwide (21–17,200 ind./10 cm^2) [22,26,27]. The composition of subtidal nematode communities in the Ba Lai estuary was more diverse than Ham Luong, but in the range of the diversity generally found in the intertidal sediments of the Mekong estuarine system (71–230 genera, 20–59 families) [57–59] and in other dam-free estuaries worldwide (106–120 genera belonged to 35–40 families) [22,26]. The generic richness was higher than that of Mondego estuary (Portugal), with only 8–19 genera [22].

According to Ngo et al. (2013), the diversity of the intertidal nematodes communities did not show any particularly common trend along the course of the Mekong estuaries in the dry season of 2009 [58]. Tran et al. (2018) reported that the Ba Lai's upstream part was characterized by significant lower density, generic richness and H' index than downstream [59]. Our results showed no trend in diversity for the natural Ham Luong estuary. In contrast, the downstream part of Ba Lai estuary was characterized by a significant higher diversity (richness and Shannon–Wiener indices) in comparison to upstream, and even to the Ham Luong estuary. The higher diversity in the downstream part of the Ba Lai is to some extent unexpected, given the presumed impact of the dam. However, the environmental analysis mainly showed that the upstream region of Ba Lai is contaminated with higher TSS and Hg, while the environment of the downstream section does not differ much from the reference estuary. It is possible that the increase in habitat heterogeneity has led to an increase in diversity [60]. The dammed downstream section is also an intermediately disturbed area compared to the reference estuary on one hand, and the more severely impacted upstream part of Ba Lai on the other [61]. Therefore, the increase in diversity in this area could be explained as conforming to the intermediate disturbance hypothesis [61]. An increased diversity, due to intermediate levels of disturbance, can be a plausible explanation here, given that the organic loading of the area initially favors the diversity and densities, but after a certain threshold impoverishes the communities.

Previous studies have documented that salinity, sediment composition and nutrient contents play major roles in driving the distribution pattern and structure of nematode communities in estuaries worldwide [27,62–64]. In some cases, the effect of salinity prevails over the role of the sediment composition, being the main limiting factor for species distribution in transitional water systems [23,26,62,65]. Nicholas et al. (1992) studied the effect of the periodic operation (opening and closing) of barrages in the mouth of the Murray River estuary, South Australia, on the survival of the nematode communities. The authors observed a higher mortality of organisms, due to the drastic decrease in salinity by a prolonged discharge of river water when the barrages were opened [32]. In the present work, salinity differed between estuarine sections in the Ba Lai estuary, but no significant correlation was found with the nematode communities. This indicates that other environmental factors played a major role in the differences observed. Our results also showed that pH positively correlated to the genera richness and the H' index, and the H' index also increased with the elevation of NO_3^-. The silt content negatively correlated with richness, despite the fact that the sediment composition was not significantly different, but the dammed upstream contained, on average, a higher silt fraction. The accumulation of fine sediment can slow down the mineralization process of organic matter and can be associated with toxic compounds, causing negative effects on organisms and their diversity [66]. The nematode density in the present research did not correlate with any environmental variables, which contrasts with the observation by Adão et al. (2009) and Nguyen et al. (2012), who found higher nematode abundances with increasing organic matter [57,62]. Similar to our findings, nematode density in the Can Gio mangrove forest, a Biosphere Reserve in southern part of Vietnam, were not significantly correlated to any measured environmental variables, such as nutrients, grain size, heavy metal, etc. [67].

The generic composition of the nematode communities was significantly affected by the interaction between nutrients (NO_3^-) and heavy metals (Hg, Fe). Other studies also reported that nutrient content is one of the environmental factors associated with the distribution pattern and structure of nematode communities in estuaries [27,62–64]. Until present, the correlation of Hg and Fe in estuarine sediment related to the nematode composition is still not well documented. Similar to our results, Alves et al. (2013) reported that *Daptonema* was the most dominant genus in subtidal soft sediments of the Mondego estuary (Portugal) [22]. Ngo et al. (2016) also found that *Parodontophora* was the main genus found in the more silty intertidal upstream parts of the Mekong estuaries. It has been noted that, in estuarine environments, nematode communities are usually comprised of a high number of genera, with only a few dominant ones [26]. In our study, even though differences in relative abundances were reported, *Parodontophora* and *Theristus* were the most dominant groups in all parts of both estuaries, reflecting their wide range in salinity tolerance. This is relatively surprising, as they were not in the list of abundant groups in many other studies from European estuaries, such as the Brouage mudflat in France [68], the Thames estuary in the United Kingdom [23] or the Mondego estuary in Portugal [22]. This may indicate that these two genera are possibly specific for tropical brackish environments. Furthermore, while *Parodontophora* was the prevailing genus in most estuarine sections, *Theristus* was the most dominant genus in the dammed upstream part, suggesting that this genus is highly tolerant to elevated heavy metal concentrations (especially Hg), low DO and high TSS. Interestingly, the genus

Theristus is normally considered as a typical 'colonizer', identified as an indicator of organic pollution in the Swartkops estuary of South Africa [69]. Moreover, *Terschellingia*, as the third most important genus in the dammed upstream part, is reported to be tolerant to pollution and anoxia [22,26].

In conclusion, the presence of the dam seems to have caused differences in the environment, as suggested by the local accumulation of contaminants and oxygen depletion in the Ba Lai estuary. The response of nematode communities to the dam effect was more subtle, with shifts in the density, diversity, presence of dominant genera and community composition. The present communities may be well adapted to the natural organic and oxygen stress in both estuaries, but potentially the dam may continue to drive the Ba Lai's ecosystem to its tipping point. Therefore, further research is needed and monitoring of the system is a major recommendation from this study.

Author Contributions: All authors have equally contributed to the conceptualization of the idea and the methods used; data acquisition (sampling and samples processing and specimens analyses), N.T.M.Y., T.T.T., N.X.Q, and T.N.B.; statistical analysis and data interpretation, N.T.M.Y., A.V., L.L. and N.X.Q.; project administration, A.V. and N.X.Q.; writing—original draft, N.T.M.Y and A.V; all authors have substantially contributed for writing—review and editing; all authors have read and agreed to the published version of the manuscript.

Funding: This research is funded by the VLIR-TEAM project: "Environmental and socio-economic impact after dam construction for local communities in the Mekong estuarine system: the case of the Ba Lai estuary".

Acknowledgments: We also would like to express our thankfulness to VLIR-UOS for supporting the funding for this research. We appreciate the support from the Marine Biology Research Group (Ghent University, Belgium) and the Department of Environmental Management and Technology (Institute of Tropical Biology, Vietnam).

Conflicts of Interest: The authors declare no conflict of interest. The funders had no role in the design of the study; in the collection, analyses, or interpretation of data; in the writing of the manuscript, or in the decision to publish the results.

Appendix A

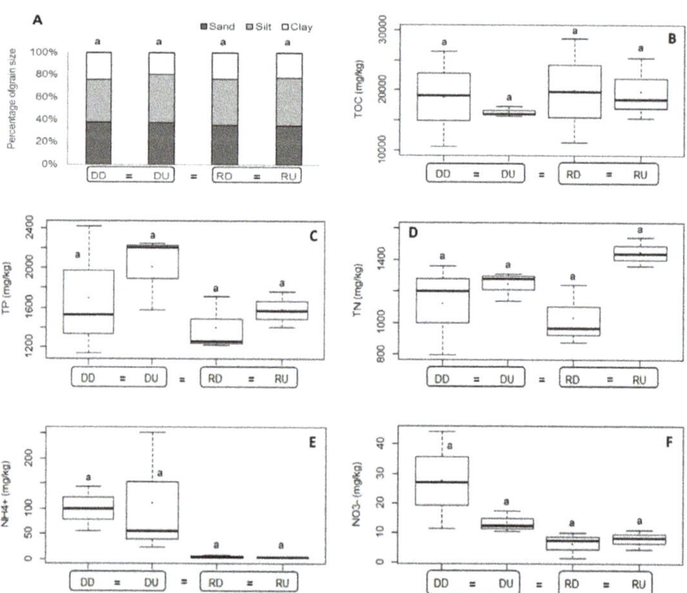

Figure A1. The grain size (sand, silt, clay in graph (**A**)) and nutrient concentrations (TOC (graph (**B**)), TP (**C**), TN (graph (**D**)), NH_4^+ (graph (**E**)) and NO_3^- (graph (**F**)) in sediment of Ba Lai and Ham Luong estuary. ▢ point to estuary level. Abbreviations: "DD" = dammed downstream, "DU" = dammed upstream, "RD" = reference downstream, "RU" = reference upstream. Letter a pointed to non-significant differences. "=" sign meant that no significant difference was found.

Table A1. The percentage of contribution of most important genera responsible for dissimilarity in relative abundance between groups based. Percentage dissimilarity (bold) between group (between the two estuaries and between each pair estuarine sections) and contributing percentage of each contributor for the dissimilarity between groups are shown together. Abbreviations: "DD" = dammed downstream, "DU" = dammed upstream, "RD" = reference downstream, "RU"= reference upstream.

Dammed Estuary and Reference Estuary (68.29%)					
Parodontophora	7.2	*Metalinhomoeus*	3.56	*Monhystrella*	2.13
Theristus	6.33	*Viscosia*	3.47	*Mononchulus*	2.12
Terschellingia	3.92	*Sphaerotheristus*	3.21	*Comesoma*	1.89
Daptonema	3.74	*Monhystera*	2.68	*Desmodora*	1.89
Rhabdolaimus	3.63	*Mesodorylaimus*	2.58	*Halalaimus*	1.86

DD and DU (71.45%)		DD and RD (70.17%)		DD and RU (71.76%)	
Theristus	6.67	*Parodontophora*	6.07	*Parodontophora*	6.59
Parodontophora	5.39	*Pseudochromadora*	3.28	*Theristus*	5.86
Rhabdolaimus	3.48	*Metalinhomoeus*	3.15	*Daptonema*	3.86
Metalinhomoeus	3.25	*Viscosia*	3.06	*Pseudochromadora*	3.38
Pseudochromadora	3.18	*Theristus*	2.7	*Terschellingia*	3.34
Terschellingia	3.02	*Comesoma*	2.5	*Sphaerotheristus*	3.27
Monhystera	2.54	*Trissonchulus*	2.48	*Viscosia*	2.42
Mesodorylaimus	2.16	*Daptonema*	2.42	*Rhabdolaimus*	2.22
Daptonema	2.14	*Marylynnia*	2.15	*Linhystera*	2.11
Halalaimus	2.11	*Terschellingia*	2.03	*Halalaimus*	2.07
Linhystera	2.05	*Amphimonhystrella*	2	*Amphimonhystrella*	2
Monhystrella	2.02	*Halalaimus*	1.99	*Cobbia*	1.77
Viscosia	1.97	*Rhabdolaimus*	1.95	*Oncholaimus*	1.75
Udonchus	1.97	*Desmodora*	1.92	*Mesodorylaimus*	1.72
Amphimonhystrella	1.9	*Sabatieria*	1.88	*Sabatieria*	1.59
Afrodorylaimus	1.76	*Mesodorylaimus*	1.86	*Microlaimus*	1.52
Sphaerotheristus	1.69	*Linhystera*	1.81	*Oxystomina*	1.49
Cobbia	1.67	*Rhynchonema*	1.75	*Desmodora*	1.49
Mononchulus	1.6	*Sphaerotheristus*	1.74	*Spilophorella*	1.44
		Cobbia	1.72	*Achromadora*	1.44

RD and RU (64.91%)		DU and RU (61.45%)		DU and RD (69.79%)	
Parodontophora	9.18	*Parodontophora*	9.05	*Theristus*	8.41
Theristus	8.3	*Theristus*	8.65	*Parodontophora*	7.34
Daptonema	5.77	*Terschellingia*	6.2	*Rhabdolaimus*	5.35
Viscosia	5.57	*Daptonema*	5.73	*Metalinhomoeus*	5.04
Sphaerotheristus	5.04	*Rhabdolaimus*	5.24	*Viscosia*	4.79
Metalinhomoeus	4.69	*Sphaerotheristus*	5.21	*Terschellingia*	4.4
Comesoma	4.34	*Metalinhomoeus*	5.08	*Comesoma*	3.67
Terschellingia	4.31	*Monhystera*	4.63	*Monhystera*	3.43
Trissonchulus	3.9	*Mesodorylaimus*	3.68	*Mesodorylaimus*	3.24
				Daptonema	3.2
				Trissonchulus	3.1

References

1. MRCS. *2017 Lower Mekong Regional Water Quality Monitoring Report*; Mekong River Commission: Vientiane, Lao PDR, 2019; p. 52.
2. Le, A.T.; Le, V.; Du, L.; Tristan, S. Rapid integrated & ecosystem-based assessment of climate change vulnerability & adaptation for Ben Tre Province, Vietnam. *J. Sci. Technol.* **2014**, *52*, 287–293.
3. VNCOLD Vietnam National Committee on Large Dam and Water Resources Development (VNCOLD). Available online: http://www.vncold.vn/en/web/default.aspx (accessed on 9 December 2018).
4. Amanda, A.B.; Bristo, D. Impact of dams on global biodiversity: A scientometric analysis. *Neotrop. Biol. Conserv.* **2016**, *11*, 101–109.
5. Diomande, D.; Kpai, N.N.; Kouadio, K.N.; Sébastino, K.; Costa, D.A.; Gourene, G. Spatial distribution and structure of benthic macroinvertebrates in an artificial reservoir: Taabo Lake (Côte d' Ivoire). *Int. J. Biol. Chem. Sci.* **2013**, *7*, 1503–1514. [CrossRef]

6. Oo, A.T.; Van Huylenbroecka, G.; Speelman, S. Differential impacts of an irrigation project: Case study of the Swar Dam Project in Yedashe, Bago region of Myanmar. *J. Dev. Agric. Econ.* **2017**, *9*, 178–189.

7. Phan, P.D. *Literature Review for the Development of Guideline for Prioritisation of Barriers to Fish Passage in Irrigation Schemes in Vietnam*; Research Institute for Aquaculture No. 3: Nha Trang, Vietnam, 2015; p. 38.

8. Wei, G.; Yang, Z.; Cui, B.; Li, B.; Chen, H.; Bai, J.; Dong, S. Impact of Dam Construction on Water Quality and Water Self-Purification Capacity of the Lancang River, China. *Water Resour. Manag.* **2009**, *23*, 1763–1780. [CrossRef]

9. Wildi, W. Environmental hazards of dams and reservoirs. *Terre Environ.* **2010**, *88*, 187–197.

10. Mattos, L.D.; Kruger, M.L.D.; Affonso, S.L.A.; Perbiche-Neves, G.; Junior, S.P. Small dams also change the benthic macroinvertebrates community in rocky rivers. *Acta Limnol. Bras.* **2017**, *29*. [CrossRef]

11. Power, M.E.; Dietrich, W.E.; Finlay, J.C. Dams and Downstream Aquatic Biodiversity: Potential Food Web Consequences of Hydrologic and Geomorphic Change. *Environ. Manag. Vol.* **1996**, *20*, 887–895. [CrossRef]

12. Lin, Q. Influence of Dams on River Ecosystem and Its Countermeasures. *J. Water Resour. Prot.* **2011**, *3*, 60–66. [CrossRef]

13. Mueller, J.S.; Grabowski, T.B.; Brewer, S.K.; Worthington, T.A.; Mueller, J.S.; Hall, A. Effects of Temperature, Total Dissolved Solids, and Total Suspended Solids on Survival and Development Rate of Larval Arkansas River Shiner. *J. Fish Wildl. Manag.* **2017**, *8*, 79–88. [CrossRef]

14. Quevedo, L.; Ibáñez, C.; Caiola, N.; Cid, N.; Hampel, H. Impact of a reservoir system on benthic macroinvertebrate and diatom communities of a large Mediterranean river (lower Ebro river, Catalonia, Spain). *Limnetica* **2018**, *37*, 209–228.

15. Ngo, X.Q.; Ngo, T.T.; Nguyen, X.D.; Vanreusel, A. Initial study on impact of Ba Lai dam construction to humanity ecosystem in Binh Dai district, Ben tre province. In Proceedings of the Human Ecology and Sustainable Development from Theory to Practice, Ha Noi, Vietnam, 13 January 2017; pp. 254–274.

16. Nguyen, S.T.; Nguyen, M.H. Study on deposition of Ba Lai estuary, Ben Tre province. *J. Sci. Vietnam Natl. Univ. Nat. Sci. Technol.* **2011**, *1S*, 211–217.

17. Bongers, T. The Maturity Index: An Ecological Measure of Environmental Disturbance Based on Nematode Species Composition. *Oecologia* **1990**, *83*, 14–19. [CrossRef] [PubMed]

18. Vanreusel, A. Ecology of the free-living marine nematodes from the Voordelta (Southern Bight of the North Sea). I; Species composition and structure of the namatode communities. *Cah. Biol. Mar.* **1990**, *31*, 439–462.

19. Moens, T.; Braeckman, U.; Derycke, S.; Fonseca, G.; Gallucci, F.; Gingold, R.; Guilini, K.; Ingels, J.; Leduc, D.; Vanaverbeke, J.; et al. Ecology of free-living marine nematodes. In *Handbook of Zoology*; De Gruyter: Berlin, Germany, 2013; pp. 109–152. ISBN 9783110274257.

20. Yeates, G.W.; Bongers, T.; De Goede, R.G.; Freckman, D.W.; Georgieva, S.S. Feeding habits in soil nematode families and genera-an outline for soil ecologists. *J. Nematol.* **1993**, *25*, 315–331.

21. Yeates, G.W.; Ferris, H.; Moens, T.; Putten, W.H. Van Der The Role of Nematodes in Ecosystem. In *Nematodes as Environmental Indicators*; CAB International: Wallingford, UK, 2009; 341p, ISBN 9781845933852.

22. Alves, A.S.; Adão, H.; Ferrero, T.J.; Marques, J.C.; Costa, M.J.; Patrício, J. Benthic meiofauna as indicator of ecological changes in estuarine ecosystems: The use of nematodes in ecological quality assessment. *Ecol. Indic.* **2013**, *24*, 462–475. [CrossRef]

23. Ferrero, T.J.; Debenham, N.J.; Lambshead, P.J.D. The nematodes of the Thames estuary: Assemblage structure and biodiversity, with a test of Attrill's linear model. *Estuar. Coast. Shelf Sci.* **2008**, *79*, 409–418. [CrossRef]

24. Ngo, X.Q.; Nguyen, N.C.; Vanreusel, A. Nematode morphometry and biomass patterns in relation to community characteristics and environmental variables in the Mekong Delta, Vietnam. *Raffles Bull. Zool.* **2014**, *62*, 501–512.

25. Ngo, X.Q.; Nguyen, N.C.; Smol, N.; Prozorova, L.; Vanreusel, A. Intertidal nematode communities in the Mekong estuaries of Vietnam and their potential for biomonitoring. *Environ. Monit. Assess.* **2016**, *188*, 91.

26. Soetaert, K.; Vincx, M.; Wittoeck, J.; Tulkens, M. Meiobenthic distribution and nematode community structure in five European estuaries. *Hydrobiologia* **1995**, *311*, 185–206. [CrossRef]

27. Smol, N.; Willems, K.A.; Govaere, J.C.R.; Sandee, A.J.J. Composition, distribution and biomass of meiobenthos in the Oosterschelde estuary (SW Netherlands). *Hydrobiologia* **1994**, *282/283*, 197–217. [CrossRef]

28. Tita, G.; Vincxy, M.; Desrosiers, G. Size spectra, body width and morphotypes of intertidal nematode: An ecological interpretation. *J. Mar. Biol. Assoc. UK* **1999**, *79*, 1007–1015. [CrossRef]

29. Tita, G.; Desrosiers, G.; Vincx, M.; Clément, M. Intertidal meiofauna of the St Lawrence estuary (Quebec, Canada): Diversity, biomass and feeding structure of nematode assemblages. *J. Mar. Biol. Assoc. UK* **2002**, *82*, 779–791. [CrossRef]

30. Wojtasik, B. The influence of water power station in Niedzica on littoral zone meiobenthos of Czorsztyński and Sromowiecki dam reservoirs (Pieniny mountains, Poland). *Teka Kom. Ochr. Kszt. Środ. Przyr.—OL PAN* **2009**, *6*, 411–423.

31. Georg, W.; Orendt, C.; Höss, S.; Großschartner, M.; Adamek, Z.; Jurajda, P.; Traunspurger, W.; de Deckere, E.; Liefferinge, C. van The macroinvertebrate and nematode community from soft sediments in impounded sections of the river Elbe near Pardubice, Czech Republic. *Lauterbornia* **2010**, *69*, 87–105.

32. Nicholas, W.L.; Bird, A.F.; Beech, T.A.; Stewart, A.C. The nematode fauna of the Murray River estuary, South Australia; the effects of the barrages across its mouth. *Hydrobiologia* **1992**, *234*, 87–101. [CrossRef]

33. Tiemann, J.S.; Gillette, D.P. Effects of Lowhead Dams on Riffle-Dwelling Fishes and Macroinvertebrates in a Midwestern River. *Trans. Am. Fish. Soc.* **2004**, *133*, 705–717. [CrossRef]

34. Montagna, P.A.; Palmer, T.A.; Beseres Pollack, J. *Hydrological Changes and Estuarine Dynamics*; SpringerBriefs in Environmental Science; Springer: New York, NY, USA, 2013; Volume 8, ISBN 978-1-4614-5832-6.

35. ISO-5667 ISO 5667-3:2018—Water Quality—Sampling—Part 3: Preservation and Handling of Water Samples. Available online: https://www.iso.org/standard/72370.html (accessed on 4 August 2019).

36. ISO-5667 ISO 5667-12:2017—Water Quality—Sampling—Part 12: Guidance on Sampling of Bottom Sediments from Rivers, Lakes and Estuarine Areas. Available online: https://www.iso.org/standard/59903.html (accessed on 4 August 2019).

37. Krumbein, W.C. Size Frequency Distributions of Sediments. *SEPM J. Sediment. Res.* **1934**, *4*, 65–77. [CrossRef]

38. U.S. EPA Method 3050B: Acid Digestion of Sediments, Sludges, and Soils. *J. Jpn. Soc. Bronchol.* **1996**, 12.

39. Brown, K.A.; McGreer, E.R.; Taekema, B.; Cullen, J.T. Determination of Total Free Sulphides in Sediment Porewater and Artefacts Related to the Mobility of Mineral Sulphides. *Aquat. Geochem.* **2011**, *17*, 821–839. [CrossRef]

40. Vincx, M. Meiofauna in marine and freshwater sediments. In *Hall, G.S. Methods for the Examination of Organismal Diversity in Soils and Sediments*; CAB International in association with United Nations Educational, Scientific, and Cultural Organization and the International Union of Biological Sciences: Wallingford, UK, 1996; pp. 187–195. ISBN 0851991491.

41. De Grisse, A.T. Redescription ou modification de quelques techniques utilisée dans l'étude des nematodes phytoparasitaires. *Meded. Rijksfac. der Landbouveten Gent* **1969**, *34*, 351–369.

42. Schmidt-Rhaesa, A. *Gastrotricha, Cycloneuralia and Gnathifera. Volume 2, Nematoda*; De Gruyter: Berlin, Germany, 2013; ISBN 9783110274257.

43. Zulini, A. Identification Manual for Freshwater Nematode Genera. In *Lecture Book for MSc*; Nematology Ghent University: Ghent, Belgium, 2010; p. 112.

44. Abebe, E.; Andrássy, I. *The Handbook on Freshwater Nematodes: Ecology and Taxonomy*; The CABI Publisher: Wallingford, UK, 2006.

45. Warwick, R.M.; Platt, H.M.; Somerfield, P.J. *Free-living Marine Nematodes Part III: Monhysterids*; Kermack, D.M., Barnes, R.S.K., Eds.; The Linnean Society of London and the Estuarine & Coastal Sciences Association: London, UK, 1998; ISBN 1 85 153 260 9.

46. Platt, H.M.; Warwick, R.M. *Free-living Marine Nematodes. Part I: British Enoplids*; Kermack, D.M., Barnes, R.S.K., Eds.; The Linnean Society of London and the Estuarine & Coastal Sciences Association: London, UK, 1983; ISBN 0 521 25422 1.

47. Platt, H.M.; Warwick, R.M. *Free-Living Marine Nematodes, Part II: British Chromadorids*; Kermack, D.M., Barnes, R.S.K., Eds.; The Linnean Society & The Estuarine & Brackish-Water Sciences Association: London, UK, 1988; ISBN 90 04 085955.

48. Nguyen, V.T. *Fauna of Vietnam. Free-living Nematodes Orders Monhysterida, Araeolaimida, Chromadorida, Rhabditida, Enoplida, Mononchida and Dorylaimida*; Science Technology: Ha Noi, Vietnam, 2007.

49. Bezerra, T.N.; Decraemer, W.; Eisendle-Flöckner, U.; Hodda, M.; Holovachov, O.; Leduc, D.; Miljutin, D.; Mokievsky, V.; Peña Santiago, R.; Sharma, J.; et al. Nemys: World Database of Nematodes. Available online: http://nemys.ugent.be/ (accessed on 4 August 2019).

50. Shannon, C.E. A mathematical theory of communication. *Bell Syst. Tech. J.* **1948**, *27*, 379–423. [CrossRef]

51. R Core Team. R: A language and Environment for Statistical Computing. Vienna, Austria. 2018. Available online: https//:www.Rproject/ (accessed on 1 July 2019).

52. Clarke, K.R.; Warwick, R.M. *Change in Marine Communities: An Approach to Statistical Analysis and Interpretation. PRIMER-E Ltd*, 2nd ed.; Plymouth. Marine Laboratory: Plymouth, UK, 2001.

53. Benjamini, Y.; Hochberg, Y. Controlling the false discovery rate: A practical and powerful approach to multiple testing. *J. R. Stat. Soc. Ser. B* **1995**, *57*, 289–300. [CrossRef]

54. Ongley, E.D. Pollution by sediments. In *Control of Water Pollution from Agriculture*; Food and Agriculture Organization of the United Nations: Rome, Italy, 1996; pp. 19–37.

55. Burton, G.A. Sediment quality criteria in use around the world. *Limnology* **2002**, *3*, 65–75. [CrossRef]

56. Phuong, P.K.; Tu, N.D.; Thanh, N.V. Heavy metals status in sediment at Can Gio mangrove, Ho Chi Minh city, Vietnam. *J. Biol.* **2011**, *33*, 81–86. [CrossRef]

57. Nguyen, V.S.; Ngo, X.Q.; Vanreusel, A.; Smol, N. The nematode community distribution in two estuaries of the Mekong delta: Cung Hau and Ham Luong, South Vietnam. *J. Biol.* **2012**, *34*, 1–12.

58. Ngo, X.Q.; Smol, N.; Vanreusel, A. The meiofauna distribution in correlation with environmental characteristics in 5 Mekong estuaries, Vietnam. *Cah. Biol. Mar.* **2013**, *54*, 71–83.

59. Tran, T.T.; Nguyen, L.Q.L.; Nguyen, T.M.Y.; Vanreusel, A.; Ngo, X.Q. Free-living nematode communities in Ba Lai river, Ben Tre province. *Vietnam J. Sci. Technol.* **2018**, *56*, 224–235.

60. Vanreusel, A.; De Groote, A.; Gollner, S.; Bright, M. Ecology and Biogeography of Free-Living Nematodes Associated with Chemosynthetic Environments in the Deep Sea: A Review. *PLoS ONE* **2010**, *5*, 16. [CrossRef]

61. Fonseca, G.; Gallucci, F. The need of hypothesis-driven designs and conceptual models in impact assessment studies: An example from the free-living marine nematodes. *Ecol. Indic.* **2016**, *71*, 79–86. [CrossRef]

62. Adão, H.; Alves, A.S.; Patrício, J.; Neto, J.M.; Costa, M.J.; Marques, J.C. Spatial distribution of subtidal Nematoda communities along the salinity gradient in southern European estuaries. *Acta Oecologica* **2009**, *35*, 287–300. [CrossRef]

63. Heip, C.; Vincx, M.; Vranken, G. The Ecology of Marine Nematodes. *Oceanogr. Mar. Biol.* **1985**, *23*, 399–489.

64. Schratzberger, M.; Warr, K.; Rogers, S.I. Functional diversity of nematode communities in the southwestern North Sea. *Mar. Environ. Res. Elsevier* **2007**, *63*, 368–389. [CrossRef]

65. Attrill, M.J. A testable in estuaries linear model for diversity trends. *J. Anim. Ecol.* **2002**, *71*, 262–269. [CrossRef]

66. Vanaverbeke, J.; Merckx, B.; Degraer, S.; Vincx, M. Sediment-related distribution patterns of nematodes and macrofauna: Two sides of the benthic coin? *Mar. Environ. Res.* **2010**, *71*, 31–40. [CrossRef] [PubMed]

67. Nguyen, D.T. *Seasonal and Spatial Patterns in Meiofauna Community Structure of the Can Gio Mangrove Forest (Vietnam) with a Focus on Nematoda and Their Role as Bioindicator*; Ghent University: Ghent, Belgium, 2009; 242p.

68. Rzeznik-Orignac, J.; Fichet, D.; Boucher, G. Spatio-temporal structure of the nematode assemblages of the Brouage mudflat (Marennes Oléron, France). *Estuar. Coast. Shelf Sci.* **2003**, *58*, 77–88. [CrossRef]

69. Gyedu-Ababio, T.K.; Furstenberg, J.P.; Baird, D.; Vanreusel, A. Nematodes as indicators of pollution: A case study from the Swartkops River system, South Africa. *Hydrobiologia* **1999**, *397*, 155–169. [CrossRef]

Article

Modern Benthic Foraminiferal Diversity: An Initial Insight into the Total Foraminiferal Diversity along the Kuwait Coastal Water

Eqbal Al-Enezi [1], Sawsan Khader [2], Eszter Balassi [3] and Fabrizio Frontalini [3,*]

[1] Kuwait Institute for Scientific Research, Safat 13109, Kuwait; eenezi@kisr.edu.kw
[2] Department of Earth and Environmental Sciences, Faculty of Science, Kuwait University, Safat 13060, Kuwait; eihabs90@gmail.com
[3] Department of Pure and Applied Sciences (DiSPeA), Urbino University, 61029 Urbino, Italy; balassieszti@gmail.com
* Correspondence: fabrizio.frontalini@uniurb.it; Tel.: +39-392-8457-666

Received: 11 March 2020; Accepted: 2 April 2020; Published: 5 April 2020

Abstract: Kuwait territorial water hosts an important part of national biodiversity (i.e., zooplankton and phytoplankton), but very limited information exists on the overall diversity of benthic foraminifera. On the basis of the integration of publications, reports and theses with new available data from the Kuwait Bay and the northern islands, this study infers the total benthic foraminiferal diversity within Kuwait territorial water. This new literature survey documents the presence of 451 species belonging to 156 genera, 64 families, 31 superfamilies and 9 orders. These values are relatively high in consideration of the limited extension and the shallow depth of the Kuwait territorial water. Kuwait waters offer a variety of different environments and sub-environments (low salinity/muddy areas in the northern part, embayment, rocky tidal flats, coral reef systems, islands and shelf slope) that all together host largely diversified benthic foraminiferal communities. These figures are herein considered as underestimated because of the grouping of unassigned species due to the lack of reference collections and materials, as well as the neglection of the soft-shell monothalamids ('allogromiids').

Keywords: benthic foraminifera; biodiversity; checklist; Kuwait; Arabian Gulf

---

1. Introduction

Checklists, lists of species, represent the baseline for the implementation of natural conservation biodiversity plans and are important inventories for wildlife management, species protection and biogeographical studies. Such inventories become even more essential in the context of global climatic change and in areas with a high degree of endemism or in those threatened by pollution.

In this context, the Arabian Gulf, an extension of the Indian Ocean, is known for the largest reservoirs of petroleum, as well as for its unique oceanographical conditions and endemic biota adapted to extreme conditions. This area has a strategic value under economic and environmental perspectives [1]. The Gulf represents the hottest marine environment and has been commonly used as a model for the 2100 climate change in tropical oceans [2]. The rapid economic development and antropogenization of the Gulf coupled with the peculiar climatic features of the region and the relatively shallow and reduced size of the basin have led to a deterioration of natural habitats [3]. In light of the present global climatic changes, an increasing of sea surface temperature (SST) of ~2.2 °C by 2100 has been recently predicted for the Gulf [4]. The same authors inferred the highest differences of mean annual warming between 2100 and 2015 along the north coast of the United Arab Emirates, the southern coastlines of the Strait of Hormuz and the northernmost part of the Gulf including Kuwait.

The Kuwait territorial waters host an important part of the national biodiversity and are also nursery grounds for a number of species [5]. A total of 235 macrofaunal taxa in the intertidal zone [6], over 340 zooplankton and phytoplankton species [7], 345 species of fish [8] and 105 species of marine plants have been recorded from Kuwait waters. Specifically, 55 tintinnid species [9]; 80 flagellate species, including benthic dinoflagellate, phototropic flagellates and euglenoids [10]; and 37 copepods species [11] were described in Kuwait. Detailed investigations on the overall diversity of phytoplankton species were provided by [7], with the identification of more than 140 diatoms, 55 dinoflagellates, 2 coccolithophorids and a cyanophyte species. The phytoplankton checklist was then updated with the recognition of over 323 species, including diatoms (202 species), dinoflagellates (108 species), silicoflagellates, naked flagellates and cyanophytes [12]. Several groups of marine organisms, including fish, are relatively well-known because of their commercial value, whereas a large number of invertebrates thriving in the benthos has been poorly considered and commonly neglected. Despite this, these invertebrates have both an ecological and a conservation relevance, as well as a scientific one.

Among them, benthic foraminifera are single-celled organisms living in a wide range of marine environments [13]. They are inferred to play a major role in the biogeochemical cycles [14] and occupy a low trophic position in the food web [15]. A limited number of foraminiferal studies has covered extensive geographical areas [16–22], and the regional-synoptic approach documenting benthic foraminiferal diversity is currently quite limited [19]. Although benthic foraminifera have been largely studied in the Arabian Gulf [23–27], a limited knowledge of their diversity exists for the Kuwait coastal water [28], as most of the available studies have focused on their application as bioindicators [29,30].

In light of this, the main objective of the present paper is to provide a checklist of recent benthic foraminiferal species and to infer the total biodiversity of this group of organisms in the Kuwait coastal marine areas on the basis of the available dataset published in the last fifty years, coupled with original data.

2. Geography, Hydrology and Study Area

The State of Kuwait lies in the northeastern part of the Arabian Peninsula, between latitudes 28°30′ and 30°05′ N and longitudes 46°33′ and 48°30′ E. The country is bordered by Iraq on the northwest, Saudi Arabia on the south and southwest, and the Arabian Gulf on the east. The total area, considering the three larger (Bubiyan, Failaka and Warba) and the six smaller (Miskan, Auha, Umm Al-Namil, Kubbar, Qaruh and Umm Al-Maradim) islands, covers about 17,818 km^2 [31]. The coastal length of Kuwait, including islands, is about 500 km, which represents about 7611 km^2 of the Kuwaiti territorial water [31]. The marine environment is situated in a transitional zone between two major geomorphic units of the northern part of the Arabian Gulf, the stable Arabian Foreland in the west and southwest, and the compound Shatt al-Arab Delta of the Mesopotamian Plain in the north. As a result, the influence of these two geomorphic units is reflected in the marine and coastal environments of the country. On the basis of environmental characteristics, the Kuwaiti coastal area can be broadly divided into northern and southern provinces. The northern province, which includes Khor Al-Sabiya, Kuwait Bay and the six northern coastal and offshore islands (Warba, Bubiyan, Miskan, Failaka, Auha and Umm Al-Namil), is characterized by extensive mudflats that are covered by a mixture of fine sand, silt and clay, bounded landward by extensive sabkhas and encrusted with algae and very rich in organisms [32]. Its clastic deposits are commonly derived from the Shatt al-Arab, the principal riverine input in the Gulf [33,34]. This river contributes to the cyclonic circulation, promotes the productivity by the discharge of nutrients and forms a plume in the northwestern part of the Gulf (i.e., Kuwait). The deflection of the Shatt al-Arab river plume due to the Coriolis effect affects the sedimentary influx in the northern area but excludes most of the Iranian coast [35]. The southern coastal province located between the coastal area of Kuwait City and the southern border with Saudi Arabia is marked by extensive rocky tidal flats masked by a relatively thick cover of sand and bounded landward by several oolitic ridges and inland sabkha flats. Coral reef systems, including platforms, patch reefs and fringing corals, are abundant in the southern area, particularly around the islands [7].

The offshore area of Kuwait is characterized by shallow depths that generally increase in a southeasterly direction, reaching a maximum depth of about 30 m. Currents along the Kuwaiti shores are generally parallel to the coast. The main hydrodynamic factors affecting Kuwait coastal areas are tidal currents and waves. Along the Kuwait coast, tides have a mean range of about 3 m (mesotidal) and are semidiurnal (with two highs and two lows) to mixed type in the north and diurnal toward south. During spring tides, tidal currents reach 0.5–0.6 m/s, with a maximum of 1 m/s at the entrance of the Kuwait Bay [7]. The shallow nature of the offshore area in general and the northern sector, in particular, limits the generation of large waves along the shoreline. Waves higher than 50 cm are rare in the northern offshore area of Kuwait. The northern coastal area is generally marked by a low-energy environment, with the currents being the prime source of energy. Currents seem to be more active along Khor Al-Sabiya, carrying tidal sediment to the deeper parts of Kuwait waters. Very extreme conditions characterize the Kuwait seawaters, with SST difference between summer and winter exceeding 20 °C, with the summer temperature reaching 36 °C and salinity over 45 [7]. The Kuwait Bay is eutrophicated and influenced by runoff, sewage discharge and several anthropogenic activities. All of these characteristics make the Kuwait water a very harsh environment for marine communities.

Five physiographic regions have been identified in Kuwait: (1) the submerged estuarine flat (shallow muddy area); (2) the submerged estuarine channel and bar system; (3) the Kuwait Bay trough; (4) the shelf slope; and (5) the islands [7,36]. The submerged estuarine flat lies off the northern coast of Kuwait and includes most of the Kuwait Bay and the area surrounding three northern islands (Failaka, Auha and Miskan Islands).

The climate of Kuwait is affected by the desert climate that is dry and hot. The annual air temperature varies between a maximum of 50 °C in summer and a minimum of 0 °C in winter. Precipitation is scarce (<10 cm/yr) and makes a negligible contribution to the freshwater budget of the Arabian Gulf. Estimates of evaporation vary from 140 to 500 cm/yr [37]. In addition to the effects of high temperatures in summer, high evaporation rates and low rain falls, winds have a pronounced influence on the oceanographic and sedimentological nature of the area. The most dominant wind direction in the Arabian Gulf is the "Shamal", a NW wind that blows the year around, and is usually associated with dust storms. The wind brings large quantities of terrigenous materials to the Kuwait's offshore zone [32].

On the basis of the above, several environments characterized by specific organisms (biotopes) were defined, such as shallow sand/mud (less than 5 m), deep sand/mud (over 5 m), coral reef, rock/algae, seagrass and productive intertidal mud/sand flats [38].

3. Previous Benthic Foraminiferal Studies within the Kuwaiti Territorial Waters

Several papers, theses and reports—the latter mainly from the Kuwait Institute of Scientific Research (KISR)—have been published on benthic foraminifera within Kuwait territorial waters from the late 1960s [23,28–30,39–59] (Figure 1).

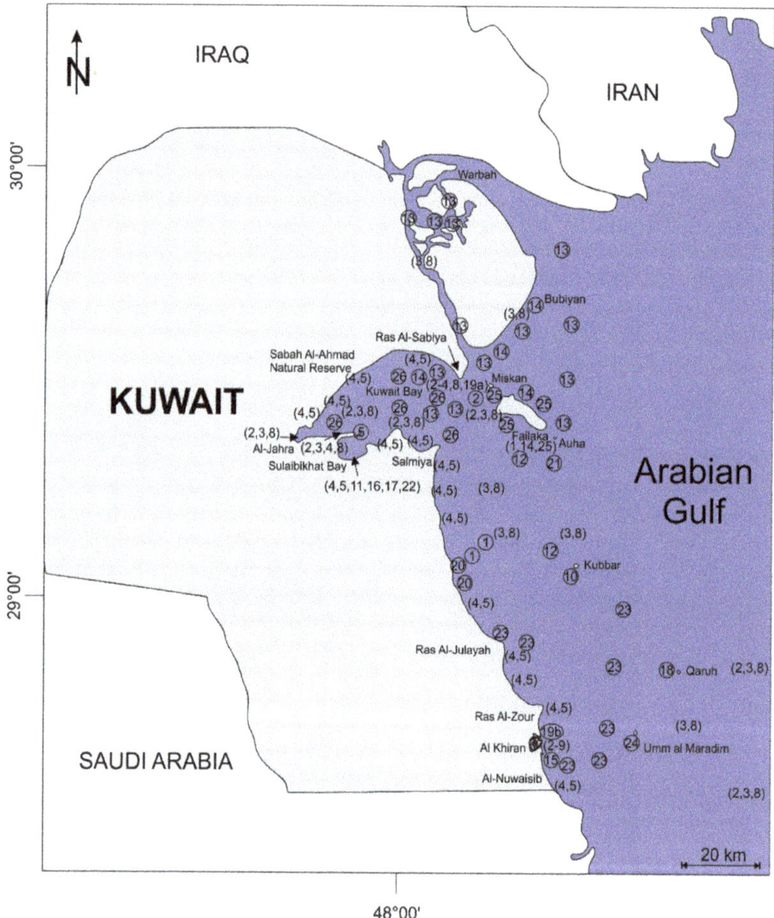

Figure 1. Location map of the Kuwait territorial waters, showing the geographical location of the investigated areas (numbers are organized by date of publication). (1) Mina Al-Ahmadi and south of Failaka Island [39]; (2) offshore area of Kuwait [40]; (3) Kuwait shoreline and the offshore area between Kuwait and Saudi Arabia [41]; (4) tidal flats from Al-Sabiya to Al-Nuwaisib, including the Sulaibikhat Bay and reefal flats of Umm Al-Namil Island and Al-Akaz [42]; (5) tidal flats from Al-Sabiya to Al-Nuwaisib, including the Sulaibikhat Bay and reefal flats of Umm Al-Namil Island and Al-Akaz [43]; (6) Khor Al-Nhaim lagoon in the Al-Khiran area [44]; (7) Al-Khiran area [45]; (8) Kuwait shoreline and the offshore area between Kuwait and Saudi Arabia [46]; (9) tidal and intertidal channels within two connected creeks (Khor Al-Mufateh and Khor Al-Mamlaha) in the Al-Khiran area [47]; (10) Kubbar Island [48]; (11) the Sulaibikhat Bay [49]; (12) offshore area of central Kuwait [23]; (13) the Shatt al-Arab Delta [50]; (14) the Western Bubiyan area and the surrounding of three islands (Failaka, Auha and Miskan) [51]; (15) Khor Iskandar, close to Al-Khiran area [52]; (16) Shuwaikh area [53]; (17) the Sulaibikhat Bay [29]; (18) Qaruh Island [54]; (19a) Ras Al-Subiya and (19b) Ras Al-Zour [55]; (20) Mina Ahmadi and Mina Abdullah [56]; (21) south of Failaka Island [57]; (22) the Sulaibikhat Bay [30]; (23) carbonate ramp transect in the southern part of Kuwait [58]; (24) Umm al Maradim Island [28]; (25) Auha, Failaka and Miskan Islands [59]; (26) Kuwait Bay: present study.

To the best of our knowledge, the first investigation of benthic foraminifera in Kuwait was carried out by [39], who identified 73 species in offshore areas around Mina Al-Ahmadi and south of Failaka Island (1 in Figure 1). A total of 120 foraminiferal species of which 100 benthic ones was recognized by [40] in the offshore area of Kuwait (2 in Figure 1). Later on, twelve localities along the Kuwait shoreline and in the offshore area between Kuwait and Saudi Arabia led to the identification of 208 species and the definition of seven new species [41] (3 in Figure 1). The relationship between sedimentology and benthic ecology (i.e., Ostracoda and foraminifera) was investigated by [42]. The authors identified 120 benthic foraminiferal species belonging to 50 genera; however, the complete list of the identified species was not provided (4 in Figure 1). The marine benthic microfauna, including benthic foraminifera, was documented in tidal flat localities from Al-Sabiya to Al-Nuwaisib, including the Sulaibikhat Bay and reefal flats of Umm Al-Namil Island and Al-Akaz [43]. Foraminifera were found to be the most abundant organisms among microfauna in the Sulaibikhat Bay, but the recognized benthic foraminiferal taxa were not provided (5 in Figure 1). A great abundance of *Peneroplis distinctiva* described as *Cribrospirolina distinctiva* from the small Khor Al-Nhaim Lagoon in the Al-Khiran area was identified by [44] (6 in Figure 1). The area around Al-Khiran was extensively sampled in 1984–1985, with the collection of 180 stations and the ostracod and foraminiferal microfauna were investigated [45]. They suggested a foraminiferal fauna typical of stress environments and identified 35 taxa (7 in Figure 1). On the basis of an extensive sampling (250 stations) in Kuwait's marine environments, coupled with the available materials [41], the distribution of 60 common benthic foraminiferal species was documented (8 in Figure 1) [46]. Forty-two benthic foraminiferal species were recognized in the total assemblages (living plus dead) of the tidal and intertidal channels within two connected creeks (Khor Al-Mufateh and Khor Al-Mamlaha) in the Al-Khiran area [47]. On the basis of the cluster analysis, the authors defined four foraminiferal assemblages, namely dry upper intertidal, wet upper intertidal, middle intertidal and lower tidal–tidal channel reflecting different environmental conditions (9 in Figure 1). The distribution of recent benthic foraminifera around the reefal Kubbar Island was documented by [48]. The benthic foraminifera assemblages were particularly diversified with 81 species, mainly represented by Miliolida and Rotaliida (10 in Figure 1). In order to identify potential bioindicators of pollution, the total benthic foraminiferal assemblages were studied in the Sulaibikhat Bay [49], leading to the recognition of 45 species (11 in Figure 1). An extensive sampling covering the Arabian Gulf led to the identification of 94 species [23]; however, only a few Kuwaiti localities were included (12 in Figure 1). An investigation in the western part of the Shatt al-Arab Delta was performed to study the total (living plus dead) foraminiferal assemblages [50]. In this study, 46 benthic foraminiferal species, mainly represented by Rotaliina, were recognized, and on the basis of the cluster analysis, three assemblages related to salinity conditions and physiographic setting were identified (13 in Figure 1). The potential impact of draining of Iraqi marshes on the sediment quality of Northern Kuwait marine area, including Khor Sabiya (western Bubiyan area), part of the Kuwait Bay and the surrounding of three islands (Failaka, Auha and Miskan), was investigated by [51] (14 in Figure 1). Additionally, 63 benthic foraminiferal species were listed. Khor Iskandar is a coastal inlet located close to the Al-Khiran area and to the border of Saudi Arabia. A total of 60 benthic foraminiferal species were identified in this area that were mainly represented by Rotaliina and Miliolina [52]. The most abundant species were *Ammonia sadoensis*, *Ammonia tepida* and *Ammonia umbonata* (15 in Figure 1). The response of benthic foraminiferal assemblages to the effect of trace elements was investigated in the eastern part of the Sulaibikhat Bay, an area near the Kuwait University in Shuwaikh [53]. The study focused on seasonal sampling in winter and summer on thirty stations and led to the identification of 54 benthic foraminiferal species (16 in Figure 1). A comprehensive investigation of the total benthic foraminifera from the polluted marine environment of the Sulaibikhat Bay was performed by [29], who identified 45 benthic foraminiferal species, of which only one is agglutinated (17 in Figure 1). The distribution of benthic foraminifera around the reefal Qaruh Island during the summer and winter seasons led to the identification of 60 benthic foraminiferal species [54] (18 in Figure 1). Benthic foraminifera were used to study the effects of two thermal pollution plants, namely Ras Al-Subiya (19a in Figure 1) and Ras Al-Zour (19a in Figure 1), and a total of 29 and 39 benthic foraminiferal species were identified, respectively [55]. Environmental baseline data, including data on benthic foraminifera, were gathered to check the feasibility to construct an offshore-treated wastewater discharge in Mina Ahmadi and Mina Abdullah refineries [56] (20 in Figure 1). In this technical report, a total of 39 benthic foraminiferal species was identified, and the assemblages were reported to be poorly diversified and characterized by high abnormalities at stations close to the shore line. The distribution, diversity and abundance of benthic foraminifera were studied in the northwestern part of the Arabian Gulf [57]. Unfortunately, as in [23], only one station falls in the Kuwait territorial waters (21 in Figure 1). In a study aimed at evaluating the environmental quality of the Sulaibikhat Bay, a total of fifty-nine benthic foraminiferal taxa were identified in the living assemblages [30]. These taxa were represented by four orders and suborders, Textulariida, Lagenida, Rotaliida and Miliolida, with the latter represented by 40 species (22 in Figure 1). Modern and relict benthic foraminiferal biofacies along a carbonate ramp transect were studied in the southern part of Kuwait [58]. In this study, a total of 141 benthic foraminiferal taxa were identified, a figure that is much higher compared to previous studies (23 in Figure 1). The benthic foraminiferal assemblages were documented around the unique and largest coral island of Kuwait, Umm al Maradim, where overall 101 and 96 species were identified in the total and living assemblages, respectively [28] (24 in Figure 1). In terms of abundance, the porcelaneous wall-type test dominated the foraminiferal assemblages, followed by hyaline and agglutinated ones. The present checklist also accounts for a recent study with the sampling of 50 stations around the three northern islands of Auha, Failaka and Miskan [59], with the recognition of 92 species (25 in Figure 1). Additionally, an extensive sediment sampling with the collection of 46 samples was performed, to study benthic foraminifera in the Kuwait Bay (26 in Figure 1).

4. Materials and Methods

A total of 50 samples were collected in May 2008, around the three northern islands (Auha, Failaka and Miskan) (Figure 2a), and 46 samples were collected in November 2018 in the Kuwait Bay (Figure 2b). These extensive surveys were carried out to study the benthic foraminiferal distribution by covering different environments, bathymetric depths and environmental conditions. Surface sediment samples were collected by using a small gravity corer. Only the uppermost part of sediments (0–1 cm) was taken from each sampling station and immediately stained with rose Bengal dye, to differentiate between living and nonliving specimens. Samples were then oven-dried, weighed and gently washed with tap water, through a 63 µm sieve, to remove clay, silt and any excess dye. The so-obtained residues were dried and weighed again. The dried samples were analyzed, using a stereo microscope in the fraction >125 µm. The specimens were picked and taxonomically identified, following the generic classifications of [60] and the specific classifications of [61–66], as well as available publications with illustrated specimens from Kuwait [41,46,47,50].

On the basis of the available foraminiferal literature survey (published papers, theses, restricted and open reports), coupled with new data from the three northern islands (Auha, Failaka and Miskan) and the Kuwait Bay, a list of all the identified benthic foraminiferal taxa in the Kuwait territorial waters has been compiled [23,28–30,39–59] (Table 1 and Supplementary Table S1).

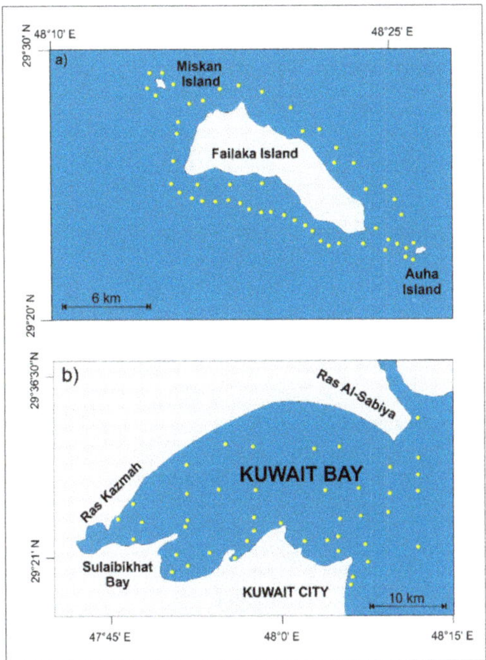

Figure 2. Geographical locations of the stations in this study: (**a**) around Auha, Failaka and Miskan Islands; and (**b**) within the Kuwait Bay.

Table 1. Benthic foraminiferal taxa (order, superfamily and family) and number of genera and species recognized in Kuwait territorial waters.

Order	Superfamily	Family	Number of Genera	Number of Species
Lituolida	Lituoloidea	Lituolidae	3	6
	Verneuilinoidea	Prolixoplectidae	1	2
		Verneuilinidae	1	1
	Trochamminoidea	Trochamminidae	3	3
Textulariida	Textularioidea	Textulariidae	4	25
		Olgiidae	1	1
	Eggerelloidea	Valvulinidae	1	1
	Hormosinoidea	Reophacidae	1	1
Miliolida	Cornuspiroidea	Cornuspiridae	1	1
	Milioloidea	Cribrolinoididae	1	10
		Hauerinidae	35	165
		Spiroloculinidae	1	18
	Miliolacea	Riveroinidae	1	1
	Nubecularioidea	Fischerinidae	2	2
		Ophthalmidiidae	3	7
		Nubeculariidae	3	3
	Soritoidea	Peneroplidae	4	7
		Soritidae	1	2
		Miliamminidae	1	2
	Squamulinoidea	Squamulinidae	1	1
Spirillinida		Spirillinidae	1	2
		Patellinidae	1	1
Lagenida	Nodosarioidea	Lagenidae	3	17
		Nodosariidae	5	5

Table 1. *Cont.*

Order	Superfamily	Family	Number of Genera	Number of Species
Vaginulinida		Vaginulinidae	1	1
Robertinida	Robertinoidea	Robertinidae	1	1
Rotaliida	Buliminoidea	Buliminidae	1	1
		Reussellidae	2	6
		Trimosinidae	2	2
	Cassidulinoidea	Bolivinitidae	7	8
		Tortoplectellidae	1	1
		Siphogenerinoididae	1	1
	Chilostomelloidea	Trichohyalidae	1	1
		Anomalinidae	1	1
	Fursenkoinoidea	Fursenkoinidae	1	2
	Turrilinoidea	Turrilinidae	1	1
		Stainforthiidae	2	2
	Bolivinoidea	Bolivinidae	1	10
	Discorboidea	Rosalinidae	3	10
		Eponididae	2	3
		Cancrisidae	2	3
		Discorbidae	1	1
		Heleninidae	1	1
	Discorbinelloidea	Discorbinellidae	2	4
		Pseudoparrellidae	1	1
	Planorbulinoidea	Cibicididae	4	7
		Planorbulinidae	2	2
		Cymbaloporidae	3	6
	Nonionoidea	Nonionidae	3	7
	Glabratelloidea	Glabratellidae	1	1
	Rotalioidea	Elphidiidae	4	21
		Elphidiellidae	1	1
		Haynesinidae	1	2
		Ammoniidae	6	25
		Notorotaliidae	1	1
	Acervulinoidea	Acervulinidae	1	2
	Asterigerinoidea	Amphisteginidae	1	1
		Epistomariidae	1	1
		Asterigerinatidae	1	1
	Nummulitoidea	Nummulitidae	2	6
	Murrayinellidae	Murrayinellidae	1	2
Polymorphinida	Polymorphinoidea	Glandulinidae	2	3
		Polymorphinidae	4	6
		Ellipsolagenidae	3	11

The largest part of these studies has been carried out in the Kuwait Bay [29,30,49,53], around the Al-Khiran area [44,45,47,55], along the Kuwaiti coastline [42,43], whereas only a few of them have focused on the islands [28,48,54,57,59], including the Bubiyan one [50] (Table 1 and Supplementary Table S1).

Most of these publications and reports are based on the total assemblages. The total assemblages combine living and dead foraminifera and include autochthonous taxa, as well as allochthonous ones that might be transported from other localities. To the best of our knowledge, the only studies that have solely considered the living (Rose-Bengal stained) part of benthic foraminifera are [28,30] and the present study in the Kuwait Bay area.

The World Register of Marine Species (WoRMS) [67] was followed for the systematic classification of the benthic foraminiferal species, genera, and the suprageneric taxonomic categories adopted in this study. The validity of the species and genera has been checked against WoRMS, and the unassigned specimens (sp.), as reported from different publications, were combined in a general category (spp.).

The presence–absence list of taxa was then used to calculate the β diversity or differentiation and γ or regional diversity. The β diversity was calculated by using the PAST (PAlaeontological STatistics) data analysis package (version 3.05) [68] and is expressed as the Whittaker's original measure (βw) [69], whereas γ diversity is calculated as the total species richness in the area. The βw represents the variation in species composition (the highest is the value of βw, the highest is the difference) and is here used to define the benthic foraminiferal species difference between selected environments within Kuwait, namely the northern islands (Failaka Island, Bubiyan Island and the Shatt al-Arab area), the southern islands (Umm Al-Maradim, Qaruh and Kubbar Islands), the southern coast (Al-Khiran, Khor Iskander and Ras Al-Zour) and the tidal flat and the Kuwait Bay (Supplementary Figure S1). Open nomenclature species (i.e., sp.) are not considered.

5. Results

A total of 451 benthic foraminiferal species were identified, represented by 8.9% agglutinated, 48.5% porcelaneous and 42.6% hyaline wall types. These taxa belong to 156 genera, 64 families, 31 superfamilies and 9 orders (Table 1). Here we recognized the following orders: Lituolida, Textulariida, Miliolida, Spirillinida, Lagenida, Vaginulinida, Robertinida, Polymorphinida and Rotaliida (Table 1 and Supplementary Table S1). The Order Allogromiida and the Suborder Globigerinina were not considered in this study, the former because all the considered studies were performed on dried sediment samples that did not allow the preservation of them, while the latter is a planktonic group.

The Order Lituolida is represented by three superfamilies (Lituoloidea, Verneuilinoidea and Trochamminoidea), four families (Lituolidae, Prolixoplectidae, Verneuilinidae and Trochamminidae) and five genera (*Ammobaculites*, *Ammomarginulina*, *Gaudryina*, *Trochammina* and *Rotaliammina*) (Table 1 and Supplementary Table S1).

Three superfamilies (Textularioidea, Eggerelloidea and Hormosinoidea), four families (Textulariidae, Olgiidae, Valvulinidae and Reophacidae) and seven genera (*Sahulia*, *Textularia*, *Siphotextularia*, *Bigenerina*, *Olgita*, *Clavulina* and *Reophax*) were identified within the Order Textulariida (Table 1 and Supplementary Table S1).

The Order Miliolida includes six superfamilies (Cornuspiroidea, Milioloidea, Miliolacea, Nubecularioidea, Soritoidea and Squamulinoidea), 12 families (Cornuspiridae, Cribrolinoididae, Hauerinidae, Spiroloculinidae, Riveroinidae, Fischerinidae, Ophthalmidiidae, Nubeculariidae, Peneroplidae, Soritidae, Miliamminidae and Squamulinidae) and 54 genera (*Cornuspira*, *Adelosina*, *Flintinoides*, *Hauerina*, *Sigmoihauerina*, *Spirosigmoilina*, *Cycloforina*, *Cribromiliolinella*, *Miliolinella*, *Miliolina*, *Lachlanella*, *Pseudolachlanella*, *Massilina*, *Pseudomassilina*, *Proemassilina*, *Pseudotriloculina*, *Sinuloculina*, *Ptychomiliola*, *Pseudopyrgo*, *Pyrgo*, *Pyrgoella*, *Quinqueloculina*, *Parrina*, *Affinetrina*, *Triloculina*, *Triloculinella*, *Biloculinella*, *Varidentella*, *Mesosigmoilina*, *Sigmoilina*, *Sigmoilopsis*, *Sigmoilinella*, *Siphonaperta*, *Articulina*, *Ishamella*, *Steigerina*, *Sigmamiliolinella*, *Spiroloculina*, *Pseudohauerinella*, *Vertebralina*, *Wiesnerella*, *Ophthalmidium*, *Edentostomina*, *Spirophthalmidium*, *Nodophthalmidium*, *Nubecularia*, *Nodobacularia*, *Peneroplis*, *Coscinospira*, *Spirolina*, *Dendritina*, *Sorites*, *Miliammina* and *Squamulina*) (Table 1 and Supplementary Table S1).

The Order Spirillinida is only composed of two families (Spirillinidae and Patellinidae) and two genera (*Spirillina* and *Patellina*) (Table 1 and Supplementary Table S1).

The Order Lagenida is represented by only one superfamily (Nodosarioidea), two families (Lagenidae and Nodosariidae) and eight genera (*Lagena*, *Procerolagena*, *Hyalinonetrion*, *Reussoolina*, *Pyramidulina*, *Dentalina*, *Nodosaria* and *Laevidentalina*) (Table 1 and Supplementary Table S1).

The Order Vaginulinida includes the family Vaginulinidae and the genus *Lenticulina* (Table 1 and Supplementary Table S1).

Similarly, the Order Robertinida is solely represented by the superfamily Robertinoidea, the family Robertinidae and the genus *Pseudobulimina* (Table 1 and Supplementary Table S1).

The Order Rotaliida encompasses 16 superfamilies (Buliminoidea, Cassidulinoidea, Chilostomelloidea, Fursenkoinoidea, Turrilinoidea, Bolivinoidea, Discorboidea, Discorbinelloidea, Planorbulinoidea, Nonionoidea, Glabratelloidea, Rotalioidea, Acervulinoidea, Asterigerinoidea, Nummulitoidea, and Murrayinellidae), 35 families (Buliminidae, Reussellidae, Trimosinidae, Bolivinitidae, Tortoplectellidae, Siphogenerinoididae, Trichohyalidae, Anomalinidae, Fursenkoinidae, Turrilinidae, Stainforthiidae, Bolivinidae, Rosalinidae, Eponididae, Cancrisidae, Discorbidae, Heleninidae, Discorbinellidae, Pseudoparrellidae, Cibicididae, Planorbulinidae, Cymbaloporidae, Nonionidae, Glabratellidae, Elphidiidae, Elphidiellidae, Haynesinidae, Ammoniidae, Notorotaliidae, Acervulinidae, Amphisteginidae, Epistomariidae, Asterigerinatidae, Nummulitidae and Murrayinellidae) and 66 genera (*Bulimina, Reussella, Fijiella, Quirimbatina, Trimosina, Parabrizalina, Sagrinella, Neocassidulina, Euloxostomum, Loxostomina, Pseudobrizalina, Bolivinellina, Tortoplectella, Rectobolivina, Buccella, Riminopsis, Fursenkoina, Floresina, Hopkinsina, Stainforthia, Bolivina, Rosalina, Neoconorbina, Gavelinopsis, Eponides, Poroeponides, Cancris, Valvulineria, Discorbis, Helenina, Discorbinella, Hanzawaia, Facetocochlea, Cibicides, Cibicidoides, Heterolepa, Lobatula, Planorbulina, Planorbulinella, Millettiana, Cymbaloporetta, Cymbaloporella, Pseudononion, Nonion, Nonionella, Pileolina, Protelphidium, Elphidium, Cribroelphidium, Porosononion, Elphidiella, Haynesina, Ammonia, Challengerella, Asterorotalia, Rotalidium, Rotalinoides Pseudorotalia, Cristatavultus, Acervulina, Amphistegina, Monspeliensina, Eoeponidella, Heterostegina, Operculina* and *Murrayinella*) (Table 1 and Supplementary Table S1).

The **Order Polymorphinida** includes only one superfamily (Polymorphinoidea), three families (Glandulinidae, Polymorphinidae and Ellipsolagenidae) and nine genera (*Glandulina, Globulotuba, Polymorphina, Guttulina, Globulina, Sigmomorphina, Oolina, Buchnerina* and *Fissurina*) (Table 1 and Supplementary Table S1).

The highest benthic foraminiferal similarity is shared between the southern coast and the southern islands ($\beta w = 0.82$), followed by the northern islands and the Kuwait Bay ($\beta w = 0.84$) (Table 2).

Table 2. βw diversity values calculated for the northern islands, southern islands, tidal flat southern coast and Kuwait Bay.

Global β Diversities			Northern Island	Southern Island	Tidal Flat	Southern Coast	Kuwait Bay
		Northern Island	0.00				
		Southern Island	0.84	0.00			
Whittaker (βw)	1.12	Tidal flat	1.80	1.69	0.00		
		Southern coast	0.89	0.82	1.79	0.00	
		Kuwait Bay	0.84	0.86	1.69	0.85	0.00

On the other hand, the highest dissimilarity is found between the tidal flat and the northern island and the southern coast $\beta w = 1.80$ and 1.79, respectively (Table 2).

6. Discussion

The integration of the recognized benthic foraminiferal taxa in the present study from Kuwait Bay and Auha, Failaka and Miskan Islands with those previously recognized in all available records within Kuwait territorial waters leads to the identification of a total of 451 species (γ diversity), belonging to 156 genera, 64 families, 31 superfamilies and 9 orders. These figures, however, would significantly underestimate the real γ diversity of benthic foraminifera in Kuwait waters for two main reasons. Most of the benthic foraminiferal studies reported numbers of taxa that were not classified at the species level (sp.) and have been here therefore grouped in a category (genus + spp.). This approximation was necessary given the absence of type materials that has prevented the comparison of the unidentified taxa. As an example, [58], which represents one of the most accurate studies of benthic foraminiferal distribution in Kuwait, reported 42 taxa of *Quinqueloculina*, of which 13 are assigned at species level, 15 are tentatively assigned (confer, cf.) and 14 are unassigned at species level, so left in open nomenclature (sp.); the latter might be higher, as in this case, than 30%. Similarly, [27] suggested that most of the taxa mainly belonging to *Quinqueloculina* and *Triloculina* genera retrieved in the Arabian Gulf were reported in open nomenclature. The lack of materials for comparison (i.e., type and reference collection) and of comprehensive taxonomical guides is not solely missing for Kuwait marine areas but also for the Arabian Gulf [27]. Therefore, it is not surprising that several new benthic foraminiferal species (i.e., *Falsonubeculina arabica* and *Pseudotriloculina hottingeri*) and genus (*Falsonubeculina*) have been recently identified in the Gulf [70–72].

Moreover, the Class Monothalamea was not here considered, though it commonly represents a largely diversified group of foraminifera [73], whose diversity is mostly neglected in Kuwait. In fact, to the best of our knowledge, no study has yet relied on or at least considered the soft-shell monothalamids ('allogromiids') in Kuwait marine areas. The only available data are the molecular ones (eDNA metabarcoding) associated with the samples collected in the Kuwait Bay in the present study. On the basis of these data, 94 Molecular Operational Taxonomic Units were recognized, using a filtering > 1000 reads [74]. The majority of sequences was assigned to two clades of monothalamous foraminifera clade Y (26.4%), abundant in many metabarcoding studies, and a new clade KUW (22%), which is reported for the first time here [74]. The third very abundant taxon is an undescribed species of Saccamminidae (20.8%), closely related to a genotype reported from the North Sea coastal water. Among other monothalamid genera, the most abundant are *Vellaria* (3.8%) and an allogromiid sp. (3.3%) [74]. Monothalamids are commonly ignored by traditional morphological studies, despite dominating the metabarcoding dataset, as in the Kuwait Bay. On the contrary, it is worth mentioning that only a very few studies, i.e., [28,30], and the present data for the Kuwait Bay area are based on the living benthic foraminiferal assemblages. The total assemblages, which are time-averaged assemblages based on the combination of living and dead foraminifera, might include allochthonous taxa that were transported from other localities.

Despite of different factors affecting positively or negatively the estimation of the benthic foraminiferal diversity in Kuwait territorial waters, the here-reported γ (452) benthic foraminiferal diversity is relatively high and can be compared to the available records and census data in other basins, such as 799 species in the Aegean Sea [21], 818 species around the Korean peninsula [22], 946 species in the Sahul Shelf and Timor Sea [75], 987 species in the Gulf of Mexico [17] and 1107 species in the northwestern European continental margin [19]. These figures are somewhat higher than those of Kuwait, but it needs to be stressed that the herein considered area, namely Kuwait territorial water, has a limited extension compared to the previous ones (i.e., the Aegean Sea or the Gulf of Mexico). Indeed, the territorial marine waters of Kuwait are characterized by shallow depths, reaching a maximum depth of about 30 m, so most of the deep-water benthic foraminifera are absent. Despite this, Kuwait waters offer a variety of different environments and sub-environments, including the relatively low salinity/muddy areas in the northern part, the embayment (Kuwait Bay), the rocky tidal flats in the southern part, the coral reef systems, the islands and the shelf slope that all together host a larger diversified biota (i.e., benthic foraminifera, macrofauna, zooplankton and phytoplankton).

The Arabian Gulf, the hottest marine environment on Earth, has been recently suggested as a model for the 2100 climate change [2], by which understanding the potential biodiversity loss triggered by the climate changes. Moreover, a differential warming within the Gulf with a more severe SST increase in the northernmost part of the Gulf, including Kuwait, was predicted [4]. These changes concurrently occur with a rapid development and antropogenization of coastal environments that ultimately lead to a deterioration of natural habitats. In light of it, the assessment of the diversity, as well as the understanding of its spatial and temporal changes, is a prerequisite and represents the baseline information for the implementation of natural conservation plans and species protection.

On the basis of the present data, we also revealed that some species, such as *Amphistegina lessonii*, have only been recently reported in Kuwait [28,58]. Amphisteginids have been reported to extend their geographical distribution in response to increased SST [76,77]. The first record of *A. lessonii* was found in two stations (15 at 14.6 m water depth and 33 at 22 m water depth), collected in 2000, around the unique island of Umm al Maradim that represents the largest and southernmost coral island in Kuwait [28]. Later on, it was documented in an offshore carbonate ramp near Qaro cay in one station (K039 at 10.5 m water depth) collected during an extensive offshore Southern Kuwait sampling campaign in October–December 2001 and in June 2002 [58,78]. Both of these localities, Umm al Maradim Island and Qaro cay are located in the southernmost part of Kuwait, away from the influence of Shatt al-Arab Delta discharge. In fact, *Amphistegina* is an algal symbiont-bearing large foraminiferal genus thriving in warm, clear, nutrient-poor, shallow environments [76,79], and its distribution within Kuwait waters is likely influenced by the plume generated by the Shatt al-Arab river in the northern area. This plume is deflected by the Coriolis effect that excludes, therefore, most of the Iranian coast [35]. Accordingly, [27] identified two species of *Amphistegina*, namely *A. lessonii* and *Amphistegina papillosa*, in several samples along the Iranian coast. Larger benthic foraminifera are an informal and polyphyletic group of calcifying foraminiferal organisms hosting symbionts (i.e., algae) and significantly contribute to $CaCO_3$ cycling in the ocean [80]. They live within the photic zone and in oligotrophic warm waters. Among the larger benthic foraminifera, only *A. lessonii, Assilina/Operculina ammonoides, Assilina/Operculina complanata, Heterostegina depressa, Coscinospira hemprichii, Peneroplis pertusus* and *Peneroplis planatus* have been reported from Kuwait [81]. In addition to them, we here also report *Operculina gaimardi, Operculina* sp., *Heterostegina* sp., *Peneroplis arietinus, Dendritina rangi, Sorites orbiculus* and *Sorites marginalis*. The most frequent recognized taxa belong to the genus *Peneroplis*, namely *P. arietinus, P. planatus* and *P. pertusus*, and *O. complanata*, followed by *O. ammonoides, O. gaimardi, S. marginalis*, and *S. orbiculus*. Most of these taxa are mainly documented in the southern islands (i.e., Umm Al-Maradim, Qaruh and Kubbar Islands) and the southern coast (Al-Khiran, Khor Iskander and Ras Al-Zour) but are basically scattered around the northern islands (i.e., Failaka Island, Bubiyan Island and the Shatt al-Arab area) or mostly absent within the Kuwait Bay (Supplementary Table S1).

The most abundant wall type, calculated based on the number of the identified species out of the γ diversity, is porcelaneous (48.5%), followed by hyaline wall type (42.6%) and only a minor percentage (8.9%) is represented by agglutinated species. The least diversified orders, in terms of genera, are Vaginulinida (one genus), Robertinida (one genus) and Spirillinida (two genera), followed by Lituolida (six genera), Lagenida (six genera) and Textulariida (eight genera). On the other hand, the highest generic diversification is found in the Order Rotaliida, represented by 66 genera, and in the Order Miliolida, including by 54 genera. The most diversified genera within Textulariida is *Textularia*, accounting for 18 species. On the other hand, *Quinqueloculina, Triloculina* and *Spiroloculina* are represented by 58, 24 and 18 species, respectively. Rotaliida are mainly denoted by *Elphidium* (13 species) and *Ammonia* (13 species).

When considering the βw diversity index comparing different environments, the highest benthic foraminiferal dissimilarity is identified between the tidal flat and both the northern islands and the southern coast. This might be explained by the different environmental and sedimentological features characterizing the areas. In fact, tidal flats are mostly located in the southern coastal province between the coastal area of Kuwait City and the southern borders with Saudi Arabia and are characterized by relatively thick cover of sand and bounded landward by several oolitic ridges and inland sabkha flats. On the other hand, the submerged estuarine area in the northern part of Kuwait, including the six northern offshore islands (Warba, Bubiyan, Miskan, Failaka, Auha and Umm Al-Namil), represents an extensive mudflat covered by a mixture of fine sand, silt and clay. This area is strongly influenced by the Shatt al-Arab river plume that affects the sedimentary influx in this area and, in turn, the thickness of the photic zone. The shelf slope is located in the central to southern part of the Kuwait coast, and according to [7], it represents a separated physiographic region mostly characterized by rocky substrates. The highest benthic foraminiferal similarity is instead documented between the southern coast and the southern islands, likely due to similar substrate (i.e., less affected by the Shatt al-Arab river plume) and more oligotrophic conditions. Three islands, namely Kubbar, Qaruh and Umm Al-Maradim are characterized by coral reef colonies in the subtidal zones that, in terms of substrate, are more similar to the southern coast than to the other physiographic areas, and this might explain the comparable benthic foraminiferal diversity.

7. Conclusions

The implementation of natural conservation plans requires an accurate knowledge of the ecosystem, biodiversity and ecosystem services. In light of the current climatic changes at the global scale, and anthropic pressure at a local one, the documentation of the diversity becomes more and more important for species and environmental protection. The Arabian Gulf is known for its unique oceanographical conditions and endemic biota adapted to extreme conditions. In the northwestern part of the Gulf lies Kuwait, which hosts an important part of biodiversity and is also a nursery ground for a number of species. Several previous investigations have documented high diversity of zooplankton and phytoplankton species thriving within the Kuwait territorial waters, but very limited information exists on the overall diversity of benthic foraminifera. On the basis of the integration of publications, reports and theses with new available data, it has been possible to infer the total benthic foraminiferal diversity within Kuwait territorial waters. This new literature survey documents the presence of 451 species belonging to 156 genera, 64 families, 31 superfamilies and 9 orders. These values are relatively high in consideration of the limited extension and the shallow depth of the Kuwait territorial waters. However, these figures are herein considered as underestimated because of the grouping of unassigned species due to the lack of reference collection and materials, as well as the neglection of the soft-shell monothalamids ('allogromiids').

Supplementary Materials: The following are available online at http://www.mdpi.com/1424-2818/12/4/142/s1. Table S1: List of modern benthic foraminiferal species recognized within Kuwait territorial water (the number within square brackets refers to the available records, as in reference; number refers to the location as in Figure 1; latitudinal and longitudinal range, depth interval, assemblages, number of samples, size fraction and sub-environments are also reported). NI: northern island, SI: southern island, TF: tidal flat, SC: southern coast, KB: Kuwait Bay, n.a.: not available, nc: not considered studies for the calculation of βw. Figure S1: Schematic map of the environments within Kuwait territorial waters: the northern islands (Failaka Island, Bubiyan Island and the Shatt al-Arab area), the southern islands (Umm Al-Maradim, Qaruh and Kubbar Islands), the southern coast (Al-Khiran, Khor Iskander and Ras Al-Zour), and the tidal flat and the Kuwait Bay (modified after [7]).

Author Contributions: Conceptualization, F.F. and E.A.-E.; methodology, F.F. and E.B.; formal analysis, F.F.; investigation, S.K. and E.A.-E.; resources, E.A.-E.; data curation, E.A.-E. and S.K.; writing—original draft preparation, F.F.; writing—review and editing, F.F. and E.B.; supervision, F.F.; project administration, E.A.-E.; funding acquisition, E.A.-E. All authors have read and agreed to the published version of the manuscript.

Funding: This research was funded by the Kuwait Foundation for the Advancement of Sciences and the Kuwait Institute for Scientific Research, grant number EM084K. The APC was funded by the Kuwait Institute for Scientific Research.

Acknowledgments: The authors are very grateful to Kuwait Institute for Scientific Research for the research support. The authors are also very grateful to four anonymous reviewers for their thoughtful and valuable comments that have greatly improved the paper. Authors thank Mohamed Ibrahim (Alexandria University, Egypt) for supervising a partial work in this paper.

Conflicts of Interest: The authors declare no conflict of interest.

References

1. Khan, N.Y.; Munawar, M.; Price, A.R.G. *The Gulf Ecosystem: Health and Sustainability*, 1st ed.; Backhuys Publishers: Leiden, The Netherlands, 2002; pp. 1–509.
2. Riegl, B.M.; Purkis, S.J. *Coral Reefs of the Gulf: Adaptation to Climatic Extremes in the World's Hottest Sea*, 1st ed.; Springer: Dordrecht, The Netherlands, 2012; pp. 1–379.
3. Sheppard, C.; Al-Husiani, M.; Al-Jamali, F.; Al-Yamani, F.; Baldwin, R.; Bishop, J.; Benzoni, F.; Dutrieux, E.; Dulvy, N.K.; Durvasula, S.R.; et al. The Gulf: A young sea in decline. *Mar. Pollut. Bull.* **2010**, *60*, 13–38. [CrossRef] [PubMed]
4. Noori, R.; Tian, F.; Berndtsson, R.; Abbasi, M.R.; Naseh, M.V.; Modabberi, A.; Soltani, A.; Kløve, B. Recent and future trends in sea surface temperature across the Persian Gulf and Gulf of Oman. *PLoS ONE* **2019**, *14*, e0212790. [CrossRef]

5.   Khan, N.Y. Physical and human geography. In *The Gulf Ecosystem: Health and Sustainability*, 1st ed.; Khan, N.Y., Munawar, M., Price, A.R.G., Eds.; Backhuys Publishers: Leiden, The Netherlands, 2002; pp. 3–21.

6.   Al-Bakri, D. *Environmental Assessment of the Intertidal Zone of Kuwait*; Kuwait Institute for Scientific Research Report; Kuwait Institute for Scientific Research: Al-Ahmadi, Kuwait, 1984; pp. 1–429.

7.   Al-Yamani, F.Y.; Bishop, J.; Ramadhan, E.; Al-Husaini, M.; Al-Ghadban, A.N. *Oceanographic Atlas of Kuwait's Waters*, 1st ed.; Kuwait Institute for Scientific Research: Al-Ahmadi, Kuwait, 2004; pp. 1–203.

8.   Bishop, J.M. History and current checklist of Kuwait's ichthyofauna. *J. Arid Environ.* **2003**, *54*, 237–256. [CrossRef]

9.   Al-Yamani, F.Z.; Skryabin, V.A. *Identification Guide for Protozoans from Kuwait's Waters. Coastal Planktonic Ciliates: I. Tintinnids*, 1st ed.; Kuwait Institute for Scientific Research: Al-Ahmadi, Kuwait, 2006; pp. 1–109.

10.   Al-Yamani, F.Z.; Saburova, M.A. *Illustrate Guide on the Flagellates of Kuwait's Intertidal Soft Sediments*, 1st ed.; Kuwait Institute for Scientific Research: Al-Ahmadi, Kuwait, 2010; pp. 1–197.

11.   Al-Yamani, F.Z.; Prusova, I. *Common Copepods of the Northwestern Arabian Gulf: Identification Guide*, 1st ed.; Kuwait Institute for Scientific Research: Al-Ahmadi, Kuwait, 2003; pp. 1–162.

12.   Al-Kandari, M.; Al-Yamani, F.Y.; Al-Rifaie, K. *Marine Phytoplankton Atlas of Kuwait's Waters*, 1st ed.; Kuwait Institute for Scientific Research: Al-Ahmadi, Kuwait, 2009; pp. 1–354.

13.   Murray, J.W. *Ecology and Applications of Benthic Foraminifera*, 2nd ed.; Cambridge University Press: Cambridge, UK, 2006; pp. 1–426.

14.   Lee, J.L.; Anderson, O.R. Symbiosis in Foraminifera. In *Biology of Foraminifera*; Lee, J.L., Anderson, O.R., Eds.; Elsevier Science Publishing Co. Inc.: San Diego, CA, USA, 1991; pp. 157–220.

15.   Moodley, L.; Middelburg, J.J.; Boschker, H.T.S.; Duineveld, G.; Pel, R.; Herman, P.M.J.; Heip, C.H.R. Bacteria and Foraminifera: Key players in a short-term deep-sea benthic response to phytodetritus. *Mar. Ecol. Prog. Ser.* **2002**, *236*, 23–29. [CrossRef]

16.   Jorissen, F.J. Benthic foraminifera from the Adriatic Sea; principles of phenotypic variation. *Utrecht Micropaleontol. Bull.* **1988**, *37*, 1–174.

17.   Sen Gupta, B.K.; Smith, L.E. Modern Benthic Foraminifera of the Gulf of Mexico: Census report. *J. Foraminifer. Res.* **2010**, *40*, 247–256. [CrossRef]

18.   Milker, Y.; Schmiedl, G. A taxonomic guide to modern benthic shelf foraminifera of the western Mediterranean Sea. *Palaeontol. Electron.* **2012**, *15*, 1–134. [CrossRef]

19.   Dorst, D.; Schönfeld, J. Diversity of benthic Foraminifera on the shelf and slope of the NE Atlantic: Analysis of datasets. *J. Foraminifer. Res.* **2013**, *43*, 238–254. [CrossRef]

20.   Martins, M.V.; Frontalini, F.; Laut, L.L.; Silva, F.S.; Moreno, J.; Sousa, S.; Zaaboub, N.; El Bour, M.; Rocha, F. Foraminiferal Biotopes and their Distribution Control in Ria de Aveiro (Portugal): A multiproxy approach. *Environ. Monit. Assess.* **2014**, *186*, 8875–8897. [CrossRef]

21.   Frontalini, F.; Kaminski, M.A.; Mikellidou, I.; du Châtelet, E.A. Checklist of benthic foraminifera (class Foraminifera: D'Orbigny 1826; phylum Granuloreticulosa) from Saros Bay, northern Aegean Sea: A biodiversity hotspot. *Mar. Biodivers.* **2015**, *45*, 549–567. [CrossRef]

22.   Kim, S.; Frontalini, F.; Martins, M.V.; Lee, W. Modern benthic foraminiferal diversity in Jeju Island and first insights into the total diversity of Korea. *Mar. Biodivers.* **2016**, *46*, 337–354. [CrossRef]

23.   Cherif, O.H.; Al-Ghadban, A.N.; Al-Rifaiy, I.A. Distribution of foraminifera in the Arabian Gulf. *Micropaleontology* **1997**, *43*, 253–280. [CrossRef]

24.   Arslan, M.; Kaminski, M.A.; Tawabini, B.S.; Ilyas, M.; Babalola, L.O.; Frontalini, F. Seasonal variations, environmental parameters and standing crop assessment of benthic foraminifera in eastern Bahrain. Arabian Gulf. *Geol. Q.* **2016**, *60*, 24–35. [CrossRef]

25.   Arslan, M.; Kaminski, M.A.; Tawabini, B.S.; Ilyas, M.; Frontalini, F. Benthic foraminifera in sandy (siliciclastic) coastal sediments of the Arabian Gulf (Saudi Arabia): A technical report. *Arab. J. Geosci.* **2016**, *9*, 7. [CrossRef]

26.   Amao, A.O.; Kaminski, M.A.; Qurban, M.A.; Thadikal, J.; Frontalini, F. A baseline investigation of benthic foraminifera in relation to marine sediments parameters in western parts of the Arabian Gulf. *Mar. Pollut. Bull.* **2019**, *146*, 751–766. [CrossRef] [PubMed]

27.   Amao, A.O.; Kaminski, M.A.; Asgharian Rostami, M.; Gharaie, M.H.M.; Lak, R.; Frontalini, F. Distribution of benthic foraminifera along the Iranian Coast. *Mar. Biodivers.* **2019**, *49*, 933–946. [CrossRef]

28.   Al-Enezi, E.; Al-Ghadban, A.N.; Al-Refai, I.; Pieretti, N.; Frontalini, F. Living benthic foraminifera around the unique Umm al Maradim Island (Kuwait). *Kuwait J. Sci.* **2019**, *46*, 59–66.

29. Al-Zamel, A.; Al-Sarawi, M.; Khader, S.; Al-Rifaiy, I. Benthic foraminifera from polluted marine environment of Sulaibikhat Bay (Kuwait). *Environ. Monit. Assess.* **2009**, *149*, 395–409. [CrossRef]

30. Al-Enezi, E.; Frontalini, F. Benthic foraminifera and environmental quality: The case study of Sulaibikhat Bay (Kuwait). *Arab. J. Geosci.* **2015**, *8*, 8527–8538. [CrossRef]

31. Neelamani, S.; Al-Salem, K.; Rakha, K. Extreme waves for Kuwaiti territorial waters. *Ocean Eng.* **2007**, *34*, 1496–1504. [CrossRef]

32. Al-Bakri, D.; El-Sayed, M.I. Mineralogy and provenance of the elastic deposits of the modern intertidal environment of the northern Arab. *Gulf. Mar. Geol.* **1991**, *97*, 121–135. [CrossRef]

33. Kassler, P. The structural and geomorphic evolution of the Persian Gulf. In *The Persian Gulf. Holocene Carbonate Sedimentation and Diagenesis in a Shallow Epicontinental Sea*, 1st ed.; Purser, B.H., Ed.; Springer: Berlin, Germany, 1973; pp. 11–32.

34. Al-Sarawi, M.A.; Gundlach, E.; Baca, B.J. Coastal geomorphology and resources in terms of sensitivity to oil spill in Kuwait. *J. Univ. Kuwait (Sci.)* **1988**, *15*, 141–184.

35. Riegl, B.; Poiriez, A.; Janson, X.; Bergman, K.L. The gulf: Facies belts, physical, chemical, and biological parameters of sedimentation on a carbonate ramp. In *Carbonate Depositional Systems: Assessing Dimensions and Controlling Parameters*, 1st ed.; Westphal, H., Riegl, B., Eberli, G.P., Eds.; Springer Science & Business Media: New York, NY, USA, 2010; pp. 145–213.

36. Khalaf, F.; Al-Bakri, D.; Al-Ghadban, A. Sedimentological characteristics of the surficial sediments of the Kuwaiti marine environment, northern Arabian Gulf. *Sedimentology* **1984**, *31*, 531–545. [CrossRef]

37. Reynolds, R.M. Physical oceanography of the Gulf, Strait of Hormuz, and the Gulf of Oman—Results from the Mt Mitchell expedition. *Mar. Pollut. Bull.* **1993**, *27*, 35–59. [CrossRef]

38. Jones, D.A.; Price, A.R.G.; Al-Yamani, F.Y.; Al-Zaidan, A. Coastal and marine ecology. In *The gulf Ecosystem: Health and Sustainability*, 1st ed.; Khan, N.Y., Munawar, M., Price, A.R.G., Eds.; Backhuys Publishers: Leiden, The Netherlands, 2002; pp. 65–103.

39. Peiris, N.I. Recent Foraminifera and Ostracoda from the Persian Gulf. Master's Thesis, University College of Wales, Aberystwyth, UK, 1969.

40. Anber, S.A. Foraminifera from the Offshore Area of Kuwait (Northern Arabian Gulf). Master's Thesis, Kuwait University, Kuwait City, Kuwait, 1974.

41. Shublaq, W. Foraminiferida from Bottom Sediments in the Northwestern Part of the Arabian Gulf (Offshore Kuwait and Saudi Arabia—Kuwait Portioned Zone). Master's Thesis, Kuwait University, Kuwait City, Kuwait, 1977.

42. Al-Abdul Razzaq, S.; Khalaf, F.; Al-Bakri, D.; Shublaq, W.; Al-Sheikh, Z. *Marine Sedimentology and Benthic Ecology of Kuwait Environment*; Report No. KISR 694; Kuwait Institute for Scientific Research: Al-Ahmadi, Kuwait, 1982.

43. Al-Abdul Razzaq, S.; Shublaq, W.; Al-Sheikh, Z. The marine benthic microfauna of the tidal flats of Kuwait. *J. Kuwait Univ.* **1983**, *10*, 101–109.

44. Al-Abdul-Razzaq, S.; Bhalla, S. On the genus *Cribrospirlina* Haman, 1972 (Foraminifera). *J. Micropaleontol.* **1987**, *6*, 63–64. [CrossRef]

45. Al-Abdul-Razzaq, S.; Bhalla, S. Microfauna from Al-Khiran area, Southern Kuwait. *Rev. Paléobiol.* **1987**, *6*, 139–142.

46. Shublaq, W. *Microfauna of Kuwait's Marine Environment (Foraminifera and Ostracoda)*; Phase I. Final Report, Restricted, EES-108; Reports No. KISR 2812; Kuwait Institute for Scientific Research: Al-Ahmadi, Kuwait, 1988.

47. Al-Zamel, A.; Cherif, O.; Al-Rifaiy, I. Tidal creeks foraminiferal distribution in Khor Al-Mufateh and Khor Al-Malaha, Khiran area, southeast Kuwait. *Rev. Micropalaeontol.* **1996**, *39*, 3–26.

48. Al-Shuaibi, A.A.; Cherif, O.H.; Al-Sarawi, M.A. Recent Subtidal Foraminiferal Distribution and Water Quality of Kubbar Island, Kuwait. Master's Thesis, Kuwait University, Kuwait City, Kuwait, 1997.

49. Khader, S. Coastal Geomorphology and Environmental Assessment of Sulaibikhat Bay. Master's Thesis, Kuwait University, Kuwait City, Kuwait, 1997, unpublished.

50. Al-Zamel, Z.; Cherif, O. Subtidal foraminiferal assemblages of the western part of the Shatt Al-Arab Delta, Kuwait, Arabian Gulf. *J. Foraminifer. Res.* **1998**, *28*, 327–344. [CrossRef]

51. Al-Ghadban, A.N. *The Potential Impact of Draining the Southern Iraqi Marshes on the Sediment Budget and Associated Pollutants in the Northern Arabian Gulf*; Kuwait Institute for Scientific Research: Al-Ahmadi, Kuwait, 2000; pp. 1–230.

52. Al-Enezi, E.H. Environmental Factors Controlling the Distribution of Foraminifera in the Umm Al-Maradem Island and Khor Iskandar Southern Part of Kuwait. Master's Thesis, Kuwait University, Kuwait City, Kuwait, 2002.

53. Al-Kandari, A.J.H. Trace Metals Concentration and Their Effects on Foraminifera at Coastal Area Adjacent to Kuwait University Campus at Shuwaikh. Master's Thesis, Kuwait University, Kuwait City, Kuwait, 2008.

54. Al-Ammar, M. Distribution of Recent Benthic Foraminifera around the Coral Reefs of Quran Island, Southern Kuwait: Variability and Biodiversity. Master's Thesis, Kuwait University, Kuwait City, Kuwait, 2011.

55. Al Theyabi, N.T. Effect of Thermal Pollution on Microfauna "Foraminifera" a Case Study of Ras Al-Zour and Ras Al-Subiya Power Stations, State of Kuwait. Master's Thesis, Kuwait University, Kuwait City, Kuwait, 2012.

56. Uddin, S.; Al-Ghadban, A.N.; Pokavanich, T.; Al Banna, K.; Al-Enezi, E.; Karam, Q.; Al-Shamroukh, D.; Al-Khabbaz, A.; Al- Yaegoub, A. *Marine Survey and Dispersion Modeling for Environmental Aspect at MAA and MAB Refineries*; Technical Report No. KISR 11496; Kuwait Institute for Scientific Research: Safat, Kuwait, 2013.

57. Nabavi, S.M.B.; Moosapanah, S.G.R.; Ghatrami, E.R.Z.; Ashrafi, M.G.; Nabavi, S.N. Distribution, Diversity and Abundance of Benthic Foraminifera of the Northwestern Persian Gulf. *J. Persian Gulf (Mar. Sci.)* **2014**, *5*, 15–26.

58. Parker, J.H.; Gischler, E. Modern and relict foraminiferal biofacies from a carbonate ramp, offshore Kuwait, northwest Persian Gulf. *Facies* **2015**, *61*, 1–22. [CrossRef]

59. Khader, S. Benthic Foraminifera Response to Environmental Stress around Three Northern Kuwaiti Islands (Failaka, Auha and Miskan Islands), NW Kuwait, Arabian Gulf. Ph.D. Thesis, Alexandria University, Alexandria, Egypt, 2020, unpublished.

60. Loeblich, A.R.; Tappan, H. *Foraminiferal Genera and Their Classification*; Van Nostrand Reinhold: New York, NY, USA, 1987; pp. 1–970.

61. Cimerman, F.; Langer, M.R. *Mediterranean Foraminifera*; Slovenska Akademija Znanosti in Umetnosti, Akademia Scientiarium et Artium Slovenica: Ljubljana, Slovenia, 1991; Volume 30, pp. 1–119.

62. Hottinger, L.; Halicz, E.; Reiss, Z. *Recent Foraminifera from the Gulf of Aqaba, Red Sea*, 1st ed.; Slovenska Akademija Znanosti Umetnosti: Ljubljana, Slovenia, 1993; pp. 1–179.

63. Sgarella, F.; Moncharmont Zei, M. Benthic foraminifera of the Gulf of Naples (Italy): Systematics and autoecology. *Bull. Soc. Paleontol. Ital.* **1993**, *32*, 145–264.

64. Jones, R.W. *The Challenger Foraminifera*, 1st ed.; Oxford University Press: Oxford, UK, 1994; pp. 1–149.

65. Debenay, J.P. *A Guide to 1,000 Foraminifera from Southwestern Pacific, New Caledonia*, 1st ed.; IRD, Ed.; Publications Scientifiques du Museum; Museum National d'Histoire Naturelle: Paris, France, 2013; pp. 1–384.

66. Amao, A.O. Benthic Foraminiferal Taxonomy, Distribution and Ecology in the Arabian Gulf. Ph.D. Thesis, King Fahd University of Petroleum and Minerals, Dhahran, Saudi Arabia, 2016.

67. World Register of Marine Species. Available online: http://www.marinespecies.org/aphia.php?p=taxdetails&id=1410 (accessed on 14 February 2020).

68. Hammer, Ø.; Harper, D.A.T.; Ryan, P.D. PAST: Paleontological statistics software package for education and data analysis. *Palaeontol. Electron.* **2001**, *4*, 9.

69. Whittaker, R.H. Vegetation of the Siskiyou mountains, Oregon and California. *Ecol. Monogr.* **1960**, *30*, 279–338. [CrossRef]

70. Amao, A.O.; Kaminski, M.A. A new foraminiferal species *Pseudotriloculina hottingeri* n. sp. from the Arabian Gulf. *J. Foraminifer. Res.* **2017**, *47*, 366–371. [CrossRef]

71. Amao, A.O.; Kaminski, M.A. *Pseudonubeculina arabica* n. gen. n. sp., a new Holocene benthic foraminifer from the Arabian Gulf. *Micropalaeontology* **2016**, *62*, 81–86.

72. Amao, A.O.; Kaminski, M.A. *Falsonubeculina*, a replacement name for *Pseudonubeculina* Amao and Kaminski 2016 (preoccupied). *Micropaleontology* **2019**, *65*, 544.

73. Lecroq, B.; Lejzerowicz, F.; Bachar, D.; Christen, R.; Esling, P.; Baerlocher, L.; Østerås, M.; Farinelli, L.; Pawlowski, J. Ultra-deep sequencing of foraminiferal microbarcodes unveils hidden richness of early monothalamous lineages in deep-sea sediments. *Proc. Natl. Acad. Sci. USA* **2011**, *108*, 13177–13182. [CrossRef]

74. Al-Enezi, E.; Al-Hazeem, S. *Benthic Foraminifera as Proxies for the Environmental Quality Assessment of the Kuwait Bay: Morphological and Environmental DNA Approaches*; Technical report KISR No. EM084C; Kuwait Institute of Scientific Research: Safat, Kuwait, 2019.

75. Loeblich, A.R.; Tappan, H. Foraminifera of the Sahul Shelf and Timor Sea. Cushm. Found. *Foram. Res. Spec. Publ.* **1994**, *31*, 1–661.

76. Langer, M.R.; Weinmann, A.E.; Lötters, S.; Bernhard, J.M.; Rödder, D. Climate-driven range extension of Amphistegina (Protista, Foraminiferida): Models of current and predicted future ranges. *PLoS ONE* **2013**, *8*, e54443. [CrossRef]

77. Guastella, R.; Marchini, A.; Caruso, A.; Cosentino, C.; Evans, J. "Hidden invaders" conquer the Sicily Channel and knock on the door of the Western Mediterranean Sea. *Estuar. Coast. Shelf Sci.* **2019**, *225*, 106234. [CrossRef]

78. Gischler, E.; Lomando, A.J. Offshore sedimentary facies of a modern carbonate ramp, Kuwait, northwestern Arabian-Persian Gulf. *Facies* **2005**, *50*, 443–462. [CrossRef]

79. Triantaphyllou, M.V.; Dimiza, M.D.; Koukousioura, O.; Hallock, P. Observations on the life cycle of the symbiont-bearing foraminifer Amphistegina lobifera Larsen, an invasive species in coastal ecosystems of the aegean sea (Greece, E. Mediterranean). *J. Foraminifer. Res.* **2012**, *42*, 143–150. [CrossRef]

80. Langer, M.R. Assessing the contribution of foraminiferan protists to global ocean carbonate production. *J. Eukaryot. Microbiol.* **2008**, *55*, 163–169. [CrossRef] [PubMed]

81. Förderer, M.; Rödder, D.; Langer, M.R. Patterns of species richness and the center of diversity in modern Indo-Pacific larger foraminifera. *Sci. Rep.* **2018**, *8*, 8189. [CrossRef] [PubMed]

MDPI

St. Alban-Anlage 66

4052 Basel

Switzerland

Tel. +41 61 683 77 34

Fax +41 61 302 89 18

www.mdpi.com

Diversity Editorial Office

E-mail: diversity@mdpi.com

www.mdpi.com/journal/diversity